11 CHEMISTRY

PNG UPPER SECONDARY

Reddy Kuama
Alex Eames
Suzanne Boniface
Terry Bunn
Mark Sayes

OXFORD

OXFORD
UNIVERSITY PRESS

Oxford University Press is a department of the University of Oxford. It furthers the University's objective of excellence in research, scholarship, and education by publishing worldwide. Oxford is a registered trademark of Oxford University Press in the UK and in certain other countries.

Published in Australia by
Oxford University Press
Level 8, 737 Bourke Street, Docklands, Victoria 3008, Australia

First published 2013
Reprinted 2014 (twice), 2016, 2017 (twice), 2018, 2019, 2021, 2022, 2023, 2024, 2025

This book was originally published by ESA Publications, Auckland, New Zealand. This edition, specially adapted for the Grade 11 syllabus in Papua New Guinea, is published by arrangement with ESA Publications. Authors of the original work were Alex Eames, Suzanne Boniface, Terry Bunn and Mark Sayes, and this edition has been produced by Reddy Kuama.

ISBN 978 0 19 5578799

Edited by Catherine Greenwood
Illustrated by Mairi Feeger
Typeset by diacriTech
Printed in China by Golden Cup Printing Co. Ltd

Oxford University Press Australia & New Zealand is committed to sourcing paper responsibly.

Contents

Unit 11.4 Energy and Reaction Rates

Unit 11.5 Metals and Non-metals

Supplementary Unit

Introduction

This book has been published to provide the information required by students in order to successfully complete the Upper Secondary course in Chemistry at Grade 11 in Papua New Guinea.

The book is written in a manner that best develops an overall understanding of the Chemistry concepts and processes set out in the PNG Grade 11 Syllabus. The order of Units in this book follows the order of Units presented in the Chemistry syllabus for Grade 11:

Unit 11.1 Application of Physical Processes

Unit 11.2 Chemical and Metallic Bonding

Unit 11.3 Types of Chemical Reactions

Unit 11.4 Energy and Reaction Rates

Unit 11.5 Metals and Non-metals

Within each Unit there are a number of Topics which follow the main subheadings and bullet points set out within each Unit in the syllabus. The aim is to provide structure and content in a concise and compact format for students to use as an effective resource to support the classroom experience. It is acknowledged that Chemistry is more interesting and meaningful when students are exposed to a variety of resources and materials and we encourage students and teachers not to rely on this book as the sole source of information.

If you have any suggestions about how this book might be improved in future editions, please make contact with Oxford University Press:

Fax: 00 61 3 9934 9100

Customer Service Email: exportsales.au@oup.com

We wish you every success in your studies.

The authors

Acknowledgments

There are many people to be thanked for their help and assistance in enabling the publication of this series to happen. First, we acknowledge the cooperation and generosity of Mark Sayes at ESA Publications in New Zealand, who responded with interest and support when the proposal to adapt his Study Guide series was put to him.

We would also like to acknowledge many other individuals who have been happy to advise and assist in different ways: Joy Sahumlal, Greg Kapanombo, Anne Sangi, Safak Deliismail, Frank Griffin, Peter Petsul and Justin Narimbi; also Catherine Greenwood, who edited the new material.

Copyright acknowledgments

The author and the publisher wish to thank and acknowledge the following copyright holders for reproduction of their material:

Shutterstock.com, p.22 (top); Science Photo Library/John McLean, p.22 (bottom).

Every effort has been made to trace the original source of copyright material contained in this book. The publisher will be pleased to hear from copyright holders to rectify any errors or omissions.

Authorship

The authors of the original editions were Alex Eames, Suzanne Boniface, Terry Bunn and Mark Sayes. This edition has been produced with the assistance of:

Reddy Kuama, lecturer in chemistry at the University of Papua New Guinea. An administrator and teacher for over twenty years for public and private institutions at secondary and tertiary levels, Mr Kuama has a passion for education and teaching young people. He has published widely, served as chief examiner and marker of the PNGHSC Year 12 Chemistry Examinations since 1998, and contributed to various school and government boards. In addition, he is one of the co-founding members and a senior fellow of the Institute of Chemists PNG (ICPNG). Mr Kuama holds a masters degree (MSc) and bachelor's degree (BSc) in chemistry from the University of Newcastle, Australia (1996, 1994), along with qualifications in administration (Dip Admin) and education (BEd, DipEd) from Pacific Adventist University (1989, 1987) and Fulton College, Fiji (1981). He resides with his wife and four children in Port Moresby.

The aim of this book is to provide Grade 11 Chemistry students in PNG with a book that they can use as a compact summary of content and skills related to the PNG Grade 11 syllabus.

Preliminary Unit
Importance of chemistry

Some of the aims of the chemistry course, clearly stated at the beginning of the Syllabus document, are to develop in students an understanding and appreciation of the methods and applications of chemistry and its development in the past, present and future contributions to life on Earth and beyond. Fundamental to this is the ability to observe, collect, analyse and interpret data to explain certain chemical principles and laws.
In this Preliminary Unit we focus on:
- Collecting information.
- Processing information by listing, sorting, collating, highlighting or summarising relevant information.
- Interpreting information by writing a report.

Introduction

To develop your capacity to observe, collect, analyse and interpret data to explain certain chemical principles and laws, you must work independently to carry out research into a relationship between chemistry and technology.

The following words and terms used in this Preliminary Unit must be understood by you.

Chemistry is:

- The branch of science concerned with the composition, properties and reactions of substances.

Technology is:

- The application of practical or mechanical sciences to industry or commerce.
- The methods, theory and practices governing such applications.
- The total knowledge and skills available to any human society for industry, art, science, etc.

Some other terms that must be understood include:

- *Collecting* – gathering together data.
- *Information* – reliable data or facts.
- *Processing* – arranging in order, eg data or a series of stages of an operation.
- *Listing* – drawing up data, etc in a table.
- *Sorting* – rearranging (eg data or stages of an operation) to improve understanding.
- *Collating* – analysing and comparing, eg data or a series of operations.
- *Summarising* – making a brief account; dispensing with needless detail.
- *Interpreting* – explaining the meaning of.

The context for investigations

Investigations involve the study of an issue or a problem. The emphasis is on the investigation of an issue in its context by researching, identifying the issues or problems, collecting, analysing and commenting on data and information. It is important to consider and explore a variety of perspectives. Some examples are in the table.

Context	Investigation
Vehicles, cities and emissions	How is humankind to control petrochemical emissions in highly populated areas?
Electricity generation and **acid rain**	Fossil fuels are burnt to produce heat energy which is then converted to electrical energy. Acid gases are produced in the burning of the fuels and these gases are responsible for acid rain.
Plastics are low density, strong and durable	How do you get rid of them, or recycle them, when they are finished with?
Water	Is there enough on planet Earth for human needs?

Sample contexts and investigations

There are many other possibilities which your teacher may make available to you.

Collecting and processing information

Taking notes

You must produce a set of notes that you can bring to the class. These notes will assist you when you write your report. These notes must be submitted along with your report.

Sources of information

You must collect appropriate information from a range of sources. You must look for information on the science and technology involved. Three different sources of information should be consulted and details of the sources of information should be given.

Selecting and collating information

You must select and collate the information relevant to your topic. Identify the links between key ideas and technologies involved. You should select relevant and useful illustrations, diagrams and graphs.

Presenting the information

You will write a report in class to show that you can process information to describe a use of chemistry knowledge. Diagrams and graphs must be hand-produced during the writing period. Your report should be one to two pages long, and should be written in your own words.

Your report will:

- State your research topic.
- Summarise, in your own words, your research findings.
- Discuss the links between the science ideas and the technology involved.
- Include a list of references, with sufficient information so that somebody else can access the information you have provided.

Example

References include businesses, tertiary education institutes, books, periodicals, websites, scientific videos, radio talks, TV programs, CD-ROMs.

Assessment criteria

The Chemistry Teacher's Guide explains the assessment criteria and the different levels of achievement: below minimum standard; low achievement; satisfactory achievement; high achievement; very high achievement.

Satisfactory achievement means	**High achievement** means	**Very high achievement** means
Processing information to *describe* a use of chemistry knowledge.	Processing information to *explain* a use of chemistry knowledge.	Processing information to *discuss* a use of chemistry knowledge.
The following may be used as guidelines of assessment:		
Gathering information: • *One* source of information. • Appropriate information collected and processed.	Gathering information: • *Three* sources of information. • Appropriate information collected and processed.	Gathering information: • *Three* sources of information. • Appropriate information collected, processed and *integrated* into the report.
Reporting information: • Information interpreted. • Report shows a link between *one* aspect of chemistry and technology of the topic.	Reporting information: • Information interpreted. • Report shows a link between *three* aspects of chemistry and technology of the topic.	Reporting information: • Information interpreted. • Report discusses link between the aspects of chemistry and technology of the topic.

Assessment criteria for the top three levels of achievement

Explanation of terms

Describe – provide characteristics of, or an account of, the scientific knowledge related to its use.

Explain – provide reasons as to how or why the scientific knowledge applies to the use.

Discuss – link ideas to integrate relevant chemistry knowledge with its use. A discussion involves elaborating, justifying, relating, evaluating, comparing and contrasting, or analysing.

Integrated – brought together to make an account complete and comprehensive.

Sample context – water

There is a great story to be told about water:

- It is so common.
- It is so unusual in its behaviour.
- It is so vital to the living world.

Here are a few questions that, when answered, will show a link between chemistry and technology. Each question illustrates an important aspect of water. When all the questions are answered, an integrated account of water can be obtained. The comments indicate interesting, important and vital features of water that may provide background information for you.

Questions	Comment
Why is water so called? Why not call it dihydrogen monoxide?	Water is historically old – its value was realised long before its composition was understood.
How much water is on planet Earth? How much water is used by one person in PNG – per day; per year; per lifetime?	There is plenty of water on planet Earth for the living world – some of the water may be of doubtful quality.
Is drinking water pure? Is it possible to get pure water? How is water purified? Are there benefits from impurities in water?	Nobody will ever drink pure water. The water we drink is fit for consumption and sometimes contains valuable minerals for human health.
Which elements react violently with water?	Only a few elements react violently with water. These elements are not familiar to most people, for obvious reasons.
Which compounds react violently with water?	Only a few compounds react violently with water. These elements are not familiar to most people, for obvious reasons.
Why is water useful as a fire extinguisher?	The ready availability of water, its stability to heat, and its ability to absorb heat (cools the fire down) make it a very good fire extinguisher.
How does water clean our bodies and our clothes?	Although water is thought by some people to be a universal solvent (ie that it will dissolve everything), it needs some help from detergents.
One of the unusual properties of liquid water is that it expands as it becomes ice, ie ice is less dense than water. What are the consequences of this behaviour of water?	Soil erosion and the breaking-up of lumps of agricultural soil are benefits of this property. In cold climates, water pipes and car radiators bursting in cold weather are a disadvantage of this property.
How important is water to the living world?	Many organisms are at least 70% water – water is vital to the living world.
How valuable are the seas as a source of minerals?	Extremely valuable. Magnesium, bromine and many other chemicals are extracted from the sea commercially.

Water – one of the most valuable chemicals

Use the information you gather to write an account of water explaining why it is considered the most important chemical to the living world.

Preliminary Unit Activity A: Processing information involving chemistry and technology

1. Explain the terms:
   a. Chemistry.
   b. Technology.
2. Explain the terms:
   a. Information.
   b. Listing.
   c. Summarising.
3. Identify the following terms:
   a. Arranging in order, eg data or a series of stages of an operation.
   b. Gathering together data.
   c. Analysing and comparing, eg data or a series of operations.
4. Using an internet search engine of your choice, find the number of websites ('hits') for the following entries:
   a. Water.
   b. Water in the world.
   c. Water in the oceans.
   d. Water in the Pacific Ocean.
   e. Volume of water in the world.
   f. State the connection between the number of hits and the entry.
5. Using an internet search engine of your choice, find the number of websites ('hits') for the following entries:
   a. Water reacting with elements.
   b. Water reacting violently with elements.
   c. State the connection between the number of hits and the entry.
6. Search the literature to find the meaning of the following terms. Write one sentence only on each term.
   a. Emissions.
   b. Acid rain.
   c. Plastic.
7. Read the following two accounts concerning water and then answer the questions below.
   1. Scientifically there is no such thing as pure water to be found on planet Earth. The early attempts, in the 19th century, to obtain pure water were by repeated distillation of samples of water; testing the electrical conductivity of the water after each distillation. When the conductivity decreased no further – it took twenty distillations for this situation to be reached – the water was judged to be pure. In practice, the water we drink is far from pure – but still fit to drink.

2. *In some Middle East countries, where petroleum is abundant and costs little, drinking water is obtained from sea water by distillation. The sea water is boiled by using heat from the combustion of petroleum products. The steam produced is then condensed and water suitable for drinking is obtained. Sea water contains many dissolved minerals, of which sodium chloride (common salt) is present in the largest quantity.*

a. Identify three chemicals mentioned in these two accounts.
b. Identify a process for judging the purity of a chemical in these two accounts.
c. Identify a process that could be described as technological in these accounts.
d. Explain why drinking water is not obtained from sea water in Papua New Guinea.

Unit 11.1 Application of Physical Processes

Topic 1: Matter and separations

Why does ice melt? Why does water evaporate? How can you separate the salt from sea water? Why does sugar dissolve easily in hot tea but not easily in cold tea? In this Unit we look at the ways in which matter is made up of particles and ways in which matter can be separated and substances changed. Topic 1 covers:

- Nature of matter.
- States of matter.
- Matter and its classification.
- Matter and its response to temperature changes.
- Matter and methods of separation.

Nature of matter

Matter is the 'stuff' around us and of what we are made. The building blocks of matter are very small **particles**.

Example

About 1 billion billion (1 000 000 000 000 000 000) carbon atoms would be visible as a full stop.

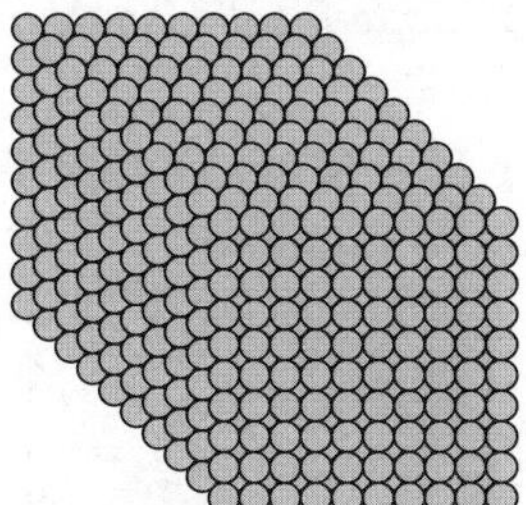

To display one million marbles, this diagram would have to be expanded by a factor of ten in all three dimensions.

A stack of one thousand marbles

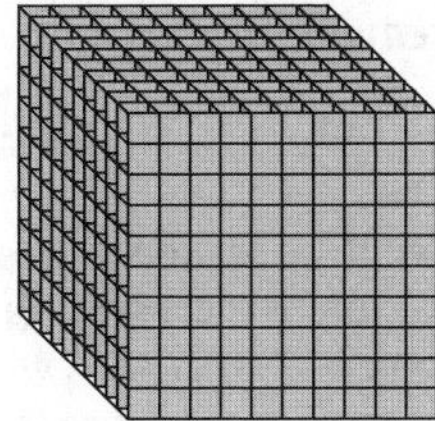

There are one thousand boxes in the stack – each box contains one million marbles, making one billion marbles altogether. One billion billion marbles would need a billion boxes!

A stack of one billion marbles

Particles may be thought of as being hard and **incompressible** (eg like pool balls).

States of matter

The three states of matter are:

- **Solid**.
- **Liquid**.
- **Gas**.

States of matter – solid, liquid and gas

The different arrangements of particles in each of the three states of matter are shown below:

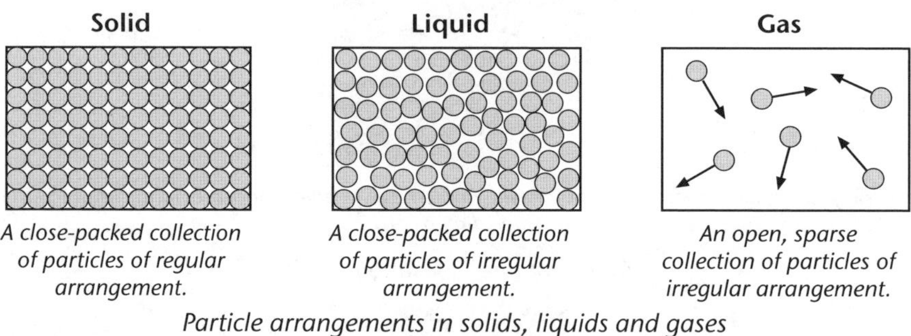

Particle arrangements in solids, liquids and gases

Solids

In the solid state, particles are attracted to each other so strongly that they are locked into a **lattice** structure. The visible part of a solid is a **crystal** – a solid with regular shape and with the **faces** at constant angles to each other. The particles within the crystal possess energy that allows them to vibrate and rotate, but they do not move from their location.

Solids are incompressible because the particles are close-packed – there is no space between particles. The arrangement of the particles is unchanged even when:

- A solid deforms (changes shape) – the particles slide over each other.
- A crystal is cut (**cleaved**) – the cleavage is between layers.

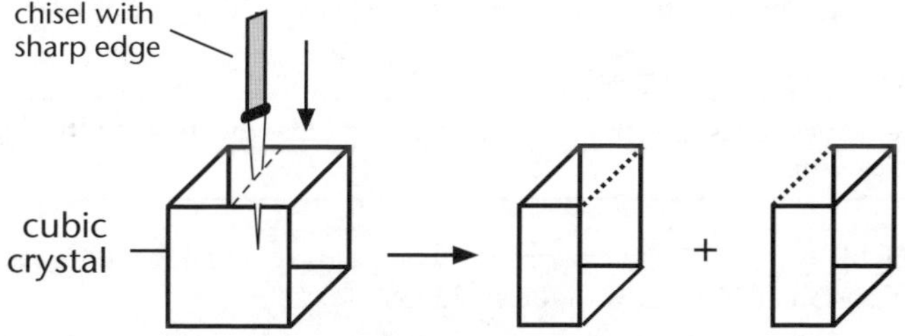

The crystal has cleaved (split) into two parts along a face to retain the crystal angles.

Cleavage of a crystal

Liquids

Liquid particles have higher energy than solid particles, so the attractive forces between liquid particles are more easily broken than those between solid particles. This means:

- Liquid particles are more mobile than solid particles – liquids are able to flow.
- The particles at the surface of a liquid can break away from the liquid and escape – the liquid evaporates.

Example

Warm windy conditions are good for drying the washing, or getting rid of puddles or floods – the liquid particles are swept away by the air movement above the liquid.

There is very little space between particles in a liquid, so liquids are incompressible.

Example

The hydraulic brakes on a car work by using oil to transfer the pressure applied by the driver on the brake pedal to the brake mechanism at the wheels. If the oil were **compressible**, all the pressure applied by the driver would be absorbed in compressing the oil – and the car would not stop!

Mixing of liquids

Some liquids:

- Mix *readily* (eg ethanol and water) – the particles in each liquid are continually breaking away from each other and can attach themselves to particles of the other liquid.
- Do *not* mix (eg water and oil) – the forces between the different liquid particles are not strong enough to hold them together and the particles 'prefer' to bond with particles of their 'own type'.

Gases

If a gas is not confined in a sealed container (eg a balloon), the gas particles will move in all directions and fill all available space.

Example

A gas **(vapour)** – eg perfume – can be smelled at one end of a large room even if the source is at the other end of the room.

Gases are compressible because there is plenty of space between the gas particles.

Example

The distance between gas particles is several hundred times greater than the diameter of the gas particles themselves.

When a gas is highly compressed, the particles become close enough to attract each other, and the liquid state can be formed, with the evolution of heat energy.

As matter is heated, the energy gained by the particles increases their **kinetic energy** and hence their **velocity**. (This happens with solids, liquids and gases but is most noticeable in a gas because the particles in a gas can move more freely.) More **collisions** of particles in the gas occur, and the impacts made by the particles on the container increase in number in a given time and thus the **pressure** of the gas increases.

Example

When a balloon is heated, the particles in the gas inside the balloon gain kinetic energy. The increase in speed and energy of the particles raises the pressure exerted by the gas, and due to the elastic nature of the rubber, the balloon expands. A point is reached where the rubber will not stretch any further and the balloon bursts.

Gases mix very quickly because:

- Gas particles are moving fast – their average speed is approximately 500 m s^{-1} (an Olympic sprinter has a speed of 10 m s^{-1}). The mass of the particle affects the speed – heavier particles move more slowly.
- There is plenty of space between the gas particles (ie there is nothing in their way).

Melting and boiling points

To change a solid to a liquid requires energy. This energy makes the solid particles vibrate and rotate to an extent that they break away from each other and the fluidity of the liquid state is achieved. All the energy being supplied is used to break the forces between the particles, so the change of state occurs at a fixed temperature – ie a solid has a precise and unique **melting point** (**mp**).

Similarly, a liquid has a **boiling point** (**bp**) – the temperature at which the liquid changes completely to a gas.

Unit 11.1 Activity 1A: Nature and states of matter

1. Draw the particle arrangement for the three states of matter. Use three boxes, one for each state.

2. Describe the behaviour of particles in:

  a. A gas.

  b. A liquid.

  c. A solid.

In your answers for questions **a.**, **b.** and **c.**, ensure you indicate:

i. The energy of the particles (low, medium or high).

ii. How a particle's movement is affected by neighbouring particles.

iii. The chances of a particle 'escaping' from its neighbours.

iv. What chances the particles have of exerting pressure on the walls of their container.

3. Explain why a hydrogen weather balloon is filled with only the minimum amount of hydrogen gas needed for it to rise. *Hint*: Consider what happens to atmospheric pressure as altitude increases above the surface of the Earth.

4. *When gas particles are pushed closer together, the particles can bond together. Heat energy is released in this process. For a liquid to become a gas or vapour, the particles have to break the bonds between each other. Heat energy is absorbed in this process.*

Using the information above, explain each of the following observations:

a. When a hand pump is used to pump up a bicycle tyre, the pump gets hot.

b. The rear of a refrigerator (where a gas is being compressed to the extent it becomes a liquid) is hot.

c. The temperature inside a refrigerator is lower than the temperature in the room where the refrigerator is standing.

d. A cigarette lighter contains a liquid fuel under pressure. When the lighter is used, the flame is due to a reaction between gas and air.

Types of matter

Element

An **element** is a type of matter that cannot be broken down into anything simpler by chemical or physical means. Only one type of **atom** is present in a given element.

Example

A block of gold of 10 cm^3 volume (1 cm × 2 cm × 5 cm) can be beaten into a sheet of gold leaf that covers 1 hectare (100 m × 100 m). The thickness of the sheet is 1 nanometre, or one billionth of a metre (0.000 000 001 m). The gold leaf contains the *same number* of the *same atoms* as the block of gold.

Compound

A **compound** is a type of matter that contains two or more elements chemically joined together in a constant ratio by **mass**, eg water.

Example

Experiment to establish composition of water by mass

The composition by mass of water, H_2O, can be determined experimentally using the apparatus:

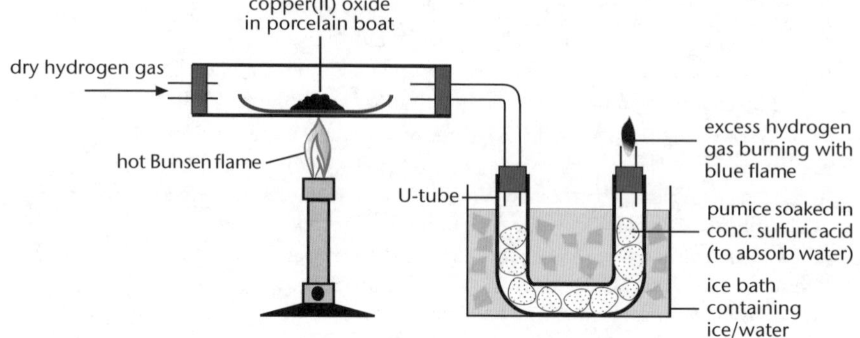

Apparatus to determine composition of water by mass

When hydrogen gas is passed over heated copper(II) oxide, the **products** of the reaction are metallic copper and steam. The steam is condensed to water and then absorbed by concentrated sulfuric acid.

If the porcelain boat is weighed empty, and then the boat and the U-tube are weighed before and after the experiment, results like the ones below can be obtained:

Mass of copper(II) oxide in the experiment	= 8.0 g
Mass of copper formed in the experiment	= 6.4 g
Mass of water formed in the experiment	= 1.8 g
Mass of oxygen in 1.8 g of water	= 8.0 – 6.4
	= 1.6 g
Mass of hydrogen in 1.8 g of water	= 1.8 – 1.6
	= 0.2 g
Ratio of mass of oxygen : mass of hydrogen in water	= 1.6 : 0.2
	= **8 : 1**

Example

Water has a constant mass composition

Any sample of water has oxygen and hydrogen present in the mass ratio of 8 : 1, ie:

- One drop of water has a mass of approximately 0.09 g and is made up of *0.08* g of *oxygen* and *0.01* g of *hydrogen*.
- The Pacific Ocean is a colossal volume of water (9×10^8 km^3). One cubic kilometre of water has a mass of 1 billion **tonnes**, so the total mass of water in the Pacific Ocean is 9×10^{17} tonnes. The water in the Pacific Ocean is made up of 8×10^{17} tonnes of *oxygen* and 1×10^{17} tonnes of *hydrogen*.

Mixture

A **mixture** is a type of matter that has any combination of elements and compounds in any ratio by mass, eg air.

Example

The composition of air

Air contains oxygen, nitrogen, argon, water vapour, carbon dioxide, dust and other materials in varying amounts.

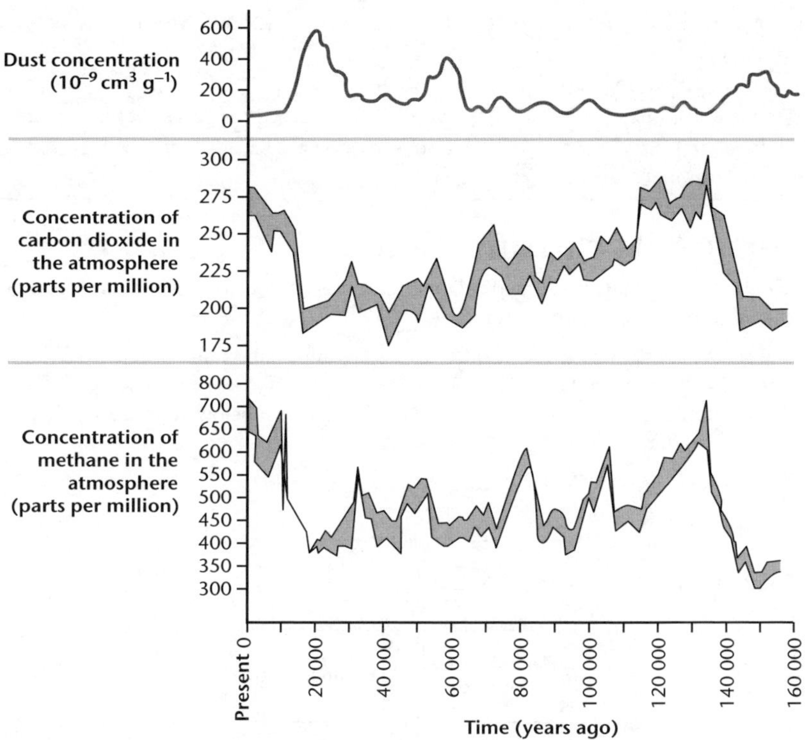

Air above industrial land – where factories are burning fuel – contains greater amounts of solid matter (eg soot) than in farming or urban areas. Air above farming land contains greater amounts of methane (resulting from the digestive processes of animals) than urban areas. Air above the sea contains salt, suspended in minute droplets of water.

Atmospheric composition of dust, carbon dioxide and methane over the last 150 000 years

Pure substance

A **pure substance** is a type of matter that is a single element or a single compound. Pure matter can be recognised by its precise physical properties.

Example

Pure water has a melting point of 0°C and a boiling point of 100°C.

Unit 11.1 Activity 1B: Types of matter

1. Use the following list of substances: ice, silver, sea water, chlorine, butter, mercury, common salt, sugar.

  a. Place each of the substances into one of the categories: Element, Compound or Mixture.

  b. Indicate which of the substances could be pure.

2. Indicate whether the following statements are true or false and give an explanation for your choice.
   a. Air is a mixture because its composition can vary.
   b. Any substance that has a sharp and precise melting point must be an element.
   c. Sea water has the same composition as river water because the water in all rivers ends up in a sea.

Matter and its classification

A simple definition of matter is something that has mass and occupies space. Matter includes everything around us, including ourselves.

One of the goals of chemistry is to organise or classify information about matter so that similarities and differences are easy to spot. There are many ways to organise matter. One way is to classify matter according to its composition. The diagram below summarises this general classification of matter.

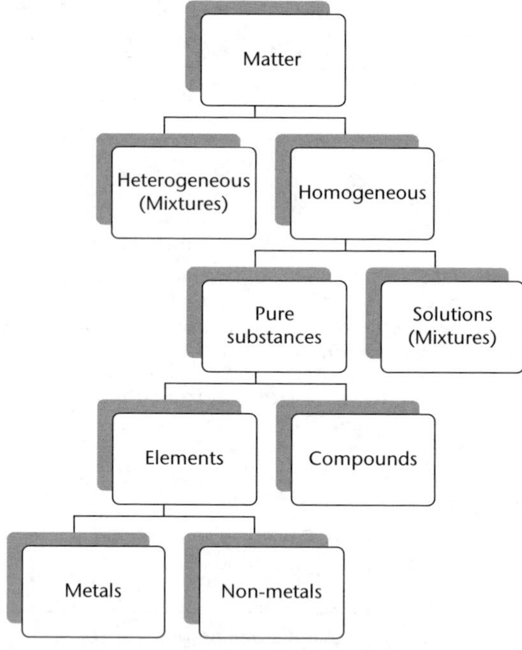

Classification of matter according to its composition. Homogeneity and heterogeneity are concepts that relate to the uniformity of a substance. If there is uniformity in the composition of a substance we say it is homogeneous. If a substance lacks uniformity in its composition, it is heterogeneous. *See p. 15 for more detail*

Heterogeneity

Heterogeneity is the state of being heterogeneous. Something is heterogeneous when it consists of many different kinds of things. For example, the population of a country can be described as heterogeneous if it is made up of many different kinds of people with different cultural backgrounds and different qualities.

In chemistry, we describe a substance as heterogeneous if it is diverse in kind or nature and composed of different kinds of substances. Heterogeneous material can consist of many states of matter or it can be a mixture of different substances in one mixture. For example, a mixture of water and petrol and grease is heterogeneous – it has different elements.

Homogeneity

Homogeneity is the state of being homogeneous; it is the opposite of heterogeneity. Something is homogeneous if it consists of things that are all the same or that are all the same type. You could say that the Western Highlands has a homogeneous population.

In chemistry, we describe a substance as homogeneous if it is something whose parts are all the same or have the same nature. Homogeneous material is not a mixture of different elements – all its elements are the same. For example, pure gold is homogeneous as all its elements are the same; so too is pure salt homogeneous.

Pure substances and solution mixtures

Homogeneous substances undergo further classification into pure substances and solutions. For example, pure gold is a pure homogeneous substance. Brass is a homogeneous solution of copper and zinc; brass is an **alloy** composed of two or more pure chemical substances, copper and zinc. Clean air is a homogeneous mixture or solution of oxygen, nitrogen and a number of other gases.

Separating mixtures

Common methods of separating mixtures involve the separation of a liquid from a solid, or a liquid from a liquid.

Decanting

A liquid and an **insoluble** solid can be separated by pouring off (**decanting**) the liquid to leave the solid in the flask. If the liquid is denser than the solid, then the solid can be skimmed off the surface of the liquid.

Example

Coffee decanting

One way of making coffee is to add boiling water to coffee beans that have been ground into small granules. After a few minutes, the soluble coffee will have dissolved and the dregs (insoluble residue) will remain at the bottom of the container. The coffee solution can be carefully decanted from the dregs to leave the dregs in the coffee pot.

Decantation

Filtration

To filter a substance, some components of the mixture must be **soluble** and the others insoluble. A **solvent** (often water, but there are many other solvents), **dissolves** some components and not others. **Porous** paper (filter paper) is the material most commonly used for separating insoluble material from soluble material.

Example

Filtering coffee

Coffee dregs (the insoluble material) can be separated from the desired coffee beverage (the **solution**) by filtering with the use of paper or a fine metal sieve.

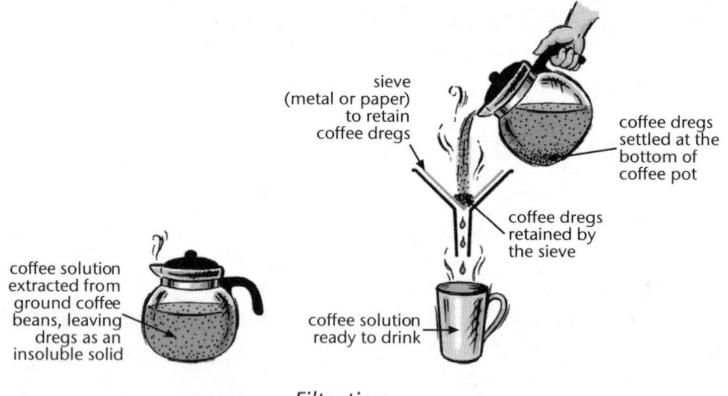

Filtration

Evaporation

Any solution contains a solvent and a **solute**. If the solvent can be removed by heating, or by allowing the solution to be in contact with the atmosphere at room temperature, the solvent will **evaporate**, leaving the solute as a residue.

Example

Evaporation – clothes drying

A liquid (eg water) attached to a porous material (eg clothing) can evaporate from the material to leave the material (clothing) dry.

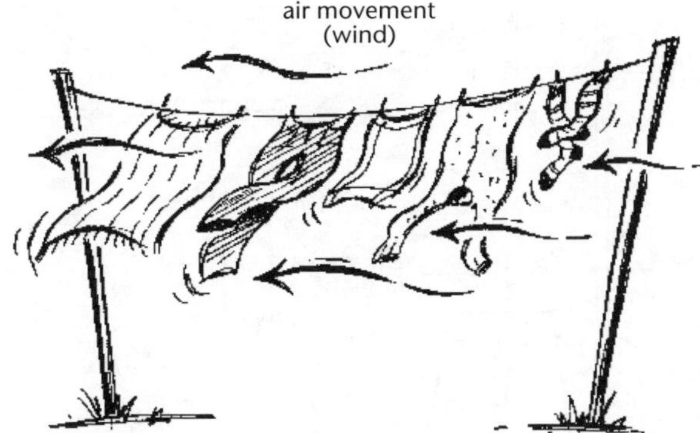

Wet clothes lose water by evaporation. The water does not boil – water particles escape from the surface of the water on the clothing and are swept away by the air movement. This allows more water particles to escape and the clothing becomes dry.

Clothes drying by evaporation

Example

Evaporation – crystals forming from solution

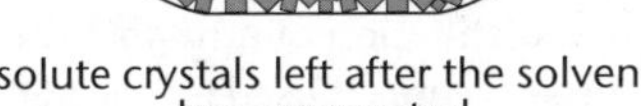

Solvent evaporates from the solution and the solute particles get closer together and form crystals.

Crystals forming from solution through evaporation

Distillation and fractional distillation

Distillation is used to separate the solvent of a solution from the solute by heating the solution so that the solvent boils. Solvent vapour is led off and condensed to produce the solvent in a pure state. The solute is left in the **distillation flask**.

Example

Distillation of a salt/water solution

When a salt/water solution is distilled using the apparatus below, the water (solvent) boils, and is led away as steam to be collected in the **conical flask**. The salt (solute) remains in the round-bottom flask.

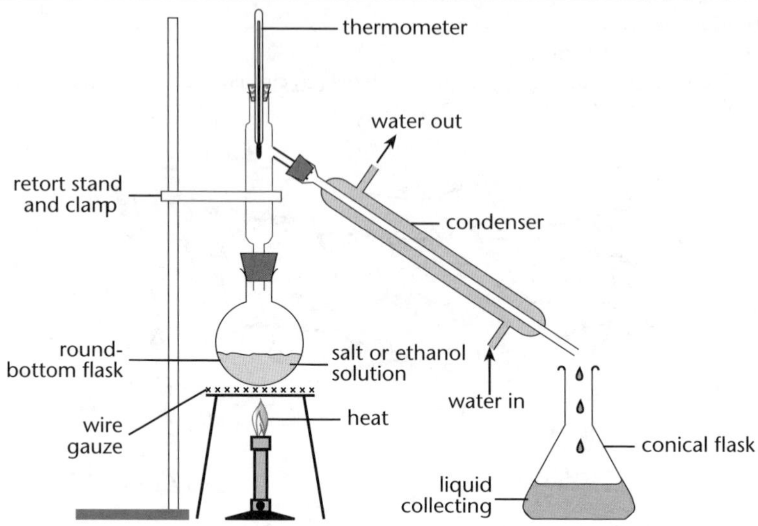

Distillation apparatus

Fractional distillation is used to separate two or more liquids whose boiling points are different – the greater the difference between the boiling points, the more effective the separation. The liquids are **vaporised** by heating, and the more **volatile** (lower boiling point) liquid or fraction is led away as a vapour and condensed to a liquid.

Example

Fractional distillation of ethanol/water solution

When the apparatus is used for the fractional distillation of an ethanol/water solution, both liquids will boil and the resulting vapour will rise. Ethanol has the lower boiling point and its vapour will rise all the way up the column and then pass into the condenser, where it will condense and be collected as a liquid in the conical flask. Water vapour condenses in the column before it can reach the condenser and is returned to the round-bottom flask.

Chromatography

Solutions that contain many solutes can be separated by applying a small quantity of the solution to a porous, absorbent material, eg paper, in an enclosed container:

- The solvent flows through the paper.
- The solutes move at different speeds along the paper.

When the solvent has reached the end of the paper, it will evaporate and eventually saturate the enclosed atmosphere. Each solute will remain in a 'fixed' position. The final appearance of the separated solute components is called a **chromatogram**. The solutes can be identified by various tests (eg each solute has a unique **colour**).

Example

Identifying the colours in 'black' ink

A spot of liquid black ink is placed on a piece of porous paper, eg filter paper. When water flows across the paper (by **capillary attraction**), several different colours will emerge – none of them black!

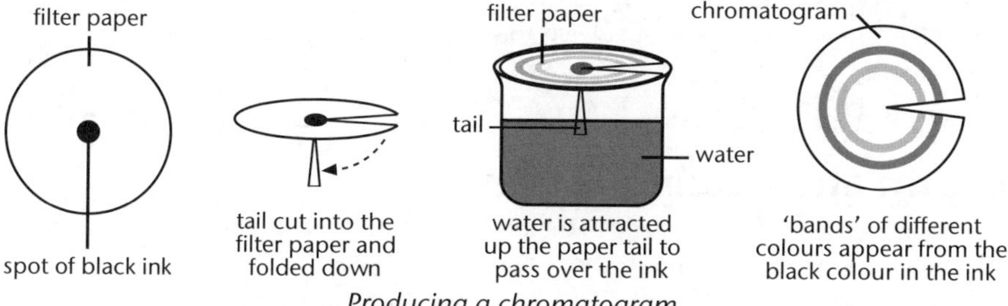

Producing a chromatogram

Best separation is achieved if the development of the chromatogram occurs in an enclosed space.

Unit 11.1 Activity 1C: Separating mixtures

1. Explain, with the aid of an example, the meaning of the following terms:

a. Solute.

b. Volatile.

c. Evaporation.

d. Insoluble.

2. *Water is placed in a beaker and a mixture containing white crystals and yellow grains of solid is added. After the mixture has been stirred, it is poured through filter paper, supported by a funnel, to obtain a clear solution – yellow grains of solid were left on the filter paper. The clear solution was left in an open dish in a warm room. Some hours later, the open dish contained dry white crystals.*

In the above description of an experiment, identify:

a. The solvent used.

b. A solute.

c. An insoluble material.

d. The name of the process used to obtain the dry white crystals.

3. Match the processes below with the problems stated. Use each process only once.

Process	Problem
a. Distillation	**i.** Investigating a blue dye to see if it was a single colour.
b. Evaporation	**ii.** Recovering petrol from fuel that has sand in it.
c. Chromatography	**iii.** Obtaining salt from sea water.
d. Decantation	**iv.** Recovering excess copper sulfate from a saturated solution
e. Filtration	**v.** Obtaining pure water from tap water.

4. Describe a process you could carry out in a school laboratory that would enable you to obtain pure samples of salt and sand from a mixture of the two substances. Include in your description a method for recovering any solvent that you may use. Draw and label diagrams of the apparatus you would use for both processes.
5. **a.** State the processes used to convert petroleum into petrol, diesel oil, fuel oil, bitumen, etc.
   b. Explain the process used to obtain common salt from sea water. Give the names of the processes used, eg melting, dissolving, filtration, condensation, etc and explain what physical changes the process involves.

Heating and cooling curves

Pure substances and mixtures respond to increases in temperature by absorbing heat. Graphs can describe what happens when substances are heated or cooled.

The graph below shows the heating curves of pure substances and mixtures.

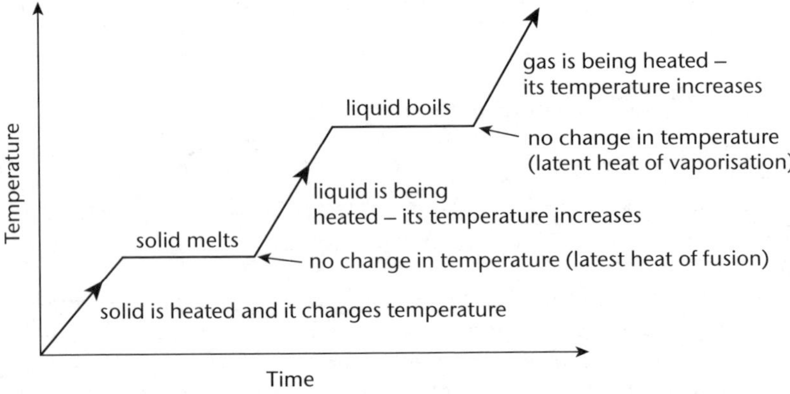

The heating curve for a solid substance

- A solid substance is heated uniformly. It absorbs heat and its temperature rises.
- The temperature of the solid substance continues to rise uniformly until it reaches the melting point of the solid.
- At the melting point, the heat being absorbed by the substance is used to melt the solid, without any further increase in temperature. All the energy is absorbed in the weakening of the forces that hold the molecules between the particles in the solid, and the solid turns to liquid. (This is called the latent heat of fusion.)
- The temperature does not increase again until all the solid has turned to liquid. The temperature of the liquid then starts to rise again, uniformly, until it reaches the boiling point of the liquid.
- At the boiling point, the heat being absorbed by the liquid is used to boil the liquid, without any further increase in temperature. All the energy is absorbed in the weakening of the forces that hold the molecules between the particles in the liquid, and the liquid turns to vapour. (This is called the latent heat of vaporisation.)

- There will be no increase in temperature until all the liquid has been turned vaporised to gas; at that point, the temperature will again increase.

Pure substances and mixtures also respond to decreases in temperature by losing heat. The graph below shows the cooling curves of pure substances and mixtures.

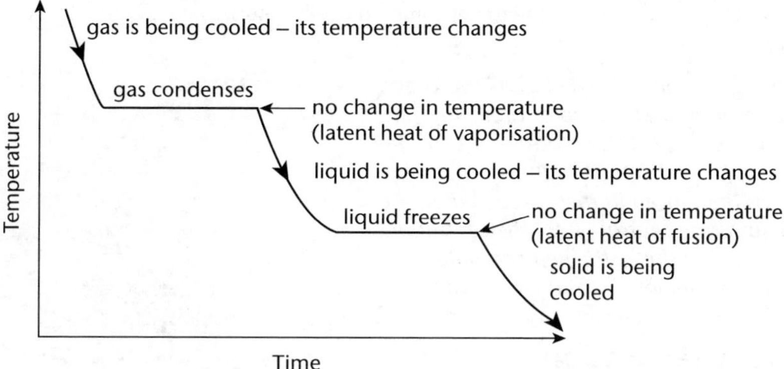

The cooling curve for a gas substance

- A gas substance is cooled uniformly. It loses heat and its temperature falls.
- The temperature of the gas continues to fall uniformly until it reaches the boiling point.
- At the boiling point, the coolness and loss of energy cause the gas to condense and the gas to turn to liquid. There is no further decrease in temperature until all the gas has been turned to liquid. In turning to liquid, the forces holding the molecules between the particles in the gas are weakened.
- The temperature does not decrease again until all the gas has turned to liquid. The temperature of the liquid then starts to reduce again, uniformly, until it reaches freezing point.
- At the freezing point, the heat being lost by the liquid strengthens the forces holding the molecules between the particles of the liquid, and the substance changes from liquid to solid. There will be no further decrease in temperature until all the liquid is turned to solid – at that point the temperature will again decrease.

The *Oxford Advanced Learner's Dictionary* defines latent as 'existing, but not yet very noticeable, active or well developed'. At the flat parts of the graphs above, there is no change in temperature because instead of changing temperature the energy is being absorbed by the change in the substance from solid to liquid (latent heat of fusion) or vice versa; or a change from liquid to vapour, or vice versa (latent heat of vaporisation).

Other methods of separation

The application of simple scientific principles and processing has lead to the application of many other separation methods that are widely used in laboratories and industries.

Centrifuging

A *centrifuge* is a machine that uses a spinning motion and centrifugal force to isolate and separate particles in a solution. Old models were spun by hand. A milk separator is a simple example of a machine that uses centrifugal force to separate milk into cream and skimmed milk.

The centrifuge works on the principle that lighter particles will tend to be moved away from the heavier particles. There are many different types of centrifuges but they all use the principle of centrifugal force which causes objects flying at speed around the centre to fly away from the centre and off a circular path.

Centrifuges (such as the one pictured) are common in laboratories and are used for separating solutions.

A centrifuge commonly used in a laboratory. It has a rotating unit, called a rotor, with fixed holes at the edge of the central plate, at an angle to the vertical. Test tubes containing solutions are placed in these slots. When the motor spins the central plate, the centrifugal force created drives the heavier material towards the bottom of the test tubes.

Magnetic separation

All materials possess magnetic properties. Some materials have strong magnetic properties, while others have weak magnetic properties. Magnetic force can be used to attract or repel substances.

Magnetic separation is a process that uses magnetic force to separate strongly magnetic material from a mixture. This separation technique is particularly useful in the mining industry. Iron, for example, is highly magnetic, and magnetic separators are often used to separate iron **ore** from the raw ore excavated from a mine.

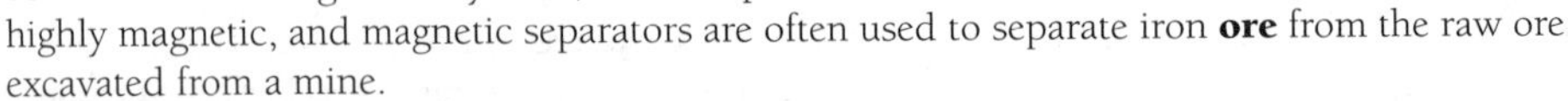

Magnetic separators are specially built machines that use magnetic force to separate mixtures. There are many different types of magnetic separators. Plate magnets, for example, are often suspended above conveyer belts to attract and remove particular materials moving along the belt.

A Bunting Suspended Plate Magnet. This is an example of a magnet that is used to remove fine particles of metal from other materials moving along a conveyor belt. Magnets like these are used in many industries such as food processing and chemical manufacturing.

Crystallisation

Crystallisation is another separation technique. It is used to separate a solid substance that has been dissolved in a liquid to form a solution. The solution is heated and the solid substance dissolves in it. Heat is maintained so that excess solvent can evaporate, leaving behind a **saturated** solution. A solution is said to be saturated when it contains the greatest possible amount of the substance that has been dissolved in it.

As the saturated solution is allowed to cool, crystals of the substance start to grow. These crystals can then be collected and allowed to dry and the excess solvent can be poured away.

Crystallisation process:

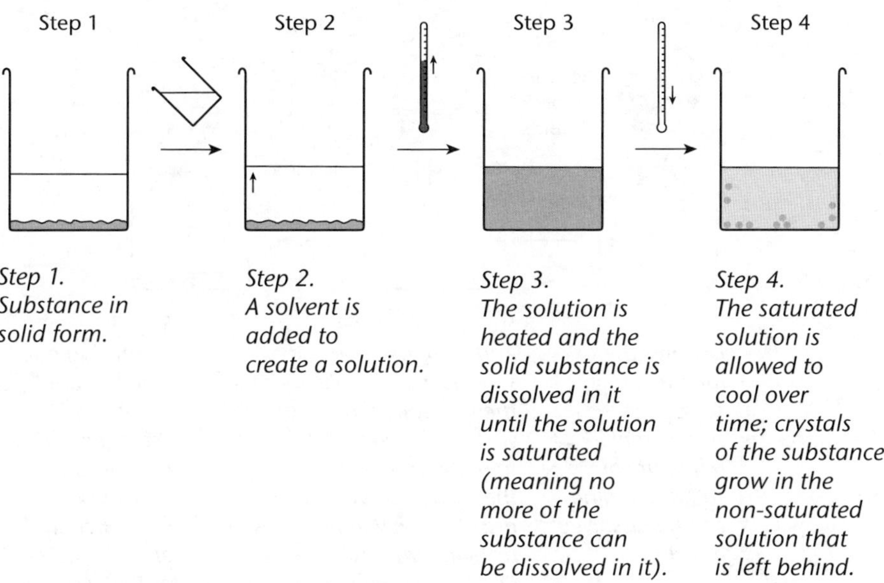

The application of this simple process has lead to the creation of industrial crystallisers that use cooling and evaporation. Cooling crystalliser tanks are used in the processing of pharmaceuticals. They are also used in the production of sugar. Evaporative crystallisers are used in various industries.

Solvent extraction

This is another separation technique, the application of which is used widely in industry. It is used in nuclear processing, mining, and the production of products such as perfumes, vegetable oils and biodiesel.

It is also known as 'liquid-liquid extraction' because the technique uses two liquids that do not dissolve in each other. (Liquids that do not dissolve in each other are called immiscible liquids.) The technique is performed using a separatory funnel.

A simple separating funnel is a glass funnel with a tap at the bottom. A mixture of liquids is placed inside the funnel and a container is placed underneath. The liquid with the higher **density** sinks to the bottom and the liquid with the lower density floats on top. When the tap is opened, the liquid with the higher density starts to flow into the container. The tap is then closed just before the liquid with the lower density starts to flow through. The liquid with the lower density remaining in the separating funnel can then be drained into a different container to separate the two liquids. Evaporation can then be used to recover solids from the two separate liquids.

Solvent extraction process:

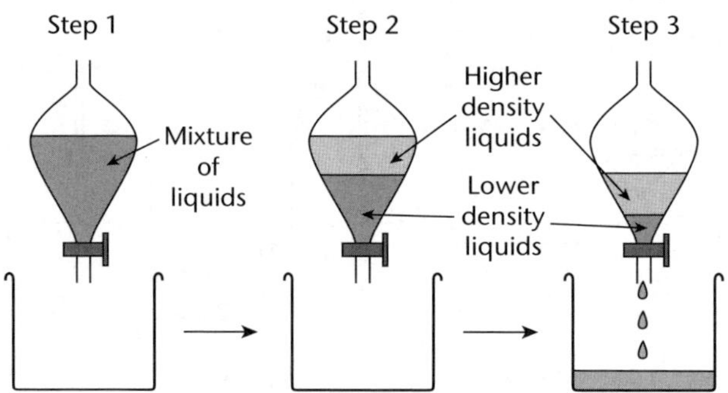

Step 1.
Dissolve the mixture in solvent. For example, a mixture of salt and iodine can be dissolved in water.

Step 2.
Add a suitable second solvent that dissolves one of the two solids. For example, we could select tetrachloromethane as the second solvent since it dissolves only iodine as well as being immiscible in water. Over time, two layers should emerge, one a colourless solution of salt in water and the other a purple solution of iodine in tetrachloromethane.

Step 3.
Decanting one layer (the bottom layer) separates it from the other. Evaporating the two layers separately will ensure recovery of the two solids.

Our lives depend on the fact that certain substances dissolve in other substances. Our bodies grow healthy because nutrients dissolve in our blood and can reach the cells and tissues. Fish can live in our rivers and oceans because oxygen dissolves in water. We enjoy certain soft drinks because the carbon dioxide dissolved in them makes them 'fizzy'. Water is a substance that dissolves many other substances, and that makes it one of our most important chemical substances.

Unit 11.1 Activity 1D: Matter and separations

1. Explain the following methods:
   a. Centrifuging.
   b. Magnetic separation.
   c. Crystallisation.
   d. Solvent extraction.

2. Which method would you use to:
   a. Separate suspension?
   b. Purify an organic compound?
   c. Process vegetable oil?

3. Classify the following substances as either homogeneous or heterogeneous:
   a. Sea water.
   b. Coppersulfate solution.
   c. Silver metal.

   Give reasons for your answer and identify whether each is an element, a compound or a mixture.

4. Explain the following terms:
   a. Evaporation.
   b. Heat of fusion.
   c. Boiling point.
   d. Intermolecular forces.

5. Explain the following terms:
   a. Freezing point.
   b. Latent heat of fusion.
   c. Super-cooling of liquids.

Unit 11.1 Application of Physical Processes
Topic 2: Diffusion of solids, liquids and gases

In line with the Syllabus p. 9, Topic 2 explains the term 'diffusion' and its significance in chemistry. Specifically, this Topic looks at:

- Diffusion of solids in water.
- Diffusion of gases.
- Graham's law.

Introduction

There are several different meanings for the word '**diffuse**'. If something is diffused, it means that it is spread out, or it becomes spread widely in all directions. If the light in a room is diffused, it is spread in many directions, which means that it shines less brightly. If people speak about the diffusion of ideas, they are describing the spread of ideas, eg 'Ideas about nationalism or racism can diffuse from propagators to others who initially may have rejected such views'.

In science, 'diffusion' describes the movement of particles from regions of higher **concentration** to regions of lower concentration. The result of diffusion is a gradual mixing of material. If temperature is kept constant and there are no other factors affecting the particles, the diffusion process will eventually result in complete mixing of particles.

If a gas or a liquid diffuses or is diffused in a substance, it means that it becomes *slowly mixed* with that substance. Diffusion in liquids and gases takes place over a relatively short period of time because the particles in liquids and gases are loosely packed together and are free to move around.

The rate of diffusion in liquids and gases is much faster than in solids. In solids, the particles are packed closely together and are not free to move around, so diffusion takes a very long period of time. For example, if gold metal and lead metal are placed in contact with each other, molecules of lead will eventually diffuse into the gold and molecules of gold will eventually diffuse into the lead. This will take thousands of years. By applying intense pressure, this diffusion can be made to occur in a few years.

Electron diffusion occurs in semi-conductors. If the electron densities in any two regions are different, then there is a net flow of electrons across the plane separating the regions.

Another form of diffusion occurs with thermal energy, between substances at different temperatures. For example, if a beaker of cold water is placed in a water bath containing water at a temperature of 90 °C, then after some time heat energy will be transferred from the hot water in the bath to the cold water in the beaker until the temperatures of the water in the bath and the water in the beaker are the same.

Diffusion is a transfer of energy and matter and is common to all particles of matter and all states of matter: solids, liquids and gases.

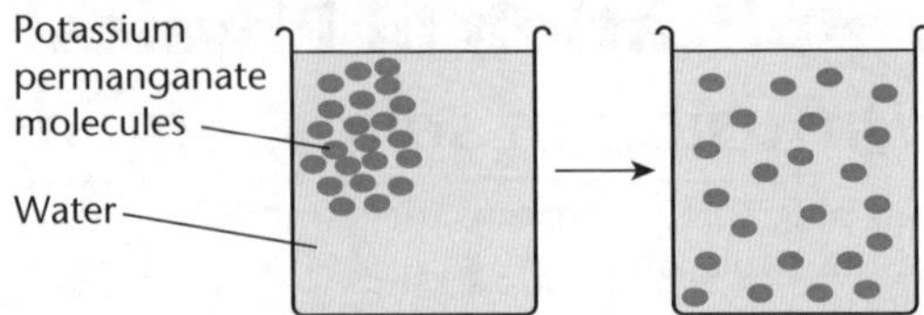

If particles of the purple salt potassium permanganate ($KMnO_4$) are dropped into a beaker of water, the particles initially remain clustered together in one corner of the beaker. Over time, the $KMnO_4$ particles diffuse in the water and become randomly and uniformly distributed, giving the solution a purple colour.

The term 'molecular diffusion' is used to describe the diffusion between two substances. It can be compared to the movement of tiny dust particles seen in a sunbeam: they seem to move under their own energy as well as by bouncing off each other. The diffusion of molecules between two substances is similar: small molecules are self-propelled by thermal energy (heat) and larger molecules are driven by collisions with other particles – but it is invisible in the diffusion of the two substances.

When the fossils of animals and leaves lie in contact with stones and rocks for a number of years, they leave an impression on the stones or rocks. This is because of the diffusion of the particles of the organic substances into the stones, which are in contact. This is especially true of a type of fossil called moulds. On decaying, the original remains completely dissolve and diffuse into the rocks, leaving behind the impression.

Diffusion in gases: Graham's law

Diffusion in gases is the rate at which two gases mix. These gases may be at the same temperature and pressure. It is not to be confused with 'effusion', which is the rate at which a gas at high pressure escapes through a pinhole into a vacuum or a region of lower air pressure.

For example when an inflated balloon is pricked, air at high pressure effuses out of the pin hole to the surrounding environment of lower air pressure.

The diffusion rate of gases is a function of temperature. The higher the temperature, the faster the gas molecules diffuse. Diffusion rate is not affected by the concentration of molecules between gases.

Graham's law of diffusion states that the absolute rate at which a gas diffuses is inversely proportional to the square root of its density.

$$\text{Rate of diffusion of gas} \propto \frac{1}{\sqrt{\text{density of gas}}}$$

Equal volumes of different gases contain the same number of particles (**Avogadro's hypothesis**), so the number of **moles** per litre at a given temperature (T) and pressure (P) is constant. Therefore, the density of a gas is directly proportional to its **molar mass** (MM) and thus the absolute rate is inversely proportional to the square root of its molar mass.

$$\text{Rate of diffusion of gas} \propto \frac{1}{\sqrt{\text{molar mass of gas}}}$$

Graham's law of diffusion can be used to compare the rates of diffusion (r) for two different gases. The following relationship holds:

$$\frac{\text{Rate of diffusion of gas A}}{\text{Rate of diffusion of gas B}} = \frac{\sqrt{\text{molar mass of gas B}}}{\sqrt{\text{molar mass of gas A}}}$$

$$\frac{r_A}{r_B} = \left(\frac{MM_B}{MM_A}\right)^{\frac{1}{2}}$$

The ratio $\frac{r_A}{r_B}$ tells us how fast gas A travels compared to gas B.

Example

If you pop a balloon of hydrogen and a balloon of carbon dioxide at the same time, which gas will reach the end of the room first?

Knowing the molar masses of H_2 (2 g/mol) and CO_2 (44 g/mol) and using Graham's law, we obtain:

$$rH_2 = rCO_2 \times \left(\frac{44}{2}\right)^{\frac{1}{2}} = 4.7 \times rCO_2$$

This means that the rate of diffusion of H_2 is 4.7 times greater than that of CO_2 and H_2 will reach the end of the room first.

Example

A sample of nitrogen and a sample of ammonia are both at the same temperature. What would their comparative rates of diffusion be?

Knowing the molar masses of N_2 (17 g/mol) and NH_3 (28 g/mol) and using Graham's law, we obtain:

$$\frac{r_{N_2}}{r_{NH_3}} = \left(\frac{28}{17}\right)^{\frac{1}{2}} = 1.28$$

This means that the rate of diffusion of N_2 is 1.28 times greater than that of NH_3.

Unit 11.1 Activity 2A: Diffusion of solids, liquids and gases

1. Which of the following best describes diffusion?

a. It only occurs in liquids.

b. It is common to all matter.

c. It is fastest in solids and liquids.

d. It occurs only in gas phase.

2. Which of the following is *not* true of diffusion?
   - **a.** It is a concept only found in science.
   - **b.** It is driven only by differences in concentration.
   - **c.** It occurs even if there is no concentration gradient.
   - **d.** It is affected by thermal energy of particles.
3. True or false? In molecular diffusion, the moving particles are small molecules that are self-propelled by thermal energy and do not require a concentration gradient to spread out through random motion.
4. True or false? When helium (4 amu) and methane (16 amu) are released at the same time in a diffusion tube 16 metres long, the helium particles will diffuse four times faster than methane.
5. True or false? From Graham's law, the correct expression relating the rate of diffusion (r_x) and the gram molar mass (MM) is: $(r_x)^2 MM = K$.
6. True or false? If gas A takes 12 minutes to diffuse through a tube and gas B takes 8 minutes, then the expression relating the gram molar mass of A to B is:

$$\frac{MM_B}{MM_A} = \left(\frac{8}{2}\right)^2$$

7. True or false? If the absolute rate of diffusion of hydrochloric acid gas is found to be 1.40×10^{-4} m/s, then the absolute rate of diffusion of nitrogen gas under the same conditions is 1.597×10^{-4} m/s (given N = 14.0067 amu, H = 1.0080 amu, Cl = 35.4530 amu).

Unit 11.1 Application of Physical Processes

Topic 3: Behaviour of gases

In line with the Syllabus p. 9, Topic 3 looks at the behaviour of gases. The Topic begins with the kinetic theory of ideal gases and then moves on to examine:

- Boyle's law.
- Charles's law.
- Combined gas law.
- Ideal gas law.
- Gas measurements.

The kinetic theory of gases

A gas is more easily compressed or expanded than a liquid or a solid. This means that changes in pressure and temperature will have a greater effect on the volume of a gas.

We shall begin our study by looking at current kinetic theory of ideal gases. The kinetic theory of ideal gases states the following:

- A gas is made up of **molecules** (eg oxygen gas is made up of molecules of the formula O_2).
- The molecules of a gas are in constant random motion and collide with each other and the side of the container.
- All collisions are perfectly elastic (the molecules rebound – they are not stuck together) and no kinetic energy is lost.
- The pressure of the gas is caused by collisions of molecules with the sides of the container.
- There are no forces (attraction) between the molecules because the molecules are too far apart.
- The volume of the gas particles is negligible (too small to observe) compared to the total volume the gas occupies.
- The absolute temperature of the gas is a measure of the average kinetic energy of the molecules.

In summary, the ideal gas model describes a system where the molecules are in constant motion, all collisions are elastic and the volume of the molecules is insignificant.

However, no gas obeys the ideal gas conditions over a large range of temperature and pressure. If they did, no gas could be liquefied by compression and cooling.

Standard conditions

Gas properties can only be compared when measured under standard conditions: standard temperature and pressure (STP) or standard laboratory conditions (SLC).

Standard temperature and pressure (STP) is:
0 °C and 1 atmosphere

or
273 K and 101.3 kPa

Standard laboratory conditions (SLC) are:
25 °C and 1 atmosphere

Boyle's law

Robert Boyle (1627–1691) was an English scientist. He found that, at a constant temperature, the volume of a fixed amount of gas decreases when the pressure on it is increased. Conversely, the volume increases when the pressure is decreased.

Boyle's law states that at a constant temperature the volume of a given mass of gas is inversely proportional to the pressure:

$$P \propto \frac{1}{V}$$

so PV = a constant, and $P_1V_1 = P_2V_2$

where: P is the pressure in kPa

V is the volume in litres

Thus for a fixed amount (moles) of a gas at constant temperature, the volume of the gas will be greater at lower pressure. Alternatively, the volume of gas will decrease as the pressure of the gas increases.

Boyle's law can be expressed graphically as follows:

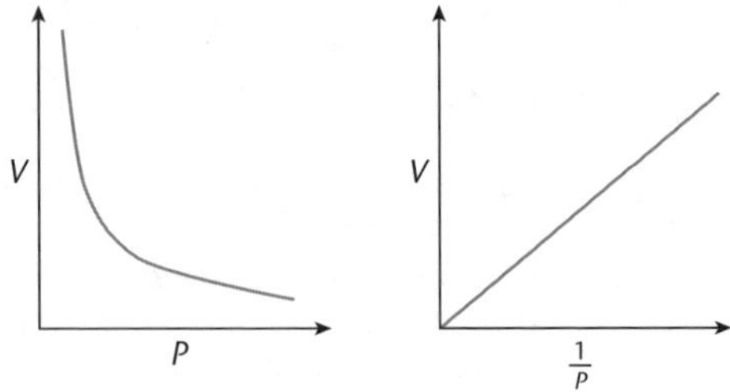

Pressure and volume relationship according to Boyle's law. A plot of pressure against volume yields a hyperbolic function. A plot of reciprocal of pressure against volume yields a linear function.

Example

A fixed amount of a gas at constant temperature has a volume of 400 mL and pressure of 1.05 atm. Calculate the volume of the same gas at 1.01 atm.

Using Boyle's law, $P_1V_1 = P_2V_2$

P_1 = 1.05 atm, V_1 = 400 mL, P_2 = 1.01 atm, V_2 = ?

1.05 atm × 400 mL = 1.01 atm × V_2

Therefore, V_2 = 415.8 mL

Thus, the gas will occupy 415.8 mL at 1.01 atm.

Charles's law

Jacques Charles (1746–1823) was a French chemist. He found that the volume of a fixed amount of gas at a fixed pressure increases when its temperature increases. Conversely, the volume decreases when the temperature is decreased.

Charles's law states that at a constant pressure the volume of a given mass of gas is directly proportional to the temperature:

$$V \propto T$$

$$\text{so } \frac{V}{T} = \text{a constant, and } \frac{V_1}{T_1} = \frac{V_2}{T_2}$$

where: V is the volume in litres

T is the absolute temperature in kelvin.

Thus for a fixed amount (moles) of a gas at constant pressure, the volume of the gas will be greater at higher temperature. Alternatively, the volume of gas will decrease as the temperature of the gas decreases.

Charles's law can be expressed graphically as follows:

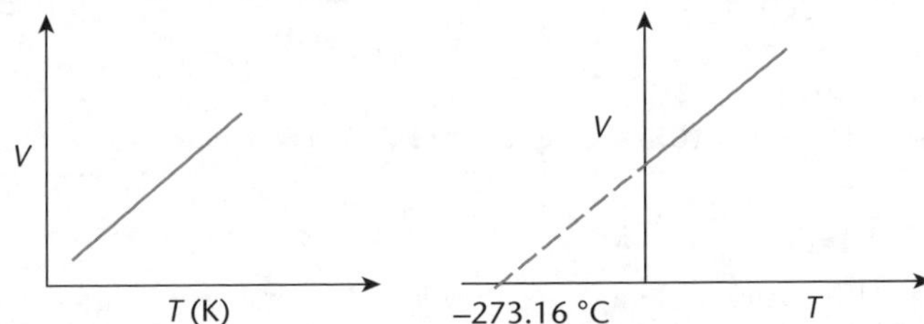

Temperature and volume relationship according to Charles's law. A plot of temperature against volume yields a linear function. By extrapolating so that the volume becomes zero, the absolute temperature can be deduced.

Example

A fixed amount of a gas at 20 °C and constant pressure has a volume of 650 mL. Calculate the volume of the same gas at standard temperature.

Using Charles's law, $\frac{V_1}{T_1} = \frac{V_2}{T_2}$

$T_1 = 20 + 273$ K, $V_1 = 650$ mL, $T_2 = 0 + 273$ K, $V_2 = ?$

$\frac{650}{293} = \frac{V_2}{273}$

Therefore, $V_2 = 605.6$ mL

Thus, the gas will occupy 605.6 mL at 0 °C.

Combined gas law

Boyle's law and Charles's law can be combined to form the combined gas law:

$$P_1V_1 = P_2V_2 = \text{a constant, and } \frac{V_1}{T_1} = \frac{V_2}{T_2} = \text{a constant}$$

$$\text{so } \frac{P_1V_1}{T_1} = \frac{P_2V_2}{T_2}$$

where: P is the pressure in kPa

V is the volume in litres

T is the absolute temperature in kelvin.

Example

Air in a weather balloon occupies 10 000 litres at STP. What volume will it occupy at a height where the pressure is 99.3 kPa and the temperature is –5 °C?

Using the combined gas law, $\frac{P_1V_1}{T_1} = \frac{P_2V_2}{T_2}\ \frac{V_1}{T_1} = \frac{V_2}{T_2}$

$P_1 = 101.3$ kPa, $P_2 = 99.3$ kPa, $T_1 = 0 + 273$ K, $V_1 = 10\ 000$ L, $T_2 = -5 + 273$ K, $V_2 = ?$

$$\frac{101.3 \times 10\ 000}{273} = \frac{99.3 \times V_2}{273}$$

Therefore, $V_2 = 10\ 014.6$ L

Thus, the air will occupy 10 014.6 L at pressure 99.3 kPa and temperature –5 °C.

Ideal gas law

Émile Clapeyron (1799–1864) was a French physicist. He combined Charles's law with Boyle's law to form the ideal gas law. The ideal gas law is a more general form of the combined gas law because it considers all the variables of temperature, pressure, volume and number of moles.

The ideal gas law can be expressed as:

$$PV = nRT$$

where: P is the pressure in kPa
V is the volume in litres
T is the absolute temperature in kelvin
n is the number of moles of gas
R is the gas constant ($R = 0.0821$ L atm mol^{-1} K^{-1} or 8.3145 J K^{-1} mol^{-1})

$R = kN_A$, where k is the Boltzmann constant and N_A is Avagadro's constant.

When pressure, volume or temperature are given in units other than atmospheres, litres or kelvin, you need to convert the units before using the value R in the ideal gas law.

Example

A sample of oxygen at 24.0 °C and 745 torr has a volume of 455 mL. How many grams of oxygen were in the sample?

We are given the volume, pressure and temperature of the gas. Use the ideal gas equation ($PV = nRT$) to calculate the number of moles, n.

To use $R = 0.0821$ L atm mol^{-1} K^{-1} we must have V in litres, P in atmospheres and T in kelvin:

$$P = \frac{745 \text{ torr} \times 1 \text{ atm}}{760 \text{ torr}} = 0.980 \text{ atm}$$

$V = 0.455$ L

$T = 24 + 273 = 297$ K

Solve for n:

$$n = \frac{PV}{RT} = \frac{(0.980 \text{ atm})(0.455 \text{ L})}{(0.0821 \text{ L atm mol}^{-1} \text{ K}^{-1})(297 \text{ K})}$$

$= 0.0183$ mol of O_2

Molecular mass of O_2 is 32.00. So 1 mol of O_2 is 32.00 g.

Convert mol O_2 to grams of O_2:

$$\frac{0.0183 \text{ mol} \times 32.00 \text{ g}}{1 \text{ mol}} = 0.586 \text{ g of } O_2$$

Thus, the sample of oxygen had a mass of 0.586 g.

Additional laws

There are other laws that apply to gases. Let us consider a few common ones.

Joseph Louis Gay-Lussac (1778–1850) was a French chemist. He formulated the general nature of the relationships of gas volumes, expressed in simple, whole number ratios.

Gay-Lussac's law states that the pressure of a fixed amount of gas held at a constant volume is directly proportional to the kelvin temperature.

$$P \propto T$$

$$V \text{ and } m \text{ are constants and } \frac{P}{T} = \text{a constant}$$

where: T is the absolute temperature in kelvin

P is the pressure.

Example

Hydrogen gas (1 volume) reacts with chlorine gas (2 volumes) to give gaseous hydrogen chloride (2 volumes) when the gases are at the same temperature and pressure.

Amedeo Avogadro (1776–1856) was an Italian chemist. He formulated a law that states that all gases of fixed number of molecules have the same temperature and pressure.

Avogadro's law states that equal volumes of all gases at the same temperature and pressure contain the same number of molecules. This means that the gas volumes and number of moles must be in the same ratio.

The molar volume is the volume occupied by 1 mole of the gas. Experimental data verifies that 1 mole of any gas occupies a volume of 22.4 L at STP (0 °C and 1 atmosphere). The following table gives the molar volumes of some gases at STP.

Molar volumes of some gases at STP		
Gas	Formula	Standard molar volume (L)
Helium	He	22.398
Argon	Ar	22.401
Hydrogen	H_2	22.410
Oxygen	O_2	22.414
Carbon dioxide	CO_2	22.414
Nitrogen	N_2	22.413

John Dalton (1766–1844) was an English scientist who is known for his work in the development of **Atomic Theory** and his research into colour blindness. Dalton discovered how gas laws apply to gas mixtures and formulated it in his law of partial pressure.

Dalton's law of partial pressure states that the total pressure of a mixture of non-reacting gases is the sum of their individual partial pressures:

$$P_{total} = P_a + P_b + P_c + \ldots$$

Example

In dry air, the pressures of oxygen, nitrogen and argon are 159.12 torr, 593.44 torr and 7.10 torr, respectively. A diver wants to fill a 4.00 L pressurised tank with 50.0 g of O_2 and 150 g of N_2. What will the total pressure have to be at 25° C? (R = 0.0821 L atm mol^{-1} K^{-1})

First calculate the number of moles of the gases ($n = \frac{m}{M}$).
Thus, $nO_2 = 1.56$ and $nN_2 = 5.36$.
Next calculate partial pressures of each gas using the ideal gas equation: $P = \frac{nRT}{V}$.
Thus, partial pressure of O_2 is 9.54 atm and partial pressure of N_2 is 32.8 atm.
Finally use Dalton's law to calculate the total pressure:

$$P_{total} = P_a + P_b = 42.3 \text{ atm}$$

Is this answer reasonable?

Check! Note total moles of gas is close to 7 mol. So the total volume is close to 7 mol × 22.4 L/mol at STP (160 L). Thus, compressing the gas from 160 L to 4 L is a factor of 1/40. So the pressure is 40 times 1 atmosphere.

Using the gas laws – a summary

Which gas laws shall I use? This is a common question students ask. The answer depends on the data supplied. First apply some reasoning to the data in the statement of problem. Is the amount of gas sample stated either in moles or in a mass unit? Or does the problem call for the calculation of the amount of gas? There is only one gas law that includes moles – the ideal gas law. So, if the amount of gas is given in moles, then that is the law to use.

If the size of the gas sample is not given, either in moles or in mass units, and it is not required, then the combined gas law is the most likely gas law to apply. This law simplifies to

Boyle's law (if the number of moles (n) and temperature (T) are constant), Charles's law (if n and pressure (P) are constant) or Gay-Lussac's law (if n and volume (V) are constant).

Determining the molar mass of a gas

When a chemist makes a new compound, its molar mass is usually determined to establish its chemical identity. If the compound is a gas, its molecular mass can be found using experimental values of pressure, volume, temperature and sample mass together with the ideal gas law.

Suppose, for example, that a gas has been trapped in a sealed bulb at known values of P, V, T and mass. The ideal gas law equation now lets us use the values of P, V and T to compute the number of moles in the sample by the equation:

$$n = \frac{PV}{RT}$$

Then we calculate the ratio of *grams to moles* to find the molecular mass.

Example

A student collected a sample of a gas in a 0.220 L gas bulb until its pressure reached 0.757 atm at a temperature of 25.0 °C. The sample weighed 0.299 g. What is the molar mass of the gas?

First substitute the data into the ideal gas law equation $PV = nRT$.

Then rearrange to solve for n.

$$n = \frac{PV}{RT}$$

$R = 0.0821$ L atm mol^{-1} K^{-1}, $V = 0.220$ L, $T = 298$ K, $P = 0.757$ atm

$$n = \frac{PV}{RT} = \frac{(0.757 \text{ atm})(0.220 \text{ L})}{(0.0821 \text{ L atm mol}^{-1} \text{ K}^{-1})(298 \text{ K})}$$

$= 6.81 \times 10^{-3}$ mol

The ratio of grams to moles $= \dfrac{0.299}{0.00681} = 43.9$ g mol^{-1}

Unit 11.1 Activity 3A: Behaviour of gases

1. Express the following gas laws in equation form:
   a. Temperature–volume law
   b. Pressure–volume law
   c. Temperature–pressure law
2. What is meant by an ideal gas?
3. State the ideal gas law in the form of an equation.
4. What is the value of the gas constant in units of L atm mol^{-1} K^{-1}?
5. What is the value of the gas constant in units of L torr mol^{-1} K^{-1}?
6. What is the value of the gas constant in units of m^3 Pa mol^{-1} K^{-1}?

7. Using oxygen as an example, carefully illustrate the following concepts:

- **a.** 1 molecule
- **b.** 1 mole
- **c.** 1 molar mass
- **d.** 1 molar volume

8. At STP how many molecules of H_2 are in 22.4 L?

9. Explain in terms of the kinetic theory how raising the temperature of a confined gas makes its pressure increase.

10. In what way does Dalton's law of partial pressure provide evidence that the molecules in a gas move and behave independently of each other?

11. What postulates of the kinetic theory of gas are not strictly true, and why?

12. At a given temperature, how is the rate of diffusion of a gas related to its molecular mass?

13. A gas has a volume of 255 mL at 755 torr. What volume will the gas occupy at 365 torr if the temperature of the gas does not change?

Unit 11.1 Application of Physical Processes

Topic 4: Solubility of solids and gases in water

In Topic 4 we examine solubility, in particular:

- Solubility of substances in water.
- Calculation of solubility.
- Solubility of gases in water.
- Factors affecting the solubility of gases in water.

What is a 'solution'?

Usually the word 'solution' means the answer to a problem, but in chemistry, the word 'solution' is used to describe a mixture in which one substance is dissolved in another substance (usually a liquid but it can be a gas or a solid).

Solutions are important in chemistry because most chemical reactions occur in solution. Thousands of important industrial processes make use of solution chemistry. All life on Earth relies on vast complex chemical processes occurring in solutions.

A **solution** is a mixture of two or more pure substances: the solvent, which is present in the major proportion, and the solute, which is present in the minor proportion.

The **solvent** is usually a liquid in which a solid, liquid or gas substance is dissolved. Water is the most common liquid solvent used in the laboratory, in industry and in the home. It is an inorganic solvent. Organic liquid solvents such as **alcohols** and **esters** are found in drycleaning chemicals, nail polish removers and paint thinners.

The **solute** is the substance (solid, liquid or gas) that is dissolved in the solvent. So a solution is a mixture obtained when a solute is dissolved in a solvent. The process of a solute dissolving in a solvent is called *dissolution*.

It is possible to have solutions composed of several solutes. Common examples are tap water, tincture of iodine, tea, coffee and air.

An *unsaturated solution* is one that can dissolve more solute. A *saturated solution* is one in which no more solute can dissolve – it is in equilibrium with an excess of the solute at a given temperature.

Kinds of solutions

Most solutions consist of a solid that has been dissolved in a liquid, but there are many other types of solutions that involve different combinations of solids, liquids and gases. Air is an example of gases dissolved in another gas – nitrogen is the solvent and oxygen, carbon dioxide, water vapour, neon and other gases are the solutes. Metal alloys, such as bronze, are examples of solids dissolved in solids.

Properties of solutions

Pure substances have a set of characteristic physical properties, such as melting and boiling points, and vapour pressure at a given temperature. Solutions also exhibit these properties, but adding a solute to a solvent causes the values to change. Melting point, boiling point, vapour pressure and osmotic pressure show a predictable change as more solute is added to a solvent. They are known as *colligative properties* because they are dependent only upon the number of solute molecules in the solution, and not on the identity of the solute.

Melting point depression

A solution will have a lower melting point than the pure solvent. The decrease in melting point is proportional to the number of solute particles in the solution. The greater the number of solute particles (that is, the higher the concentration), the greater the melting point depression.

Boiling point elevation

A solution will have a higher boiling point than the pure solvent. The increase in boiling point is proportional to the number of solute particles in the solution. The greater the number of solute particles, the greater the boiling point elevation.

Vapour pressure depression

The vapour pressure of a liquid is the pressure of the vapour above the liquid that results from evaporation. The vapour pressure of a solution is less than that of the pure solvent. When water boils at 100 °C, its vapour pressure is equal to the atmospheric pressure allowing bubbles of steam to escape from the liquid state. If salt or sugar is added to the boiling water, it will stop boiling because the solution has a lower vapour pressure than pure water.

Osmosis

Osmosis is a colligative property of solutions that is important in biology, medicine and related areas. *Osmosis* is the diffusion of molecules through a semipermeable membrane – a membrane that allows one type of molecule to pass through but not others. So solvent molecules can pass through, but not solute molecules. Diffusion occurs from a solution with a low solute concentration to a solution with a higher solute concentration until there is an equal concentration on both sides of the membrane. Sometimes different kinds of molecules can pass through at different rates.

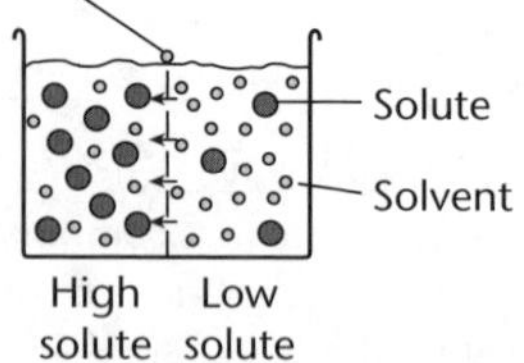

Osmosis is the diffusion of solvent molecules across a semipermeable membrane.

Osmosis is important in the transfer of molecules through animal cell membranes and plant cell walls. Osmosis plays a role in seed germination and the rising of sap into the branches and leaves of trees.

Concentration

The **concentration** of a solution is defined as the amount of solute in a given quantity of solvent. In science, we often speak of **concentrated** or **dilute** solutions. A concentrated solution contains more solute than a dilute solution.

The concentration of a solution is usually expressed in terms of the mass of solute per unit mass of the solvent. It does not imply anything about the amount of solute or solvent present; rather it gives the ratio of solute to solvent in terms of some convenient (and arbitrary) units.

Units of concentration include:

- Mass/mass (g of solute/kg of solvent).
- Molarity (moles of solute/L of solvent).
- Molality (moles of solute/kg of solvent).
- Mole fraction (moles of solute/moles of solvent).

Other units include grams of solute per 100 grams of water (g/100 g) and grams of solute per 100 mL of water (g/100 mL).

In the laboratory, it is usually more convenient to determine the volume of a sample of a liquid rather than its mass, so a more practical unit of concentration is the mass of solute in a given volume of the solution. For example, a sugar solution may contain 50 g of sugar per 100 mL of solvent.

Example

A salt solution is prepared by dissolving 2.85 grams of sodium chloride (NaCl) in 50 grams of water. The concentration of this solution could be expressed as 0.057 g NaCl/1 g water, 57.0 g NaCl/1000 g water or 57.0 g NaCl/1000 mL water.

Solubility

Solubility is a measure of the maximum amount of solute that can be dissolved in a given amount of solvent to form a stable solution. There is a limit to how much solute can be dissolved in a liquid solvent at a particular temperature. When the maximum amount of solute has been added and no more solute will dissolve at that temperature, the solution is said to be saturated. However, up to the solubility limit, solutions can be produced in any desired proportion.

Solubility varies for different substances. At 20 °C, 100 mL of water will dissolve up to 170 grams of the very soluble salt barium iodide (BaI), 35.7 grams of the moderately soluble sodium chloride (NaCl) but only 0.0014 grams of calcium carbonate ($CaCO_3$). Other salts, such as silver chloride (AgCl), are insoluble – they do not dissolve at all.

Sometimes insoluble substances appear to dissolve in a solvent but are actually taking part in a chemical reaction. For example, zinc is insoluble in hydrochloric acid. When zinc is added to hydrochloric acid, they react to form hydrogen and zinc chloride, which is then soluble in hydrochloric acid.

Some solvents and solutes can mix in all proportions, with no saturation point being reached. These substances are said to be **miscible**. For example, water and ethanol can mix in all proportions, as can a mixture of any two gases. Completely miscible substances mix together to give *homogeneous* solutions.

In a saturated solution, there is an **equilibrium** between the solute in solution and the undissolved solute. Solute particles (molecules, atoms or ions) are continuously going into solution and leaving the solution. The solution is saturated, so the total number of dissolved solute molecules cannot increase. This means dissolving must be balanced by an equal number of molecules of the solute leaving the solution and redepositing as excess solid solute.

At equilibrium these two processes occur at the same rate, so the amount of solute in solution stays constant. Therefore, we can say that, at a given temperature, a saturated solution is in equilibrium with an excess of the solute.

It is possible to make a *supersaturated solution* – one that contains an excess of the equilibrium amount of solute. This can be done by heating a saturated solution, adding more solute until no more will dissolve, and letting the solution cool again. Supersaturated solutions are unstable but can stay in solution if left undisturbed. Once disturbed, by agitation or the addition of a solid particle, the excess solute can be brought out (precipitate) of solution. The solution returns to its saturated state.

Unit 11.1 Activity 4A: Solubility

Use copper sulfate as a solute and water as a solvent to make up three different solutions of copper sulfate: unsaturated, saturated and supersaturated. Label the solutions accordingly.

Measuring solubility

Solubility is usually expressed as a concentration.

The disadvantage of using concentration to measure solubility is that it can be affected by the presence of other solutes in the solvent (eg the common ion effect).

Conditions affecting solubility

The solubility of a solute in a solvent is affected by pressure, temperature and the nature of the solute and solvent.

Pressure

Pressure does not have a great effect on the solubility of solids or liquids in liquid solvents. However, the solubility of a gas at a constant temperature is directly dependent on pressure. This is known as *Henry's law,* after William Henry (1774–1836), the English chemist who formulated it:

$$P = kc$$

where: P is the partial pressure of the gas in atmospheres
k is a constant
c is the concentration of the dissolved gas in mol/L.

The effect of pressure on a gaseous solute can be seen when you remove the top off a bottle of carbonated soft drink. Soda water is made by dissolving carbon dioxide gas in water under pressure. Removing the bottle top results in **effervescence**, or fizzing, as carbon dioxide leaves the solution and the pressure of the gas comes to equilibrium with atmospheric pressure.

Unit 11.1 Activity 4B: Conditions affecting solubility

1. Open a bottle or can of carbonated soft drink that has been kept in a refrigerator for at least a day.
2. Put about 125 mL of the soft drink into each of three microwave-safe containers.
3. Leave container 1 at room temperature. Stir this sample 10 times using a circular motion.
4. Put container 2 in a microwave oven and heat it for 30 seconds. Stir this sample 10 times using a circular motion.
5. Put container 3 in a microwave oven and heat it for 60 seconds. Stir this sample 10 times using a circular motion.
6. After 5 minutes, or when each sample is cool enough to drink, place your tongue in each solution and taste each one to assess the degree of fizziness. Record your observations.
7. Repeat step 6, but taste the samples in the reverse order.
8. Compare your observations with those of another student. Try to account for any differences.

Temperature

The solubility of gases in water is inversely proportional to temperature – as temperature increases, solubility decreases. There is no general relationship for the solubility of liquids in liquids: solubility can increase or decrease with temperature, and sometimes there is little effect at all. Generally, the solubility of solids in liquids is directly proportional to temperature – as temperature increases, solubilty increases. This increase in solubility can be large or very small.

The effect of temperature on the solubility of a salt in water can be graphed as a *solubility curve* – a graph of temperature against solubility. Temperature has a very large effect on the solubility of potassium nitrate (KNO_3). Increasing the temperature from 20 °C to 40 °C nearly doubles the solubility. However, the solubility of sodium chloride (NaCl) does not increase very much over a temperature range of 100 °C. A few salts, such as cerium(III) sulfate ($Ce_2(SO_4)_3$), become less soluble in water as temperature increases. This is known as retrograde or inverse solubility.

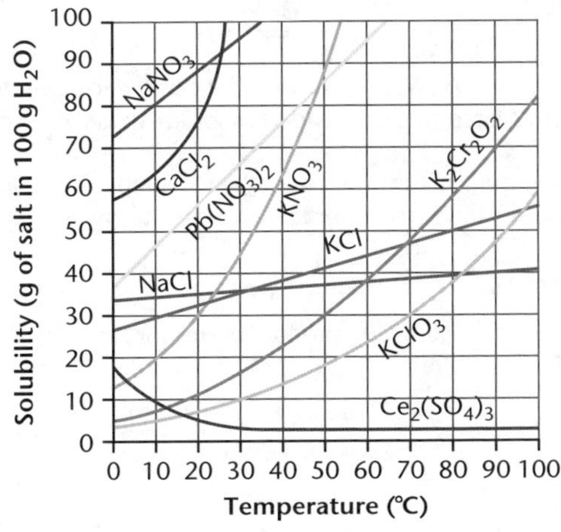

Solubility curves for some inorganic salts in water

This difference in solubility in hot and cold solvents is used in the process of purifying substances by *recrystallisation*. A saturated solution is allowed to cool until crystals of the solute deposit. These crystals can then be collected and dried.

The solubility of **organic compounds** nearly always increases with temperature.

Nature of solute and solvent

A useful guide for predicting solubility is 'like dissolves like', which means that a solute will dissolve best in a solvent that has a similar chemical structure to itself. So a polar solvent such as water will dissolve polar solutes and a non-polar solvent such as petrol will dissolve non-polar solutes such as petoleum jelly. *Liquid–liquid extraction* is a technique that uses differences in solubilities to separate and purify chemicals.

Ionic crystalline substances are held together by electrostatic forces between the positive and **negative ions**. When the crystal dissolves, these ions break apart and become surrounded by polar water molecules. Whether the crystal dissolves depends on which force is stronger: the force of attraction between the crystal ions or the force of attraction between the water molecules and the ions.

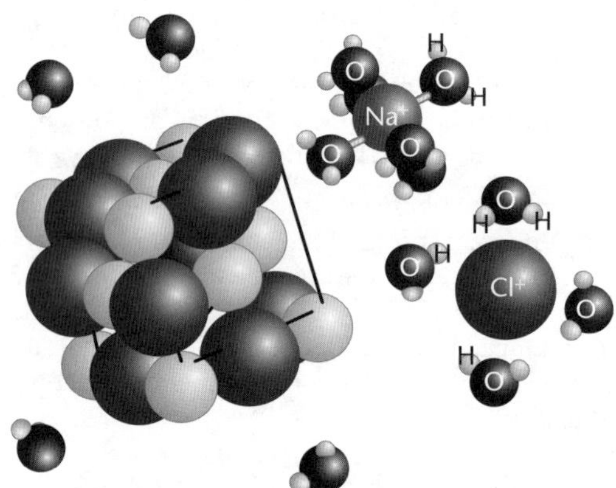

When an ionic salt such as NaCl dissolves, the positive and negative ions break apart and become surrounded by polar water molecules.

The rate of solubility is affected by the surface area of the solute. For example, smaller crystals of salt or sugar dissolve faster than larger crystals because they have a larger surface area.

Solubility products

The solubility product (K_{sp}) of a salt is the product of the concentrations of ions in a saturated solution. For example, for silver chloride (AgCl) at 25 °C:

$$AgCl(s) \rightleftharpoons Ag^{+}(aq) + Cl^{-}(aq)$$

$$K_{sp} = [Ag^{+}] \times [Cl^{-}] = 1.8 \times 10^{-10}$$

In contrast, the K_{sp} of sodium chloride (NaCl) in water is about 36, so it is much more soluble.

Solubility products can be used to predict how much salt will dissolve in a given volume of water or whether a salt will precipitate out of solution. The value depends on the type of salt, temperature and the common ion effect.

Example

At 25 °C, the K_{sp} for silver chloride (AgCl) is 1.8×10^{-10}. Calculate how much AgCl will dissolve in 1000 mL of water.

$$AgCl(s) \rightleftharpoons Ag^+(aq) + Cl^-(aq)$$

$$K_{sp} = [Ag^+] \times [Cl^-] = 1.8 \times 10^{-10}$$

$[Ag^+] = [Cl^-]$, in the absence of other silver or chloride salts.

$[Ag^+]^2 = 1.8 \times 10^{-10}$

$[Ag^+] = 1.34 \times 10^{-5}$

Thus, 1000 mL of water can dissolve 1.34×10^{-5} moles of AgCl(s) at room temperature.

To convert to grams, mass = moles × molar mass

Solubility rules

The solubility rules are a set of guidelines to help predict whether an ionic solid will dissolve in water.

Negative ion	Solubility rule
NO_3^-	All nitrates are soluble.
ClO_4^-	All chlorates are soluble.
Cl^-	All chlorides are soluble, except AgCl, Hg_2Cl_2 and $PbCl_2$.
I^-	All iodides are soluble, except AgCl, Hg_2Cl_2 and $PbCl_2$.
SO_4^{2-}	All sulfates are soluble, except $CaSO_4$, $BaSO_4$, $PbSO_4$, Ag_2SO_4, $HgSO_4$ and $SrSO_4$.
CO_3^{2-}	All carbonates are insoluble, except $(NH_4)_2CO_3$ and those of the **group** 1 elements.
PO_4^{3-}	All phosphates are insoluble, except $(NH_4)_3PO_4$ and those of the group 1 elements.
OH^-	All hydroxides are insoluble, except those of the group 1 elements and $Ba(OH)_2$ and $Sr(OH)_2$. $Ca(OH)_2$ is slightly soluble.
S^{2-}	All sulfides are insoluble, except those of the groups 1 and 2 elements and $(NH_4)_2S$.

Solubilty of oxygen gas

Oxygen gas (O_2) has a solubility of 40 mg/L of water at 20 °C. The low solubility of oxygen is because the oxygen molecules only interact weakly with the water molecules. The solubility of oxygen gas is strongly dependent on temperature. As temperature increases, its solubility decreases.

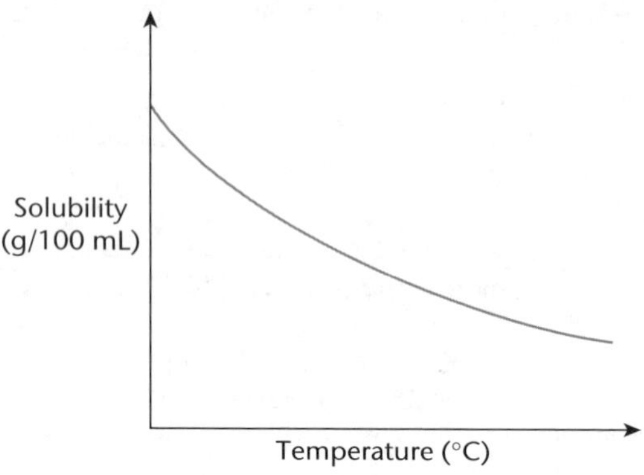

Solubility curve for oxygen gas

Aquatic organisms such as snails and fish depend on oxygen dissolved in water. Some power stations use water from rivers as a coolant and may return warm water to the environment. Increasing the temperature of the river water reduces the amount of dissolved oxygen, which can threaten the survival of the aquatic life.

Oxygen is more soluble in fresh water than in sea water because its solubility decreases with increasing concentrations of dissolved solids.

Solubilty of carbon dioxide gas

Carbon dioxide gas (CO_2) has a solubility of 1.8 g/L of water at 20 °C. Carbon dioxide is much more soluble than oxygen in water because the carbon dioxide molecules interact strongly with the water molecules. The **electronegative** oxygen atoms of CO_2 from strong hydrogen bonds with the H atoms of water. The carbon dioxide and water molecules also react to produce H+ ions:

$$CO_2 + H_2O \leftrightarrow HCO_3^- + H^+$$

This is a slow reaction. At equilibrium, less than 1% dissolved CO_2 is converted to H_2CO_3. Most of the CO_2 remains as dissolved molecular CO_2.

Calculating solubility

Solubility can be calculated from data in solubility tables or solubility curves.

Example

Use the solubility curve for potassium chloride (KCl) to calculate how much KCl can be dissolved in 200 g of water at 80 °C.

From the solubility curve, we can see that at 80 °C, 50 g of KCl will dissolve in 100 g of water.

So 2 × 50 g will dissolve in 200 g of water.

Thus, 100 g of KCl will dissolve in 200 g of water at 80 °C.

Example

A saturated solution of potassium chlorate ($KClO_3$) is formed from 100 g of water at 80 °C. If the solution is cooled to 50 °C, how many grams of $KClO_3$ will precipitate out of solution?

From the solubility curve, we can see that at 80 °C, 40 g of $KClO_3$ will dissolve in 100 g of water to form a saturated solution, and at 50 °C, 20 g of $KClO_3$ will dissolve. Decreasing the temperature from 80 °C to 50 °C will result in a supersaturated solution at the lower temperature and the excess will precipitate out, leaving 20 g dissolved.

Thus, 40 – 20 = 20 g of precipitate will be formed.

Unit 11.1 Activity 4C: Solubility of substances

1. Name the solutes found in air.
2. What solutes are contained in a bottle of SP beer?
3. Why do solids tend to become more soluble as temperature increases while gases tend to become less soluble?
4. Which of the following substances would you expect to be soluble in water?
   - **a.** Sodium carbonate
   - **b.** Barium sulfate
   - **c.** Calcium carbonate
5. State Henry's law for the effect of pressure on the solubility of gases in liquids.
6. Which of the following is a unit of solubility?
   - **A.** g/mol
   - **B.** g/mL
   - **C.** atm
   - **D.** K
7. True or false? A solution will have a lower boiling point than the pure solvent.

Unit 11.2 Chemical and Metallic Bonding

Topic 1: Structure of the atom

Do you ever think about why some substances behave differently from other substances? Why some conduct electricity and others don't? Why water is so different from its constituent elements of oxygen and hydrogen?
In Lower Secondary you learned about atoms and chemical changes. In this Unit you learn that the combination of atoms of various elements leads to the formation of new substances. Topic 1 describes atomic structure and bonding by explaining:

- Atomic mass, mass number, atomic number.
- Electron configuration of the first 20 elements.
- Isotopes.

Introduction

The term **atom** is part of our language. It took scientists many years to provide evidence that convinced the world that atoms exist.

Date	Discovery	Scientist and his contribution to present-day understanding of atoms
1803–1808	Atomic Theory	J. Dalton's scientific work convinced the world that the **atom** existed. He pictured atoms to be hard and indivisible (like billiard balls).
1869 1913–1914	Periodic Table	D. Mendeleev created the **periodic table** based on the chemical similarity of some elements when arranged in atomic mass order. H. Moseley established **atomic number** for each element, confirming Mendeleev's periodic table.
1896	Radioactivity	J. Becquerel discovered that uranium was **radioactive** – the element released invisible radiation, indicating that some atoms were not stable.
1897	Electron	J.J. Thomson discovered the first subatomic particle (the electron) and showed it was present in all matter.
1911	Nuclear atom and proton	E. Rutherford showed that an atom was mainly space, with an extremely small **nucleus** where all the mass was concentrated. The existence of the **proton** was predicted to explain the positive nucleus of an atom.
1914	Electrons in energy levels	N. Bohr produced evidence to show that very low-mass electrons were in **energy levels** outside the nucleus.
1932	Neutron	J. Chadwick showed that a neutral particle existed in the nuclei of atoms. It took a long time to identify the **neutron**, because of its lack of charge.

The development of the periodic table was an essential part in confirming the understanding of atoms and atomic structure. *A periodic table is located on page 301 of this book.*

Scientific discoveries in the development of the modern understanding of the atom

Atomic structure

There are about 10^{18} (one billion billion) carbon atoms in the full stop at the end of this sentence. The nucleus of the atom is $\frac{1}{100\,000}$ of the size of the atom. There are three types of **subatomic particle**:

- **Protons** (positive charge) – found inside the nucleus.
- **Neutrons** (neutral) – found inside the nucleus.
- **Electrons** (negative charge) – arranged in energy levels around the **nucleus**.

All atoms contain these particles – except the lightest atom, hydrogen, which contains no neutrons.

Subatomic particle	Mass on the atomic scale	Charge
Proton	1	1+
Electron	0.0005	1–
Neutron	1	Zero

Properties of subatomic particles

An atom can be represented in the following manner:

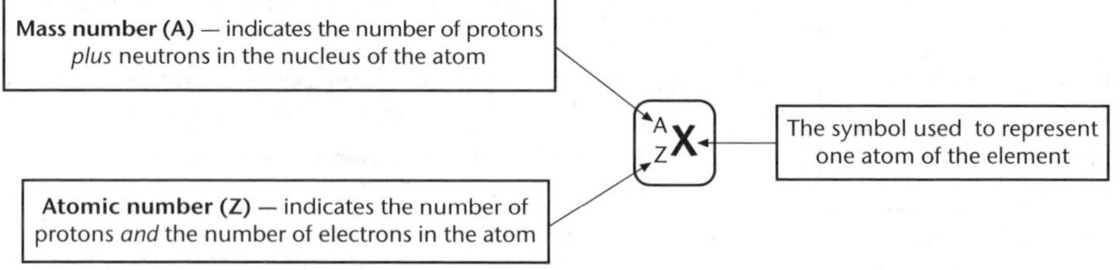

Atomic symbol, mass number (A) and atomic number (Z)

Example

The simplest atom is hydrogen, represented by $^{1}_{1}H$.

This representation means the element hydrogen (atomic symbol H) has:

- Atomic number (Z) of one – indicating that the hydrogen atom contains one proton and one electron.
- Mass number (A) of one – indicating there are no neutrons in the atom.

Example

Atomic structure

A sodium atom has 11 protons, 12 neutrons and 11 electrons.

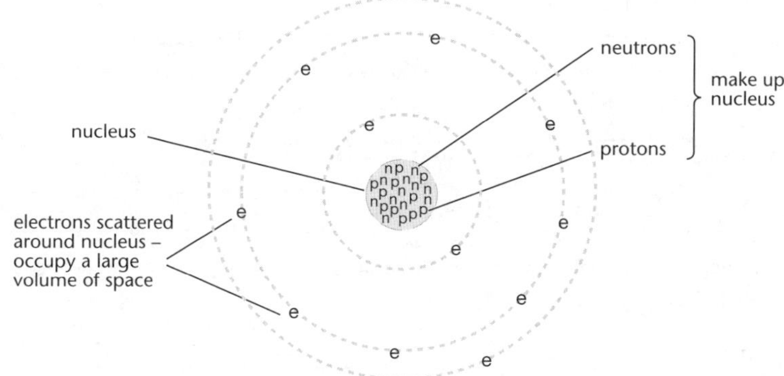

The nucleus is drawn much bigger than scale.

Structure of the sodium atom

Unit 11.2 Activity 1A: Atomic theory

1. **a.** State who developed the atomic theory.

b. State which was the first subatomic particle discovered.

c. State which was the most difficult subatomic particle to discover and why it was difficult to discover.

d. State what phenomenon was discovered that indicated some atoms were unstable.

e. State what Rutherford's contribution was to the understanding of the modern atom.

2. Match the following statements with the subatomic particles:

Statement	Subatomic particle
a. Two thousand of these would balance one hydrogen atom on a set of scales.	**i.** Proton
b. These particles are the reason why atoms of an element can have different masses.	**ii.** Electron
c. The particle that exists in the nucleus of every atom.	**iii.** Neutron

3. Complete the following account concerning atomic theory using the words supplied (in alphabetical order) at the end of the account. The dashes indicate the number of letters in the missing word.

The _ _ _ _ _ _ _ of an atom contains nearly _ _ _ the mass of the atom and is surrounded by _ _ _ _ _ _ _ _ _ in energy _ _ _ _ _ _. Although it is not possible to see atoms – clusters of about one hundred atoms can be seen under an electron _ _ _ _ _ _ _ _ _ _ – every student of chemistry believes that they _ _ _ _ _.

_ _ _ _ _ _ bombs have been exploded. So this is the atomic _ _ _ or the _ _ _ _ _ _ _ age; or maybe by now we have moved on to the _ _ age!

Word list: age; all; Atomic; electrons; exist; IT; levels; microscope; nuclear; nucleus.

4. Explain the meaning of the X, A and Z in the representation $^{A}_{Z}\mathbf{X}$.

Isotopes

Every element has:

- A unique atomic number.
- The same number of protons and electrons in every one of its atoms.

Isotopes are atoms of an element with different numbers of neutrons in the nucleus of the atom.

The *only* difference between isotopes of the same element is the number of *neutrons* in the nucleus of the atom – isotopes of the same element have the same number of protons and electrons.

Neutrons do not alter the chemical behaviour of the atoms – thus, the chemistry of isotopes of the same element is the same.

Example

If two isotopes of a given element undergo the same chemical reaction, the only detectable difference in the chemical behaviour is in the rates of their reaction.

Some common isotopes

Most elements have more than one isotope – some isotopes are more abundant than others. Some common isotopes are:

- *Carbon-14*, $^{14}_{6}C$ – a radioactive isotope of carbon-12 that is produced in the upper atmosphere when nitrogen atoms are bombarded by cosmic rays. Carbon dioxide is then formed – thus carbon-14 gets into plant material by **photosynthesis**. The age of dead carbon material can be dated (**carbon dating**) by measuring its radioactivity.
- *Iodine-131*, $^{131}_{53}I$ – one of the products of nuclear fission (eg an atomic bomb explosion). This radioactive isotope, if ingested, would end up in the thyroid gland, where it can cause cellular damage.
- *Strontium-90*, $^{90}_{38}Sr$ – another fallout product of an atomic bomb explosion. This radioactive isotope is chemically similar to calcium – if it falls on paddocks, cows may ingest it, causing the radioactive strontium to be found in the milk. Humans who consume this milk may have bones and teeth containing radioactive strontium-90 (instead of calcium), causing cellular damage.
- *Hydrogen-2*, $^{1}_{2}H$ – this isotope, commonly called **deuterium** (symbol D), is not radioactive. Chemically, deuterium behaves like hydrogen and forms deuterium oxide, D_2O (similar to water, H_2O) or '**heavy water**'. Heavy water is 11% heavier than ordinary water – ordinary water contains 0.015% heavy water.

Atomic mass

The atomic mass of any element is the average of the atomic mass of all its isotopes.

Example

Atomic masses of copper and chlorine

The atomic masses of copper and chlorine are, respectively, 63.5 and 35.5. Copper and chlorine each contain two isotopes:

- $^{63}_{29}Cu$ and $^{65}_{29}Cu$ for copper.
- $^{35}_{17}Cl$ and $^{37}_{17}Cl$ for chlorine.

Isotope	Symbol	Number of protons	Number of electrons	Number of neutrons
Copper-63	$^{63}_{29}Cu$	29	29	34
Copper-65	$^{65}_{29}Cu$	29	29	36
Chlorine-35	$^{35}_{17}Cl$	17	17	18
Chlorine-37	$^{37}_{17}Cl$	17	17	20

Isotopic data for copper and chlorine atoms

For both copper and chlorine, the isotope with the smaller atomic mass is present in the ratio of 3:1 compared with the isotope with the larger atomic mass:

- For copper, $3 \times 63 + 1 \times 65 = 254$.
- The average copper atom has a mass of $254 \div 4 = 63.5$.
- For chlorine, $3 \times 35 + 1 \times 37 = 142$.
  The average chlorine atom has a mass of $142 \div 4 = 35.5$.

Note: The most accurate value for the atomic mass of copper is 63.6, indicating the ratio of isotopes is not 3:1 (exactly).

For most elements, the isotopic composition is dominated by one isotope – the atomic mass quoted for the element tends to be the mass of the most common isotope.

Element	For every 10 000 atoms			Atomic mass, A
Oxygen (in air)	9 976 are *oxygen-16*	4 are *oxygen-17*	20 are *oxygen-18*	16
Hydrogen (in water)	9 999 are *hydrogen-1*	1 is *hydrogen-2*	–	1
Sodium (in common salt)	10 000 are *sodium-23*	–	–	23

Isotopic composition of oxygen, hydrogen and sodium

Unit 11.2 Activity 1B: Isotopes

1. a. Name the only atom that contains no neutron(s).

b. Explain why atoms of $^{14}_{6}C$ and $^{14}_{7}N$ have the same mass.

c. The most abundant isotopes of carbon, iodine and strontium are $^{12}_{6}C$, $^{127}_{53}I$, $^{88}_{38}Sr$, respectively. Draw up a table to compare these isotopes with carbon-14, iodine-131 and strontium-90. In the table, list the isotopes by symbol and indicate the number of protons, electrons and neutrons in each isotope.

2. a. The first atomic bomb used uranium-235 as its nuclear fuel. Natural uranium consists of 99.3% uranium-238 and 0.7% uranium-235. To make the bomb, approximately 1 kg of pure uranium-235 was needed. Calculate the mass of natural uranium needed to produce 1 kg of uranium-235.
   b. If the symbol for uranium is U and its atomic number is 92, write the symbols for the two isotopes of uranium.
3. Consider the three atoms, $^{1}_{1}H$, $^{2}_{1}D$, $^{3}_{1}T$, where D stands for deuterium and T stands for **tritium**. Give as much information as you can about the relationship between these three atoms.

Electron configuration

Once scientists had accepted the structure of the nuclear atom, where all the positive charge was in the nucleus, a major problem remained – how did the (negatively charged) electrons avoid being attracted into the nucleus? New physics was developed and indicated the atoms could be stable if:

- The electrons existed in energy levels.
- The energy levels were located at certain, defined distances from the nucleus.
- Each energy level had a maximum occupancy of electrons.

The **electron configuration** describes how electrons are arranged outside the nucleus of an atom. For the first 20 elements, the electron arrangement follows the pattern:

Energy level	Number of electrons
1	Maximum of 2
2	Maximum of 8
3	Maximum of 8
4	Two electrons (up to element 20) – elements with atomic number above 20 require the energy level to contain more electrons.

Electron configuration in energy levels

Electron configuration of elements 1–20

Element	Atomic number	Electron configuration of atom
Hydrogen	1	1,
Helium	2	2,
Lithium	3	2, 1
Beryllium	4	2, 2
Boron	5	2, 3
Carbon	6	2, 4
Nitrogen	7	2, 5
Oxygen	8	2, 6

Element	Atomic number	Electron configuration of atom
Fluorine	9	2, 7
Neon	10	2, 8
Sodium	11	2, 8, 1
Magnesium	12	2, 8, 2
Aluminium	13	2, 8, 3
Silicon	14	2, 8, 4
Phosphorus	15	2, 8, 5
Sulfur	16	2, 8, 6
Chlorine	17	2, 8, 7
Argon	18	2, 8, 8
Potassium	19	2, 8, 8, 1
Calcium	20	2, 8, 8, 2

Electron configuration of elements 1–20

For the elements hydrogen, helium, lithium and beryllium, two electrons (or none in the case of hydrogen) in the outer energy level of the particle of the element produces a stable arrangement of electrons.

Atoms of the **inert** gas elements have a complete energy level. With the exception of helium (which has two electrons in the outer energy level), the atoms of the inert gas elements have eight electrons in the outer energy level. Particles with eight electrons in their outer energy levels are stable (chemically unreactive) – this is the **octet rule**.

Unit 11.2 Activity 1C: Electron configuration

1. Complete the table below identifying the elements and stating their atomic numbers.

Electron configuration	Element	Atomic number
a. 2		
b. 2, 2		
c. 2, 8, 2		
d. 2, 8, 8, 2		
e. 2, 5		
f. 2, 8, 5		

2. Below are listed the electron configurations of the first 20 elements in numerical order. Rearrange the twenty boxes so that a pattern of numbers is displayed. *Hint*: Place all the boxes with the same last number in the same vertical column.

1	2	2, 1	2, 2	2, 3	2, 4	2, 5	2, 6	2, 7	2, 8
2, 8, 1	2, 8, 2	2, 8, 3	2, 8, 4	2, 8, 5	2, 8, 6	2, 8, 7	2, 8, 8	2, 8, 8, 1	2, 8, 8, 2

Atom models

Atoms can be represented by 'pictures' or atom models.

Example

Atom model

In the atom model for the carbon atom:

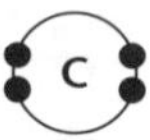

- The **'C'** represents the nucleus of a carbon atom plus the inner energy level that contains two electrons.
- The four electrons in the outer energy level are placed in a circle and displayed in pairs.

Atom	Atom model showing number of electrons in outer energy level	Description
$^{12}_{6}C$ Carbon-12	C	**'C'** represents the nucleus of a carbon atom containing 6 protons and 6 neutrons, plus 2 electrons in the first energy level.
$^{16}_{8}O$ Oxygen-16	O	**'O'** represents the nucleus of an oxygen atom containing 8 protons and 8 neutrons, plus 2 electrons in the first energy level.
$^{24}_{12}Mg$ Magnesium-24	Mg	**'Mg'** represents the nucleus of a magnesium atom containing 12 protons and 12 neutrons, plus 2 electrons in the first energy level, and 8 electrons in the second energy level.
$^{27}_{13}Al$ Aluminium-27	Al	**'Al'** represents the nucleus of an aluminium atom containing 13 protons and 14 neutrons in the first energy level, plus 2 electrons in the first energy level, and 8 electrons in the second energy level.
$^{35}_{17}Cl$ Chlorine-35	Cl	**'Cl'** represents the nucleus of a chlorine atom containing 17 protons and 18 neutrons in the first energy level, plus 2 electrons in the first energy level, and 8 electrons in the second energy level.

Atom models for carbon-12, oxygen-16, magnesium-24, aluminium-27 and chlorine-35

Unit 11.2 Activity 1D: Atom models

1. Draw atom models to represent the following atoms:

Atom	Mass number	Atomic number	Atom model
H	2	1	
Be	9	4	
N	14	7	
S	32	16	

2. Complete the table for **b.** to **e.** in the style of **a.**, which has been done for you. For each question, identify the element (W, X, Y) and/or establish the values of mass number A, and atomic number Z. You may need to refer to the periodic table at the back of the book to identify an element from its atomic number.

	Atom	Atom model showing number of electrons in outer energy level	Description
a.	$^{4}_{2}He$ Helium-4	He	'**He**' represents the nucleus of the helium atom containing 2 protons and 2 neutrons.
b.	$^{20}_{10}W$ _ _ _ _ - __	__	'**W**' represents the nucleus of the _ _ _ _ atom containing __ protons and __ neutrons, plus 2 electrons in the first energy level.
c.	$^{A}_{Z}Mg$ _ _ _ _ _ _ _ _ _ _ - __	Mg	'**Mg**' represents the nucleus of the _ _ _ _ _ _ _ _ _ _ atom containing 12 protons and 12 neutrons, plus 2 electrons in the first energy level, and 8 electrons in the second energy level.
d.	$^{A}_{5}X$ _ _ _ _ _ - __	__	'**X**' represents the nucleus of the _ _ _ _ _ atom containing 5 protons and 4 neutrons, plus __ electrons in the first energy level.
e.	$^{35}_{Z}Y$ _ _ _ _ _ _ _ _ _ - __	__	'**Y**' represents the nucleus of the _ _ _ _ _ _ _ _ atom containing __ protons and 18 neutrons, plus 2 electrons in the first energy level, and 8 electrons in the second energy level.

Unit 11.2 Chemical and Metallic Bonding

Topic 2: Covalent bonding

Usually we use the word 'bonding' to describe the formation of a special relationship, but in chemistry, bonding is the process of atoms joining together. The formation of new substances is made possible by bonding. Chemical bonding focuses on the formation of ionic and covalent compounds. Topic 2 describes atomic structure and bonding by explaining:

- Covalent bonding.
- Molecules.
- Single and multiple covalent bonds.
- Lewis diagrams.

Introduction

Atoms can **bond** (join together) by sharing electrons in their outer energy levels (**covalent bonding**) to form **molecules**. By sharing the electrons in a covalent bond, the atoms in the molecule can achieve eight electrons in their outer energy level – the **octet rule**. An exception is the hydrogen atom which requires only two electrons in its outer energy level when it bonds with other atoms to form a molecule.

Atoms of the *same* element or of *different* elements can form covalent bonds.

Example

Hydrogen atoms can bond with each other to form a hydrogen molecule, H_2, or with an oxygen atom to form a water molecule, H_2O.

Covalent bonds in elements 1–20

The first 20 elements in the **periodic table** are:

H							He
Li	Be	B	C	N	O	F	Ne
Na	Mg	Al	Si	P	S	Cl	Ar
K	Ca						

The first 20 elements of the periodic table

The shaded elements in the diagram above have atoms that form covalent bonds when they combine with themselves (eg H_2, Cl_2, O_2, N_2), and when they combine with each other (eg H_2O, PCl_3, CH_4).

Molecules

Molecules can be represented by:

- A **molecular formula**, eg Cl_2, O_2.
- A **structural formula** that shows the bonding present.

Example

The chlorine molecule can be represented by Cl–Cl, where the single line indicates a single covalent bond. The oxygen molecule, O=O, has two covalent bonds (referred to as a **double bond**).

- A 3-D picture.

Example

3-D models of molecules

The atoms in the following molecules are represented by spheres which 'fuse' together due to the electrons used for bonding.

water molecule, H_2O

methane molecule, CH_4

ammonia molecule, NH_3

3-D models of water, methane and ammonia

Lewis diagrams

Lewis diagrams (also called **electron dot diagrams**) show models of atoms, or models of molecules, with the correct number of electrons in the outer energy level(s).

Example

Lewis diagram

The Lewis diagram of the water molecule, H_2O, is:

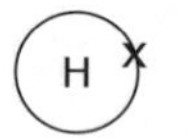

+

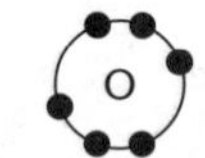

+

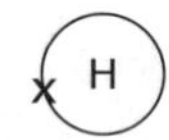

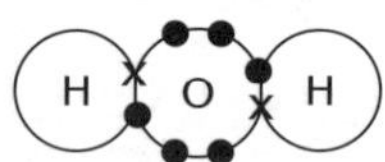

two separate hydrogen atoms, each with one electron in its outer energy level *and* one oxygen atom with six electrons in its outer level

a molecule of water in which each hydrogen atom shares one electron with the oxygen atom – both hydrogen atoms achieve a stable outer energy level of two electrons *and* the oxygen atom achieves a stable outer energy level of eight electrons

Single covalent bonds

A single covalent bond forms when one electron in the outer energy level of each atom is shared.

Example

The hydrogen molecule

The hydrogen atom is the simplest *atom* and the hydrogen molecule is the simplest *molecule*. The formation of the hydrogen molecule involves the proton in one hydrogen atom attracting the electron in the other hydrogen atom, and vice versa. These attraction forces are stronger than both:

- the repulsion forces between the protons in the nuclei of the two atoms, and
- the repulsion forces between the electrons outside the nuclei of the two atoms.

This means the molecule is more stable than the separate atoms.

The hydrogen and chlorine molecules are **diatomic** – ie they contain two atoms. The formulae for these molecules, H–H and Cl–Cl, show that a single covalent bond forms between the atoms, involving two shared electrons – one electron provided by each atom.

The formation of the hydrogen molecule, H_2

two separate hydrogen atoms, each with one electron in its outer energy level

a molecule of hydrogen in which each atom shares one electron with the other atom – both atoms achieve a stable outer energy level of two electrons

The formation of the chlorine molecule, Cl_2

two separate chlorine atoms, each with seven electrons in their outer energy level

a molecule of chlorine in which each atom shares one electron with the other atom – both atoms achieve a stable outer energy level of eight electrons

Covalent bonding in molecules of hydrogen and chlorine

In addition to diatomic molecules, molecules can also be:

- Triatomic – contains three atoms, eg H_2O.
- Tetratomic – contains four atoms, eg PCl_3.
- Pentatomic – contains five atoms, eg CH_4.

The formation of the phosphorus trichloride molecule, PCl_3

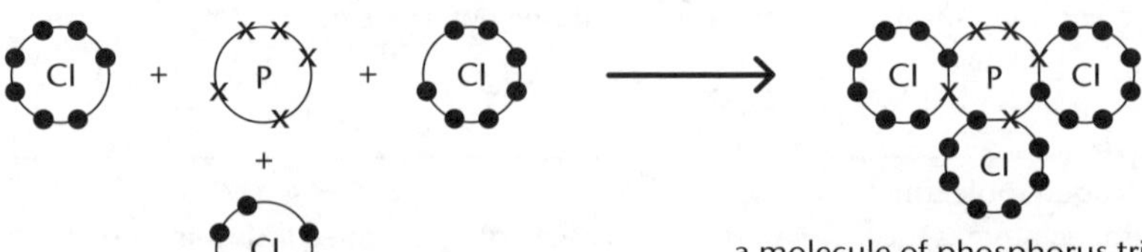

three separate chlorine atoms, each with seven electrons in its outer energy level, *and* one phosphorus atom with five electrons in its outer energy level

a molecule of phosphorus trichloride in which each chlorine atom shares one electron with the phosphorus atom – the three chlorine atoms achieve a stable outer energy level of eight electrons *and* the phosphorus atom achieves a stable outer energy level of eight electrons

The formation of the methane molecule, CH_4

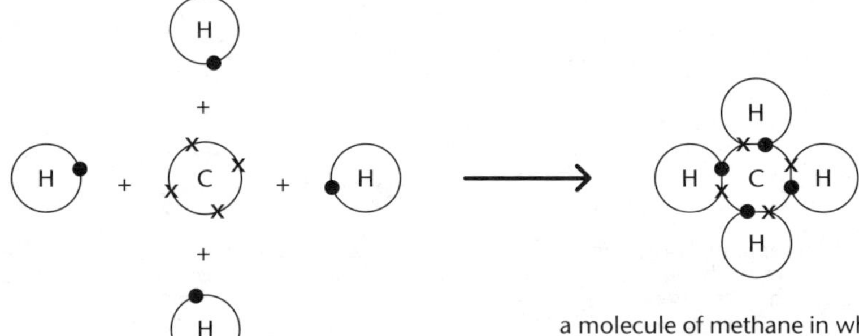

four separate hydrogen atoms, each with one electron in its outer energy level, *and* one carbon atom with four electrons in its outer energy level

a molecule of methane in which each hydrogen atom shares one electron with the carbon atom – the four hydrogen atoms achieve a stable outer energy level of two electrons *and* the carbon atom achieves a stable outer energy level of eight electrons

Covalent bonding in molecules of phosphorus trichloride and methane

Molecules with single bonds can be represented in 'shorthand'.

Example

Shorthand diagrams of molecules with single bonds

```
H—O—H        Cl—P—Cl          H
                |             |
                Cl         H—C—H
                              |
                              H
```

H_2O
a triatomic molecule containing two single covalent bonds

PCl_3
a tetratomic molecule containing three single covalent bonds

CH_4
a pentatomic molecule containing four single covalent bonds

Multiple covalent bonds

Double bonds

An atom can achieve eight electrons in its outer energy level – the **octet rule** – by sharing *two* electrons in its outer energy level with *two* electrons in the outer energy level of another atom. This creates a **double bond**.

Oxygen, O_2, and carbon dioxide, CO_2, both contain covalent double bonds.

The oxygen molecule, O_2

two separate oxygen atoms, each with six electrons in its outer energy level

a molecule of oxygen in which each atom shares two electrons with the other atom – both atoms achieve a stable outer energy level of eight electrons

The carbon dioxide molecule, CO_2

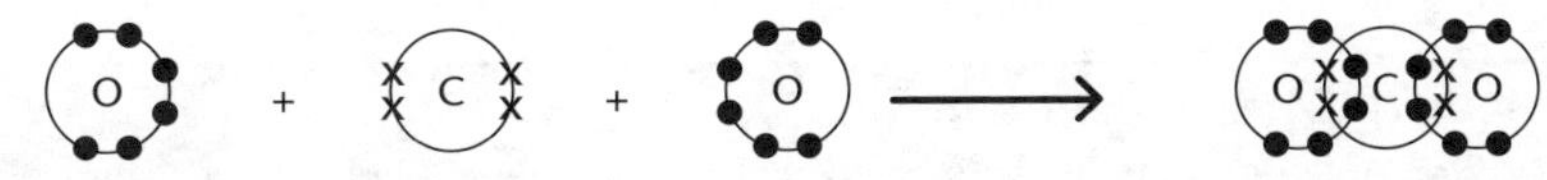

an oxygen atom with six electrons in its outer energy level

a carbon atom with four electrons in its outer energy level

an oxygen atom with six electrons in its outer energy level

molecule of carbon dioxide in which each oxygen atom shares two of its electrons with two electrons from the carbon atom – each atom achieves a stable outer energy level of eight electrons

Covalent double bond formation

Molecules with double bonds can be represented in shorthand.

Example

Shorthand diagrams of molecules with double bonds

O=O	O=C=O
O_2	CO_2
a diatomic molecule containing one double covalent bond	a triatomic molecule containing two double covalent bonds

Triple bonds

Triple bonds are rare, but occur when two nitrogen atoms bond to each other to form a diatomic nitrogen molecule, N_2.

Example

The nitrogen molecule, N_2

Each nitrogen atom has five electrons in its outer energy level. If each atom shares three of its electrons with the other atom in a covalent triple bond, then each atom achieves a stable octet of electrons in its outer energy level.

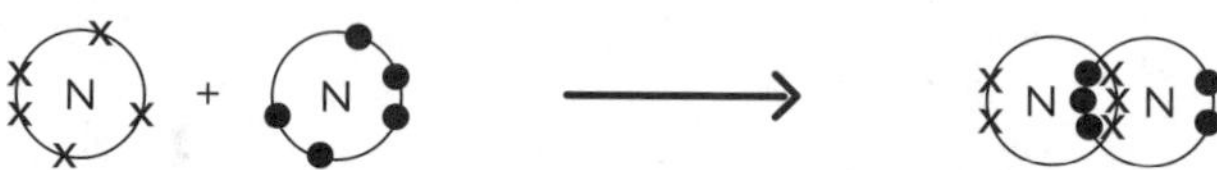

two separate nitrogen atoms, each with five electrons in its outer energy levels

a molecule of nitrogen in which each atom shares three electrons with the other atom – both atoms achieve a stable outer energy level of eight electrons

Triple covalent bond formation

The nitrogen molecule can be represented as N≡N.

Unit 11.2 Activity 2A: Covalent bonding

1. On the section of the periodic table provided below, use different-coloured pens, or different shading, to indicate:
   a. Which elements have atoms with a complete outer energy level of electrons.
   b. Which elements have atoms that form covalent bonds with atoms of the same element, or with atoms of another element.

H							He
Li	Be	B	C	N	O	F	Ne
Na	Mg	Al	Si	P	S	Cl	Ar
K	Ca						

2. Complete the following account of covalent bonding using the words listed (in alphabetical order) below. The number of letters in the missing word is indicated by the number of spaces.

   _ _ _ _ _ _ _ _ *is an element that has an atom with one electron in its outer* _ _ _ _ _ _ *level. By* _ _ _ _ _ _ _ *this electron with another* _ _ _ _ _ _ _ _ *atom, both atoms obtain a stable outer energy level of* _ _ _ *electrons. The particle formed is called a* _ _ _ _ _ _ _ _. *All atoms involved in forming a covalent bond have a* _ _ _ _ _ _ _ _ *outer energy level once the bond is formed.*

   Word list: complete; energy; hydrogen; hydrogen; molecule; sharing; two.
3. Draw Lewis diagrams of the atoms and the molecule to show the formation of covalent bonding in a molecule of:
   a. Hydrogen.
   b. Fluorine.

- **c.** Hydrogen sulfide.
- **d.** Nitrogen trichloride, NCl_3.
- **e.** Silicon tetrahydride, SiH_4.

4. Draw Lewis diagrams of the atoms and the molecule to show the formation of a molecule that contains:

- **a.** One double bond (*Hint*: Draw ethene, C_2H_4).
- **b.** Two double bonds (*Hint*: Draw carbon disulfide, CS_2).
- **c.** One triple bond (*Hint*: Draw ethyne, C_2H_2).

5. *F–F represents a molecule of fluorine which contains a single covalent bond.*

Write similar sentences that describe what the following structures represent:

- **a.** Hydrogen fluoride, H–F.
- **b.** Carbon disulfide, S=C=S.
- **c.** Phosphorus trihydride, $\begin{matrix} H-P-H \\ | \\ H \end{matrix}$.
- **d.** Ethene, $\begin{matrix} H & & H \\ | & & | \\ C & = & C \\ | & & | \\ H & & H \end{matrix}$.
- **e.** Hydrogen peroxide, H—O—O—H.
- **f.** Ethyne, H—C≡C—H.

Unit 11.2 Chemical and Metallic Bonding

Topic 3: Ionic bonding

Topic 3 continues our exploration of chemical bonding and the formation of new substances. The previous Topic looked at covalent bonding; now Topic 3 examines atomic structure and bonding by explaining:

- Ions.
- Ionic compounds.
- Ionic bonding.

Introduction

Elements react in either of two ways:

- **Covalent bonding** – atoms share electrons in their outer energy levels.
- **Ionic bonding** – electrons are transferred between atoms of different elements. Electron(s) are lost by an atom of one element and gained by an atom of another element.

Ions

Atoms have different abilities to lose electrons from or add electrons to their outer energy level:

- Elements whose atoms lose electrons readily form **positive ions** (**cations**) – these elements are commonly called **metals**. One definition for a metal is 'the atom of the element forms positive ions'. (An exception to this definition of a metal is the hydrogen atom, which forms a positive ion, but is *not* metallic.)
- Elements whose atoms gain electrons readily form **negative ions** (**anions**) – these elements are commonly called **non-metals**.
- The atoms of some non-metallic elements – the **inert gases** – do not gain, lose or share electrons. These atoms already possess stable octets.
- The atoms of other non-metallic elements (eg carbon, C, and silicon, S) do not gain or lose electrons and thus do not form ions – instead, the atoms form covalent bonds.

An atom will generally gain or lose electron(s) to form a stable octet (*eight* electrons) in its outer energy level. The exceptions are:

- Hydrogen atoms – have *no* electrons in their outer energy level when forming the positive **hydrogen ion**.
- Lithium and beryllium atoms – have *two* electrons in their outer energy level when forming positive ions.

Atoms that lose electrons to form cations

Atoms that gain electrons to form anions

H							He
Li	Be	B	C	N	O	F	Ne
Na	Mg	Al	Si	P	S	Cl	Ar
K	Ca						

Positive and negative ion formation for the elements 1–20

Ionic compounds

Compounds that contain ions are called **ionic compounds** – the term *molecule* does not apply to ionic compounds. Each ion is a separate particle that strongly attracts ions of opposite charge by **electrostatic charge**, forming an **ionic bond**.

Example

Sodium chloride

The ionic compound sodium chloride has the formula NaCl – this indicates the ratio of sodium ions, Na^+, to chlorine ions, Cl^-, is 1 : 1. A **lattice** structure exists where each sodium ion is surrounded by six chlorine ions and each chlorine ion is surrounded by six sodium ions, resulting in a 1 : 1 ratio of ions. It is not possible to associate one sodium ion with one chlorine ion, ie one sodium ion does not 'belong' to any particular chlorine ion.

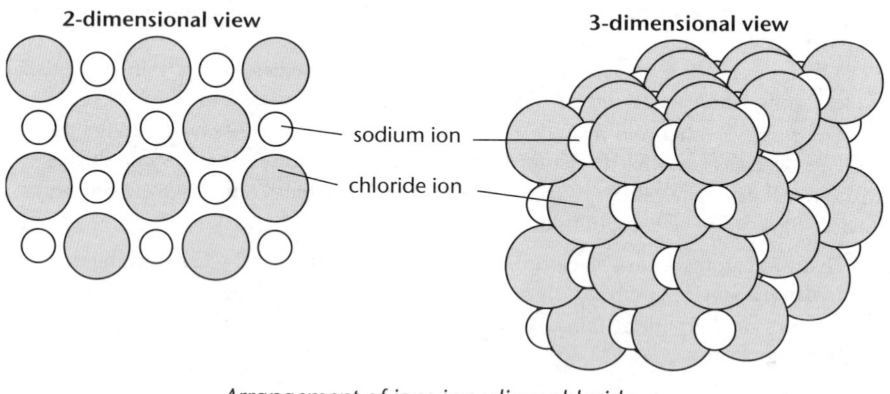

Arrangement of ions in sodium chloride

Ionic bonding

In ionic bonding, there is a *transfer* of electrons, so that both atoms achieve an octet of electrons in their outer energy levels:

- Atoms of metals have 1, 2 or 3 electrons in their outer energy level and can *donate* these electrons to atoms of non-metals.
- Atoms of non-metals have 5, 6 or 7 electrons in their outer energy levels and can *accept* electrons from metals.

Example

Formation of single charge ions

In the formation of sodium fluoride, $Na^{+}F^{-}$, the sodium atom loses one electron and the fluoride atom gains one electron:

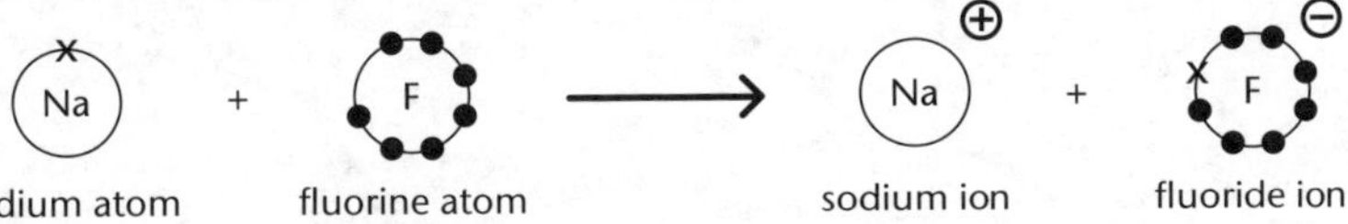

Example

Formation of double charge ions

Double positive charge and double negative charge ions are formed by a metal atom donating two electrons to a non-metal atom, eg magnesium sulfide, $Mg^{2+}S^{2-}$ (a lattice of double positive ions balancing an equal number of double negative ions):

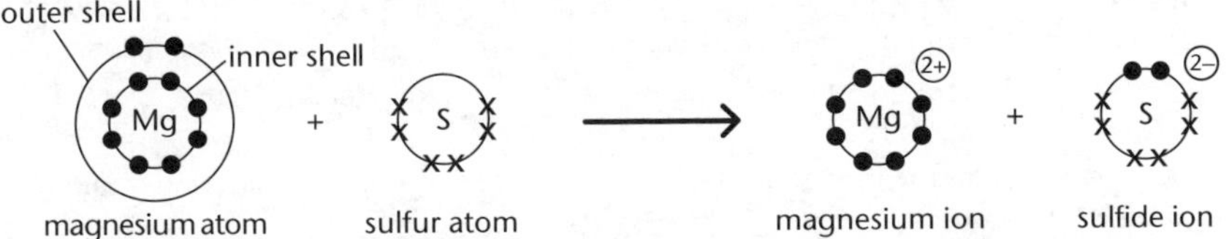

Example

Formation of aluminium (triple charge) and oxide (double charge) ions

Triple positive charge and double negative charge ions are formed by two metal atoms each donating three electrons to three non-metal atoms which each receive two electrons, eg aluminium oxide, $(Al^{3+})_2\,(O^{2-})_3$ (a lattice of triple positive ions and double negative ions – the overall charge must balance, hence 2 × $^{3+}$ balances 3 × $^{2-}$):

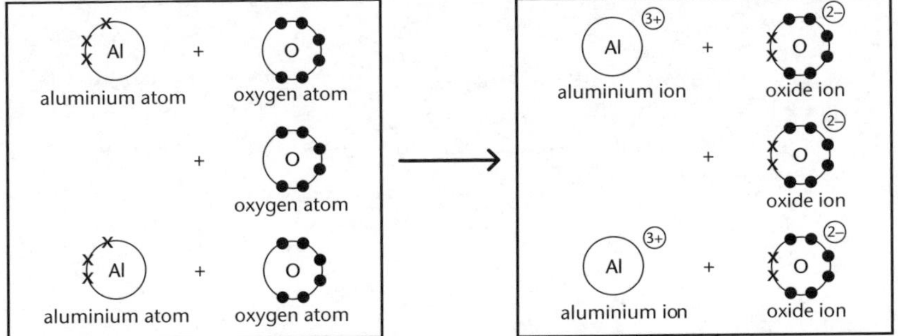

Unit 11.2 Activity 3A: Ions, ionic compounds and ionic bonding

1. By using different-coloured pens, or by different shading, indicate on the section of the periodic table given below:

a. Which elements have atoms that can form positive ions.

b. Which elements have atoms that can form negative ions.

H							He
Li	Be	B	C	N	O	F	Ne
Na	Mg	Al	Si	P	S	Cl	Ar
K	Ca						

2. Complete the following account of ionic bonding, using the words listed (in alphabetical order) below. The number of letters in the missing word is indicated by the number of spaces.

Metals have one, two or three _ _ _ _ _ _ _ _ _ in the outer energy level of their _ _ _ _ _. By _ _ _ _ _ _ _ _ this/these electron(s) to another _ _ _ _, the metal atom becomes stable because it has _ _ _ _ _ (or none, or two in a few cases) electrons in its outer energy level. The atom that _ _ _ _ _ _ _ the electron(s) is from a non-metallic element. The atom accepting the electron(s) gains a _ _ _ _ _ _ _ _ charge. All _ _ _ _ are charged and attract each other _ _ _ _ _ _ _ _. An ionic compound is electrically neutral – thus the ions must _ _ _ _ _ _ _ their charges. The ions are arranged in a three-dimensional lattice where a particular positive ion is surrounded by negative ions and _ _ _ _ _ _ _ _ _. The overall ratio of ions is _ _ _ _ _ _, eg 1 : 1, 1 : 2, 2 : 1, etc. The term _ _ _ _ _ _ _ _ should not be used when referring to ionic compounds – the term _ _ _ _ _ _ _ _ is preferred.

Word list: accepts; atom; atoms; balance; donating; eight; electrons; ions; molecule; negative; particle; simple; strongly; vice versa.

3. Identify which of the following species have the electron configuration of an inert gas atom (ie 2 or 2, 8 or 2, 8, 8).

a. Al.

b. Al^{3+}.

c. O^{2-}.

d. O_2^{2-}.

e. H^-

f. Be^{2+}.

g. K.

h. Ca^{2+}.

4. From the first twenty elements, identify as many species as you can with the electron configuration of 2, 8, 8. *Hint*: There are six, only one of which is an atom.

5. The following diagrams illustrate the formation of ions from atoms. Complete the diagrams on the right-hand side to show:

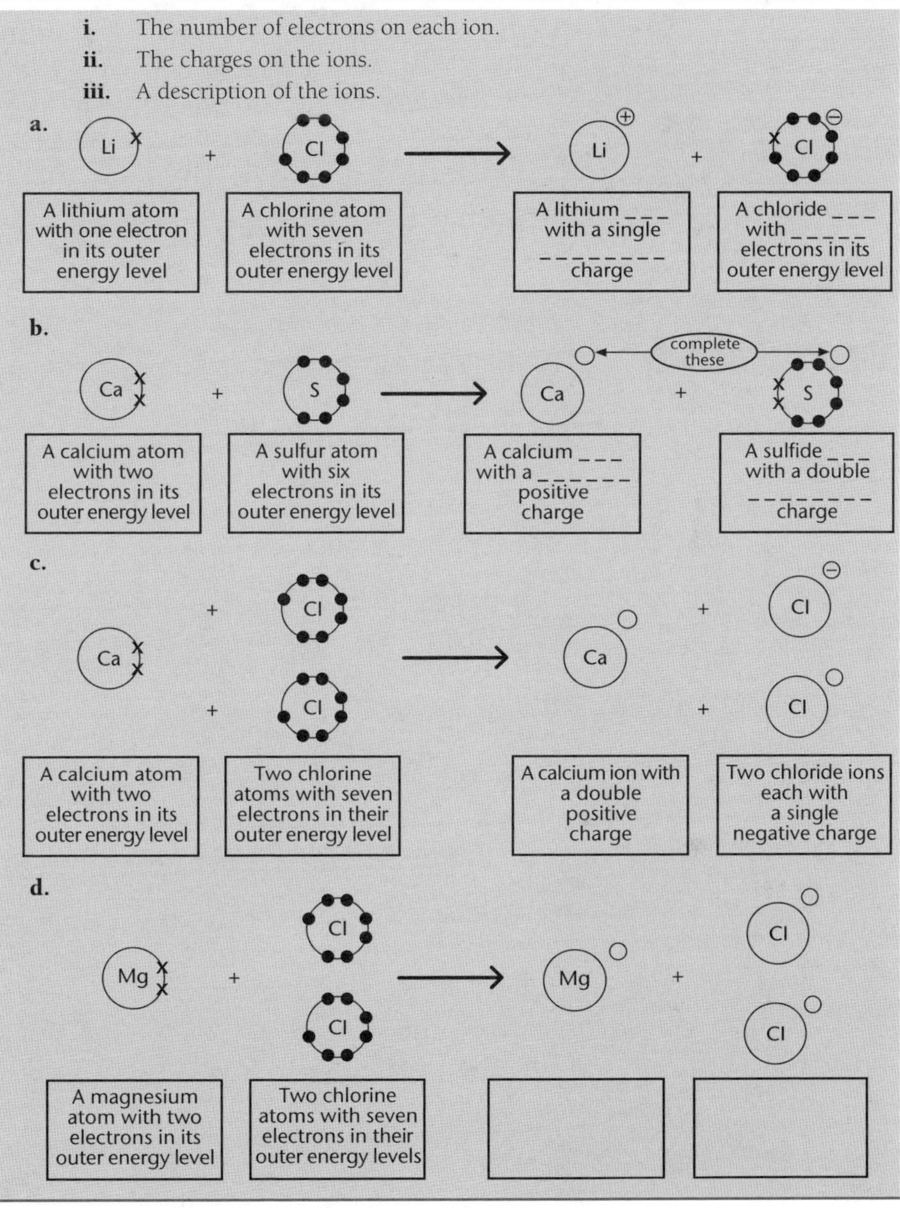
i. The number of electrons on each ion.
ii. The charges on the ions.
iii. A description of the ions.
a.
Li
Cl
Li
Cl
A lithium atom with one electron in its outer energy level
A chlorine atom with seven electrons in its outer energy level
A lithium ___ with a single ________ charge
A chloride ___ with _____ electrons in its outer energy level
b.
complete these
Ca
S
Ca
S
A calcium atom with two electrons in its outer energy level
A sulfur atom with six electrons in its outer energy level
A calcium ___ with a ______ positive charge
A sulfide ___ with a double ________ charge
c.
Ca
Cl
Cl
Ca
Cl
Cl
A calcium atom with two electrons in its outer energy level
Two chlorine atoms with seven electrons in their outer energy level
A calcium ion with a double positive charge
Two chloride ions each with a single negative charge
d.
Mg
Cl
Cl
Mg
Cl
Cl
A magnesium atom with two electrons in its outer energy level
Two chlorine atoms with seven electrons in their outer energy levels

Unit 11.2 Chemical and Metallic Bonding

Topic 4: Ionic compounds

In Topic 4 we continue our examination of ions and ionic compounds, and precipitates, which is the term used to describe a solid substance that has been separated from a liquid in a chemical process. This Topic deals with:

- Writing balanced equations for the formation of precipitates and complex ions.

Introduction

Ionic compounds are made up of positive ions (**cations**) and negative ions (**anions**).

- A cation is usually a metal atom that has lost an electron or electrons.
- An anion is formed when a non-metal atom gains an electron or electrons (**monatomic**) or when two or more atoms bond covalently with the addition of an electron or electrons (**polyatomic**).

Common ions

Cations	Symbol
Sodium	Na^+
Potassium	K^+
Silver	Ag^+
Ammonium	NH_4^+
Copper(I)	Cu^+
Hydrogen	H^+
Lithium	Li^+
Calcium	Ca^{2+}
Magnesium	Mg^{2+}
Barium	Ba^{2+}
Mercury	Hg^{2+}
Lead	Pb^{2+}
Tin	Sn^{2+}
Zinc	Zn^{2+}
Copper(II)	Cu^{2+}
Iron(II)	Fe^{2+}
Cobalt	Co^{2+}
Aluminium	Al^{3+}
Iron(III)	Fe^{3+}
Chromium	Cr^{3+}

Anions	Symbol
Chloride	Cl^-
Bromide	Br^-
Iodide	I^-
Fluoride	F^-
Sulfide	S^{2-}
Oxide	O^{2-}
Nitride	N^{3-}
Phosphide	P^{3-}
Hydroxide	OH^-
Nitrate	NO_3^-
Nitrite	NO_2^-
Hydrogencarbonate	HCO_3^-
Hydrogensulfate	HSO_4^-
Hypochlorite	OCl^-
Thiocyanate	SCN^-
Permanganate	MnO_4^-
Carbonate	CO_3^{2-}
Sulfate	SO_4^{2-}
Sufite	SO_3^{2-}
Chromate	CrO_4^{2-}
Dichromate	$Cr_2O_7^{2-}$
Phosphate	PO_4^{3-}

Ionic compounds are neutral – the ratio of positive and negative ions ensures that the charges will balance (cancel).

Example

Formulae of ionic compounds

Magnesium fluoride Ions are Mg^{2+} and F^-.

Since the magnesium ion has a +2 charge, two fluoride ions are needed to balance the positive charge:

Mg^{2+} with F^- and F^-.

The formula for magnesium fluoride is MgF_2 – this shows that there are twice as many fluoride ions as magnesium ions.

Aluminium oxide Ions are Al^{3+} and O^{2-}.

To balance the charge, two aluminium ions and three oxide ions are required:

Al^{3+}, Al^{3+} and O^{2-}, O^{2-}, O^{2-}.

The formula for aluminium oxide is Al_2O_3.

Brackets are used in ionic formulae containing more than one polyatomic ion to avoid confusion when showing the relative number of polyatomic ions present.

Example

	Polyatomic ion	
	Zinc hydroxide	**Ammonium sulfate**
Ions present	Zn^{2+} and OH^-	NH_4^+ and SO_4^{2-}
Charges balanced	Zn^{2+} and OH^- and OH^-	NH_4^+ and NH_4^+ and SO_4^{2-}
Formula	$Zn(OH)_2$	$(NH_4)_2SO_4$

When an element can form ions of variable charge (eg iron can exist as Fe^{2+} or Fe^{3+}), the ion in the formula is identified using roman numerals.

Example

Iron(II) oxide is made up of the ions Fe^{2+} and O^{2-}, so the formula is FeO.

Iron(III) oxide is made up of the ions Fe^{3+} and O^{2-}, so the formula is Fe_2O_3.

Unit 11.2 Activity 4A: Formulae of ionic compounds

1. Name the following ionic compounds:

a. $ZnCO_3$
b. K_3PO_4
c. $Ca(HCO_3)_2$
d. Cu_2O
e. $(NH_4)_2Cr_2O_7$
f. KSCN
g. NaOCl
h. $Co(NO_3)_2$
i. $Fe_2(SO_4)_3$
j. $Pb(OH)_2$
k. $CrCl_3$
l. $NaMnO_4$

2. Write formulae for the following ionic compounds:

a.	Magnesium hydroxide	**k**.	Potassium sulfite
b.	Mercury nitrate	**l**.	Aluminium sulfate
c.	Iron(II) oxide	**m**.	Cobalt bromide
d.	Lead iodide	**n**.	Sodium chromate
e.	Copper(II) bromide	**o**.	Copper(I) chloride
f.	Lithium hydrogencarbonate	**p**.	Iron(III) fluoride
g.	Ammonium carbonate	**q**.	Potassium permanganate
h.	Silver nitrite	**r**.	Tin oxide
i.	Barium nitride	**s**.	Silver sulfide
j.	Calcium phosphide	**t**.	Ammonium chloride

Soluble ionic compounds

Some ionic compounds are soluble in water. When ionic solids dissolve in water, they completely dissociate to form **aqueous** ions.

In equations representing chemical reactions, the state (or phase) of the **reactants** and products is shown using the following symbols:

(*g*) gas, (*ℓ*) liquid, (*s*) solid, (*aq*) dissolved in water.

Example

A solution of sodium chloride

Sodium chloride, a soluble ionic compound, dissolves in water to form a solution containing sodium ions, $Na^+(aq)$, chloride ions, $Cl^-(aq)$, and a large amount of water molecules, $H_2O(\ell)$. The dissolving process is represented by the equation:

$$NaCl(s) \xrightarrow{H_2O} Na^+(aq) + Cl^-(aq)$$

Sodium chloride solution

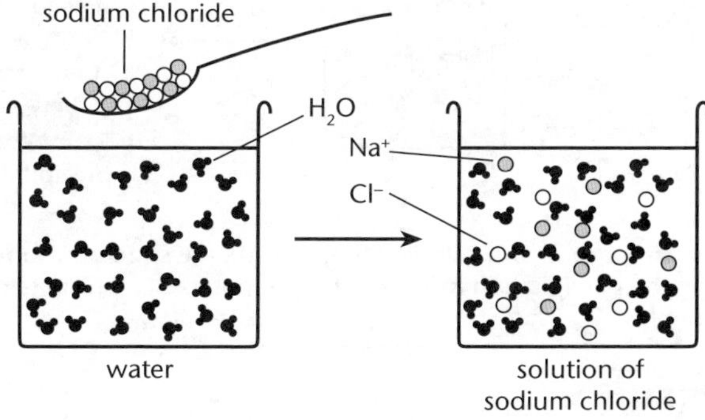

Solutions are clear. They may be colourless (eg $NaCl(aq)$) or coloured (eg $CuSO_4(aq)$).

Insoluble ionic compounds

Insoluble ionic compounds do not readily form solutions. However, even the most insoluble compound will dissolve to some extent if left long enough in water.

Soluble and insoluble compounds are the extremes of a range of solubilities. 'In-between' compounds may be described as *slightly* or **sparingly soluble**.

Whether a substance is defined as being soluble or insoluble also depends on the field of science in which it is used. To a geologist, the mineral calcium carbonate (from limestone rocks) is regarded as being soluble, since it will dissolve in water over millions of years, leaving caves and tunnels in rock formations. However, to a chemist, calcium carbonate is insoluble, because so little would dissolve during a typical experimental time frame of interest.

Precipitation

Insoluble ionic solids can form when two solutions of soluble compounds are mixed. Cations from one solution and anions from the other solution react to form an insoluble solid or **precipitate**. This process is called **precipitation**.

Example

A precipitate of calcium carbonate

When a calcium chloride solution is mixed with a solution of sodium carbonate, calcium ions, $Ca^{2+}(aq)$, combine with carbonate ions, $CO_3^{2-}(aq)$, to form a precipitate of insoluble calcium carbonate, $CaCO_3(s)$.

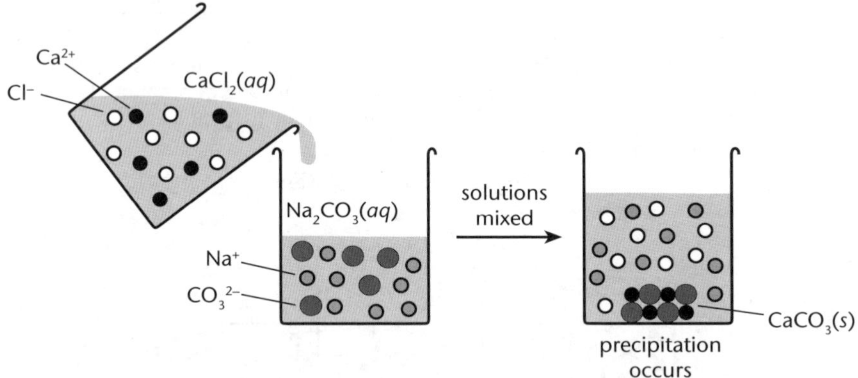

The sodium ions, $Na^+(aq)$, and chloride ions, $Cl^-(aq)$, are **spectator ions** since they are *not* involved in the precipitation reaction. They do not form a precipitate because sodium chloride is soluble.

Solubility rules

Solubility rules describe which compounds are soluble and which are insoluble. The solubility rules can be used to predict the formation of precipitates when solutions are mixed.

Ion(s) present in compound		Solubility	Exceptions	
Sodium, Na^+; Potassium, K^+; Ammonium, NH_4^+		All are soluble		
Nitrate	NO_3^-	All are soluble		
Chloride	Cl^-	All are soluble	*except*	silver chloride and lead chloride
Sulfate	SO_4^{2-}	All are soluble	*except*	lead sulfate, calcium sulfate and barium sulfate
Carbonate	CO_3^{2-}	All are insoluble	*except*	those of Group 1 (sodium carbonate, potassium carbonate) and ammonium carbonate
Hydroxide	OH^-	All are insoluble	*except*	those of Group 1 (sodium hydroxide, potassium hydroxide)
Iodide	I^-	All are soluble	*except*	silver iodide and lead iodide

Precipitates

Most precipitates are *white*. Those containing **transition metal** ions, however, often have a characteristic colour; eg *blue* copper hydroxide, $Cu(OH)_2$, and *green* iron(II) hydroxide, $Fe(OH)_2$.

Some precipitates appear as small, separate particles in solution, while others, especially those containing hydroxide ions, may appear **gelatinous** or 'jelly-like'.

Precipitates can be removed from the remaining solution either by **filtration** or by allowing them to settle before **decanting** off the unwanted solution.

Formation of precipitates

To decide if a precipitate forms, the ions involved must be known and the solubility rules must be applied.

Solubility grids are useful for predicting what precipitates (if any) will form when two solutions are mixed.

Only combinations of cations with anions are considered:

✓ represents a combination giving an insoluble compound (precipitate).

✗ represents a combination giving a soluble compound.

Combinations of 'anion with anion' and 'cation with cation' are ignored and marked with a dash, –.

Example

Predicting precipitation

In the Example on page 76, $CaCl_2$ solution mixed with Na_2CO_3 solution, the ions in one solution, Ca^{2+} and Cl^-, are put across the top of the grid; ions in the second solution, Na^+ and CO_3^{2-}, are written down the grid:

Ions from $CaCl_2$

Ions from Na_2CO_3		Ca^{2+}	Cl^-
	Na^+	–	✗
	CO_3^{2-}	✓	–

From the solubility rules, it can be predicted that a precipitate of calcium carbonate will form when the two solutions are mixed.

So, a ✓ is placed on the grid where Ca^{2+} and CO_3^{2-} ions 'intersect'. Sodium chloride is soluble, so Na^+ and Cl^- will *not* form a precipitate, hence an ✗ is placed on the grid where they 'intersect'. Cation/cation and anion/anion combinations are ignored, so given a –.

Ions in solution are in pairs.

- Because all nitrates are soluble, if a metal ion is needed in solution, the metal nitrate should be used.
- All sodium, potassium and ammonium compounds are soluble. To obtain an anion in solution, it is common to use the sodium compound of that anion (although the potassium or ammonium ion compound could be used).

Appearance and solubility of compounds formed from cations and anions						
	Cl^-	I^-	OH^-	NO_3^-	CO_3^{2-}	SO_4^{2-}
Ag^+	White solid insoluble	Yellow solid insoluble	Brown solid* insoluble	White solid soluble	Yellow solid insoluble	White solid soluble
Na^+	White solid soluble	White solid soluble	White solid soluble	White solid soluble	White solid soluble	White solid soluble
NH_4^+	White solid soluble	White solid soluble	White solid soluble	White solid soluble	White solid soluble	White solid soluble
Ba^{2+}	White solid soluble	White solid soluble	White solid insoluble**	White solid soluble	White solid insoluble	White solid insoluble
Cu^{2+}	Green solid soluble	DNE	Blue solid insoluble	Blue solid soluble	Green solid insoluble	Blue solid soluble
Fe^{2+}	Green/ yellow solid soluble	Grey solid soluble	Green solid insoluble	DNE	Grey solid insoluble	Green solid soluble
Mg^{2+}	White solid soluble	White solid soluble	White solid insoluble	White solid soluble	White solid insoluble	White solid soluble
Pb^{2+}	White solid insoluble	Yellow solid insoluble	White solid insoluble	White solid soluble	White solid insoluble	White solid insoluble
Zn^{2+}	White solid soluble	White solid soluble	White solid insoluble	White solid soluble	White solid insoluble	White solid soluble
Al^{3+}	White solid soluble	White solid soluble	White solid insoluble	White solid soluble	DNE	White solid soluble
Fe^{3+}	Yellow-brown solid soluble	DNE	Red-brown solid insoluble	Light purple solid soluble	DNE	Yellow solid soluble

* The brown solid is silver oxide; silver hydroxide is unstable.
** Barium hydroxide is sparingly soluble.
DNE = does not exist.

Compounds listed as *insoluble* will form a precipitate if two solutions, each of which contains one of the ions in the precipitate, are mixed.

Unit 11.2 Activity 4B: Insoluble compounds and precipitates

1. Below is a list of ionic compounds. Decide whether each compound is soluble or not. If insoluble, write down the formula of the solid. If soluble, write the formulae of the ions present in a solution of the compound.

- **a.** Sodium chloride.
- **b.** Barium chloride.
- **c.** Ammonium nitrate.
- **d.** Lead sulfate.
- **e.** Lead hydroxide.
- **f.** Silver chloride.
- **g.** Sodium sulfate.
- **h.** Magnesium nitrate.
- **i.** Lead carbonate.
- **j.** Aluminium hydroxide.
- **k.** Zinc chloride.
- **l.** Copper(II) carbonate.
- **m.** Barium sulfate.
- **n.** Iron(III) chloride.
- **o.** Zinc hydroxide.
- **p.** Potassium chloride.
- **q.** Sodium carbonate.
- **r.** Sodium iodide.
- **s.** Iron(II) sulfate.
- **t.** Ammonium carbonate.
- **u.** Calcium carbonate.
- **v.** Aluminium chloride.
- **w.** Silver iodide.
- **x.** Lead iodide.
- **y.** Copper(II) hydroxide.

2. The compounds below dissolve in water to form solutions containing ions. For each compound, write an equation representing the dissolving process.

- **a.** Potassium hydroxide.
- **b.** Sodium nitrate.
- **c.** Magnesium chloride.
- **d.** Copper(II) sulfate.
- **e.** Sodium carbonate.
- **f.** Aluminium nitrate.
- **g.** Sodium bromide.
- **h.** Potassium iodide.
- **i.** Sodium hydroxide.
- **j.** Ammonium sulfate.
- **k.** Zinc chloride.
- **l.** Magnesium nitrate.

3. Write the name and formula of the precipitate that would be formed (if any) when the following solutions are mixed. If no precipitate forms, write 'None'.

- **a.** Sodium carbonate and calcium chloride.
- **b.** Lead nitrate and potassium iodide.
- **c.** Potassium hydroxide and copper sulfate.
- **d.** Magnesium sulfate and sodium chloride.
- **e.** Silver nitrate and zinc chloride.
- **f.** Barium nitrate and aluminium sulfate.
- **g.** Ammonium sulfate and copper chloride.
- **h.** Iron(II) chloride and sodium hydroxide.

Ionic equations

Equations for precipitation reactions show only the ions involved in precipitation. Spectator ions are not shown. Such equations are called **ionic equations**.

Example

Ionic equations for precipitation of $CaCO_3$

When calcium chloride solution, $CaCl_2(aq)$, is added to sodium carbonate solution, $Na_2CO_3(aq)$, calcium ions, Ca^{2+}, combine with carbonate ions, CO_3^{2-}, to form a precipitate of $CaCO_3$. This reaction is written:

$$Ca^{2+}(aq) + CO_3^{2-}(aq) \rightarrow CaCO_3(s)$$

The spectator ions Na^+ and Cl^- are *not* shown in the equation.

Formal equations are sometimes written when quantitative considerations are involved. A formal equation is required when *amounts* of *substances* reacting are required.

Example

Formal equations

For the Example on page 76, the formal equation is:

$$CaCl_2(aq) + Na_2CO_3(aq) \rightarrow CaCO_3(s) + 2NaCl(aq)$$

Unit 11.2 Activity 4C: Ionic equations

1. Write an ionic equation for the formation of iron(II) hydroxide when sodium hydroxide is added to freshly prepared iron(II) sulfate solution.

2. Write ionic equations for the precipitation of:

a. Copper(II) hydroxide. **b.** Copper(II) carbonate.

c. Barium sulfate. **d.** Silver chloride.

3. Write balanced ionic equations for the formation of the precipitate formed when the following solutions are mixed:

a. Sodium sulfate, Na_2SO_4, and Calcium nitrate, $Ca(NO_3)_2$.

b. Potassium hydroxide, KOH, and Magnesium chloride, $MgCl_2$.

c. Lead nitrate, $Pb(NO_3)_2$, and Potassium chloride, KCl.

d. Aluminium sulfate, $Al_2(SO_4)_3$, and Sodium hydroxide, NaOH.

e. Sodium carbonate, Na_2CO_3, and Magnesium nitrate, $Mg(NO_3)_2$.

f. Silver nitrate, $AgNO_3$, and Sodium iodide, NaI.

4. Write ionic equations for the formation of the precipitate when the following pairs of solutions are mixed:

a. Sodium carbonate and calcium chloride.

b. Lead nitrate and potassium iodide.

c. Potassium hydroxide and copper sulfate.

d. Silver nitrate and zinc chloride.

e. Barium nitrate and aluminium sulfate.

f. Iron(III) chloride and sodium hydroxide.

Complex ions

Complex ions consist of a metal ion and a ligand – the ion or molecule bonded to the metal. Square brackets are used in the formula, eg $[Cu(NH_3)_4]^{2+}$.

When metal ions dissolve in water, the water molecules bond to the ions to form a complex ion, ie the water molecules act as ligands.

Example

When copper sulfate, $CuSO_4(s)$, is dissolved in water, aqueous copper ions and sulfate ions are present:

$$CuSO_4(s) \xrightarrow{H_2O} Cu^{2+}(aq) + SO_4^{2-}(aq)$$

Aqueous copper(II) ions, Cu^{2+} (aq), are better represented as $[Cu(H_2O)_4]^{2+}$, where each Cu^{2+} is bonded to four water molecules:

$$Cu^{2+}(aq) + 4H_2O(\ell) \rightarrow [Cu(H_2O)_4]^{2+}(aq)$$

Complex ions can also be formed when some metal ions are dissolved in solutions containing ammonia, NH_3, or hydroxide ions, OH^-.

Complex ions formed with $NH_3(aq)$ and $OH^-(aq)$					
	Cu^{2+}	Zn^{2+}	Al^{3+}	Ag^+	Pb^{2+}
$NH_3(aq)$	$[Cu(NH_3)_4]^{2+}$ royal blue	$[Zn(NH_3)_4]^{2+}$ colourless		$[Ag(NH_3)_2]^+$ colourless	
$OH^-(aq)$		$[Zn(OH)_4]^{2-}$ colourless	$[Al(OH)_4]^-$ colourless		$[Pb(OH)_4]^{2-}$ colourless

Another useful complex ion is formed between iron(III) ions and thiocyanate ions, SCN^-; the complex ion that forms, $[Fe(SCN)]^{2+}$, is a blood-red colour:

$$Fe^{3+}(aq) + SCN^-(aq) \rightarrow [Fe(SCN)]^{2+}(aq)$$

Complex ions may be formed when a precipitate dissolves in excess of solution. Complex ions may form when excess of **reagent** causes an insoluble salt to dissolve.

Example

When a few drops of sodium hydroxide solution are added to a solution of zinc chloride, a white precipitate, $Zn(OH)_2$, forms, which dissolves when excess sodium hydroxide is added:

$$Zn^{2+}(aq) + 2OH^-(aq) \rightarrow Zn(OH)_2(s)$$

$$Zn(OH)_2(s) + 2OH^-(aq) \rightarrow [Zn(OH)_4]^{2-}(aq)$$

The second reaction is sometimes written as:

$$Zn^{2+}(aq) + 4OH^-(aq) \rightarrow [Zn(OH)_4]^{2-}(aq)$$

Unit 11.2 Activity 4D: Complex ions

1. Complete the following equations for the formation of complex ions:

a. ________ + ________ $\rightarrow [Cu(NH_3)_4]^{2+}$

b. $Pb(OH)_2(s) + 2OH^- \rightarrow$ ________

c. $Zn^{2+}(aq) + 4NH_3(aq) \rightarrow$ ________

d. ________ + ________ $\rightarrow [AlOH)_4]^-$

e. $Fe^{3+}(aq) + SCN^{-}(aq) \rightarrow$ ________

f. ________ + ________ $\rightarrow$ $[Ag(NH_3)_2]^+$

2. a. Sodium hydroxide is added to a precipitate of zinc hydroxide. The precipitate disappears. Write a balanced equation for the reaction occurring.

b. A solution of copper sulfate has a small amount of ammonia solution added to it. A blue precipitate forms. When excess ammonia solution is added, the precipitate disappears. Write balanced equations for the reactions occurring.

c. A solution of lead nitrate has a small amount of aqueous sodium hydroxide added to it. A precipitate forms which disappears when excess of the hydroxide solution is added. Write balanced equations for the reactions occurring.

d. When silver nitrate is added to aqueous potassium chloride, a white precipitate forms. When aqueous ammonia is added, the precipitate disappears. Write balanced equations for the reactions occurring.

Unit 11.2 Chemical and Metallic Bonding

Topic 5: Metallic bonding

Topic 5 continues our exploration of chemical bonding and the formation of new substances. In previous Topics in the Unit we looked at covalent bonding and ionic bonding; now we cover:

- Describing molecular, ionic, metallic and covalent network solids, their melting points, polarity and solubility.

Introduction

Matter refers to everything in and around us – water, carbon dioxide in the air, enamel on our teeth, interstellar dust and haemoglobin in our blood.

Matter can be classified by its state or type.

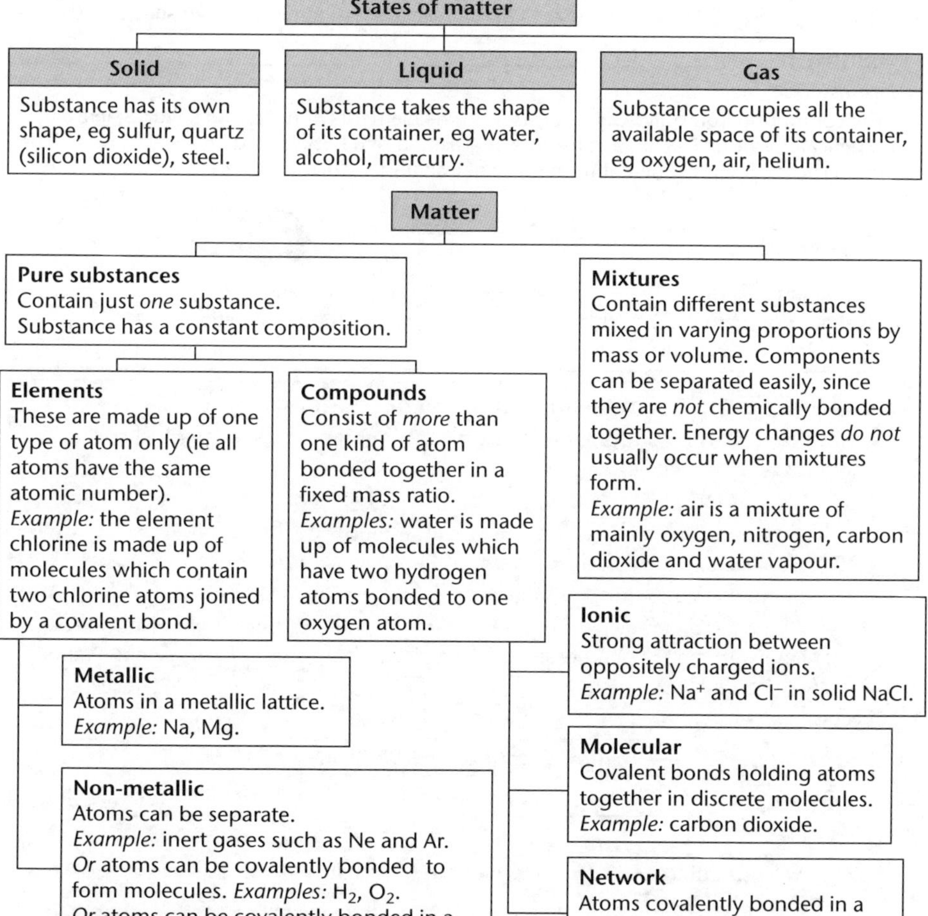

Chemical and physical changes

A substance can undergo **physical changes** or **chemical changes**.

Physical changes are easily reversed, for example melting ice, dissolving copper sulfate in water.

Chemical changes are more difficult or impossible to reverse.

Chemical changes include processes such as:

- Elements reacting together, eg magnesium metal reacts with oxygen in the air – difficult to reverse.
- Compounds reacting together, eg the starch and sugars in toast reacting with oxygen in the air as the toast burns – impossible to reverse.

When compounds form from elements, a chemical change takes place:

- An energy change occurs.
- The compound 'formed' has *constant* composition.
- The properties of the compound are *different* from the properties of its constituent elements.

Example

Iron sulfide

The compound iron sulfide, FeS, forms from the elements iron, Fe, and sulfur, S, in a chemical reaction. FeS has a constant composition of 63.6% Fe and 36.4% S by mass. The properties of FeS are different from those of iron and sulfur.

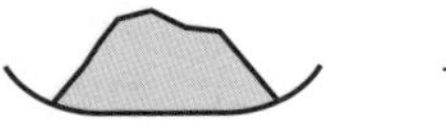

+

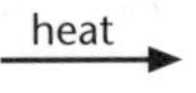

Sulfur powder (yellow, non-magnetic) | Iron filings (black, magnetic) | Iron sulfide powder (black, non-magnetic)

Unit 11.2 Activity 5A: Chemical and physical changes

1. Classify the following processes as physical or chemical changes. For each change, give one reason for your choice.
   - **a.** Boiling water.
   - **b.** Chips frying.
   - **c.** Lighting a match.
   - **d.** Sugar dissolving in water.
   - **e.** Mixing sand and common salt.
   - **f.** A steak being grilled.
   - **g.** Lithium metal reacting with water.
   - **h.** Ice cubes forming.
2. When solid magnesium metal is burnt in oxygen, a white powder forms. This powder has a high melting point and doesn't conduct electricity. Account for these observations.
3. Classify the statements **a.** to **g.** below as features of physical or chemical changes. Record your answers in a table like the one below.

Physical change	Chemical change

   - **a.** No new substances form.
   - **b.** Difficult or impossible to reverse.
   - **c.** Involves significant energy changes.
   - **d.** Occurs when substances dissolve.
   - **e.** Easily reversed.
   - **f.** New substances form.
   - **g.** Occurs when living things die.

States of matter

Particles in the solid state are close-packed, in a regular arrangement (often called a lattice).

In the liquid state, particles are not as closely packed and are irregular in their arrangement.

In the gas state, particles are widely separated.

Forces between particles are, on average, constant.

- In the solid state, the forces are strong enough to keep the particles in a fixed position (although the particles do vibrate and rotate about their fixed positions).
- In the liquid state, the greater energy of the particles enables the disruption of the lattice and particles are free to move around.
- In the gaseous state, the even higher energy of the particles means the particles can break away completely from each other.

Solids

Solids have the following properties:

- **Incompressible**.
- Fixed shape and volume.

There are four main types of solid. Within each type, there can be further groupings.

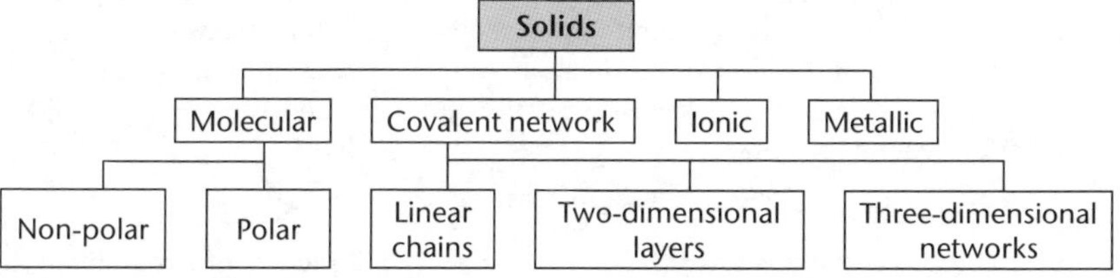

Different types of solid have different physical properties.

Property	Type of solid			
	Metallic	**Ionic**	**Covalent network**	**Molecular**
Electrical conductivity	Conduct	Conduct in the molten state or when dissolved in water	Usually non-conducting	Non-conducting
Melting point	Usually high (some have low melting points, eg mercury, lithium)	High	High	Usually low
Solubility in water	Always insoluble unless react with water (eg lithium and calcium)	Variable solubility (eg NaCl soluble, AgCl insoluble)	Always insoluble (eg quartz)	Variable solubility (eg ammonia is very soluble, CO_2 moderately soluble, and methane insoluble)
Hardness	Usually hard (eg iron), but some are soft (eg lead and lithium)	Hard	Usually hard (eg quartz)	Usually soft

Physical properties are determined by the types of particles present in the solid and the consequent forces of attraction between the particles. Particles are atoms, ions or molecules.

Forces of attraction between the particles can be strong (eg metallic bonds, ionic bonds or covalent bonds) or weak (such as intermolecular forces).

- If a solid has a high melting point, then there will be strong forces of attraction between the particles. These forces of attraction will only be broken when there is a large input of energy. Low melting point solids have weak intermolecular forces between the particles.
- If a solid conducts electricity, it must have charged particles (ions or electrons) that are free to move around (ie are **delocalised**).
- If a solid is **malleable** (able to be hammered) or **ductile** (able to be drawn into a wire), it must have layers that can slide over each other without disrupting the lattice.

Metallic solids

Metals are three-dimensional network structures. Metals consist of:

- A three-dimensional array of metal atoms.
- Delocalised or loosely-held **valence** electrons, moving freely through the lattice.

The metal atoms are arranged in a regular 3-dimensional lattice. The valence electrons from each atom are delocalised, ie they are free to move between the atoms. When electrons move away from the atom, a cation is left behind. The metallic bond is the **electrostatic attraction** between these cations and the electrons. (Metals are sometimes described as *positive ions* (nuclei and inner shell electrons) locked into a lattice and surrrounded by a 'sea' of valence electrons that move among them.)

Metallic lattice

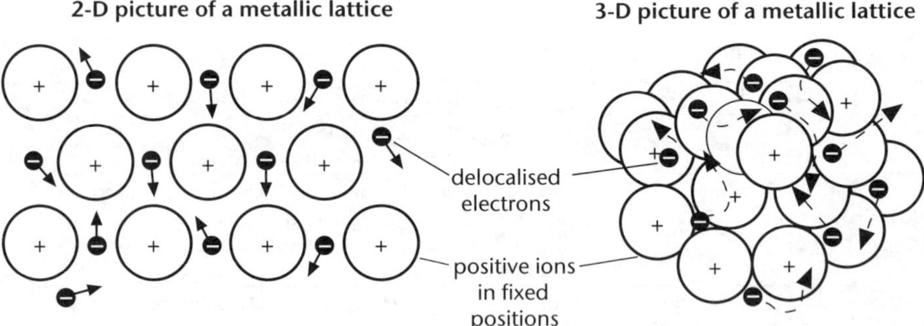

Solubility of metallic substances

Metals are not soluble in water. Some react with water, producing metal ions, hydroxide ions, OH^-, and hydrogen gas:

	metal	+	water	→	metal hydroxide	+	hydrogen gas
eg:	sodium	+	water	→	sodium hydroxide	+	hydrogen
	2Na	+	$2H_2O$	→	$2Na^+ + 2OH^-$	+	H_2

Properties of metals

Since delocalised electrons are continuously moving through the lattice of positive ions, the attractive forces have no set direction, and so:

- The solid lattice is easily deformed – metals are malleable (bend easily) and ductile (can be drawn out into thin wire). The layers of atoms can slide over each other and still remain bonded by the moving electrons.
- Metals conduct electricity – due to the mobility of the delocalised electrons.
- Metals conduct heat – due to the vibration of the atoms/positive ions transferring heat energy.
- **Metallic bonds** are very strong and metals usually have high melting points.

Ionic solids

Ionic substances consist of cations and anions.

The strong force of attraction between the positive and negative ions is known as the ionic bond.

Simple molecules do not exist in ionic compounds. This is because each cation is surrounded by anions, which in turn are surrounded by more cations to form a giant 3-D lattice.

Ionic compounds are described by their **empirical formula**, the simplest ratio of different ions present in the extended lattice.

Example

The formula of magnesium chloride

Different samples of magnesium chloride, $MgCl_2$, will have different *numbers* of ions present, depending on the mass of the sample, but all samples will have the same *ratio* of one magnesium ion to two chloride ions.

In ionic solids, ions attract each other electrostatically, so that each is surrounded by a group of oppositely charged ions, forming a stable three-dimensional lattice.

Example

The sodium chloride lattice

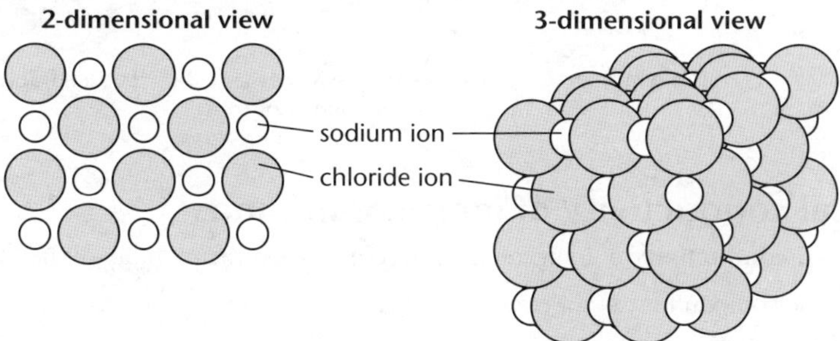

The chloride ions are larger than the sodium ions because chloride ions have their valence electrons in the *third* level, whereas the valence electrons in sodium ions are in the *second* energy level.

Melting and boiling points of ionic substances

Most ionic compounds are solids at room temperature, with high melting and boiling points, because considerable energy is needed to break the strong ionic bonds in the lattice.

Example

Sodium chloride

Sodium chloride is a typical ionic solid. Its high melting point (801 °C) and boiling point (1465 °C) show considerable energy is needed to break the ionic lattice.

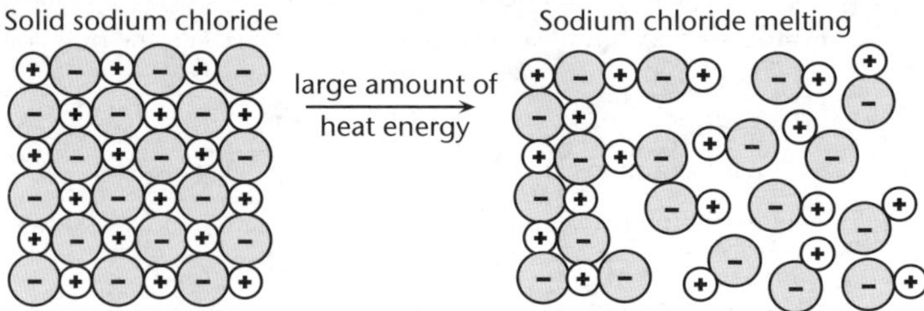

Crystal hardness of ionic solids

The strong electrostatic forces hold ions in a rigid lattice, producing a hard structure.

If an attempt is made to distort a crystal, the ions move so that anions are next to anions and cations are next to cations – the like charges repel each other, and the crystal lattice becomes **brittle** and fractures:

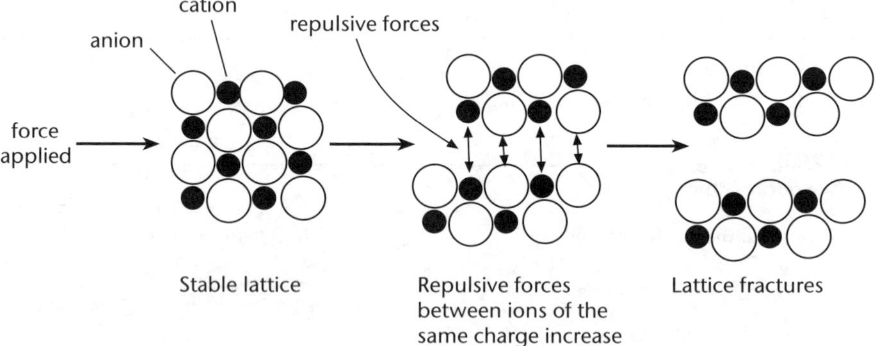

Electrical conductivity of ionic compounds

The **electrical conductivity** of ionic compounds depends on the mobility of their ions:

- Solid ionic compounds do not conduct – the ions are in fixed positions and are unable to move freely.
- **Molten** and dissolved ionic solids do conduct – the ions are able to move freely.

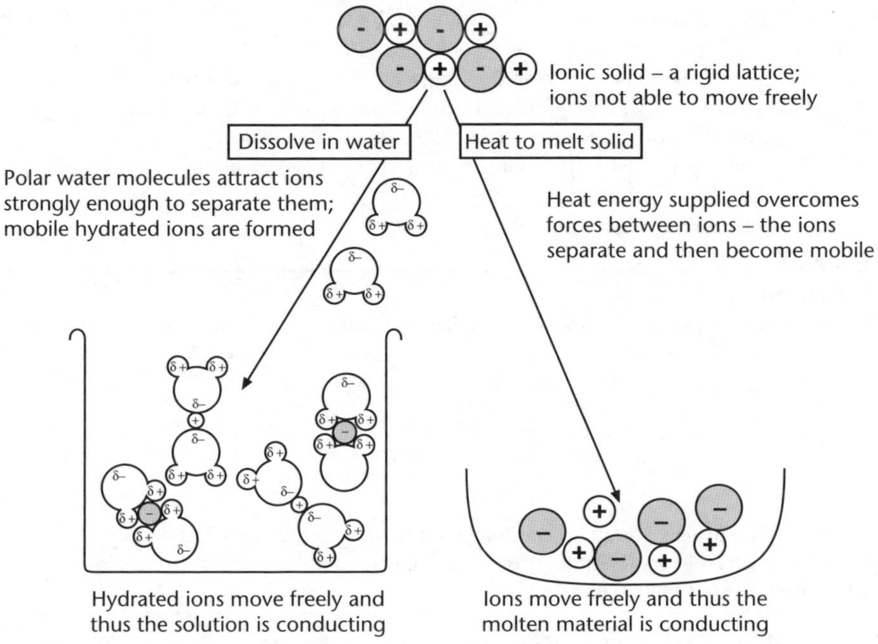

Mobile ions move towards oppositely charged **electrodes**, carrying an electric charge.

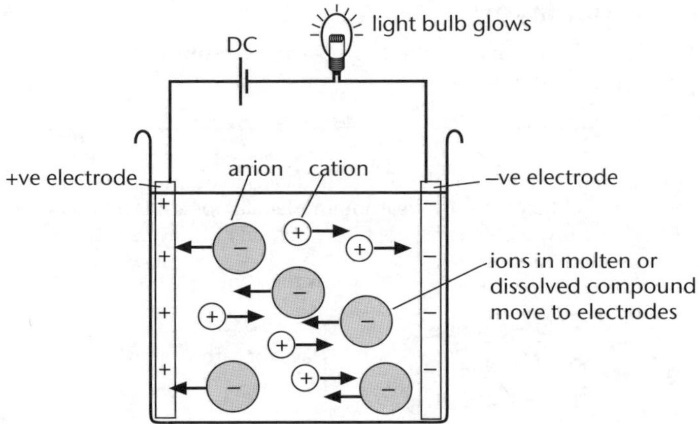

Covalent network solids

Compounds that consist of large networks of atoms covalently bonded to each other are called **covalent network** or **macromolecular** solids. Covalent network substances exist as either linear chains, two-dimensional layers, or three-dimensional networks.

Linear chains

The atoms in linear chains are formed into 'infinitely' long, one-dimensional chains. As solids, they have the following properties:

- Low melting points – since covalent bonding exists in only one dimension. However, the longer a molecule is, the higher its melting point – as the length of a chain increases, more sites exist for intermolecular forces to occur between adjacent chains.
- Low electrical conductivity – no charged particles (electrons or mobile ions) are present to carry charge.
- Exist as soft, flexible substances.

Example

Polyethene

Polyethene (or polythene) is made of covalently bonded carbon and hydrogen atoms in long, two-dimensional chains.

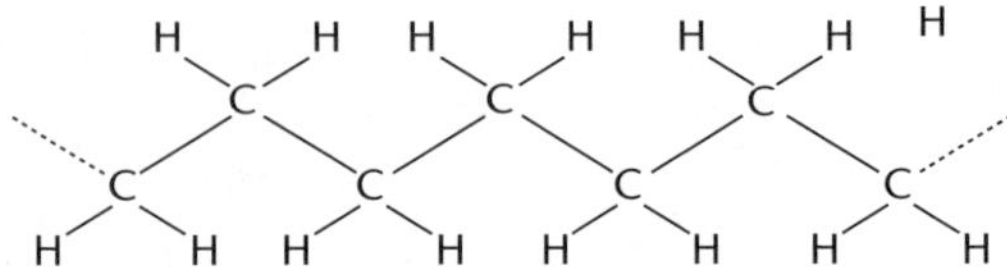

The chains are usually curled up and folded rather than being stretched out.
Polyethene stretches much more lengthwise than crosswise, as lengthwise chains slip past each other much more easily.

Rubber and **plastic** form extended linear chains. One physical form (**allotrope**) of sulfur, called **plastic sulfur**, also exists in a one-dimensional chain.

Two-dimensional layers

Two-dimensional network solids have unusual properties. **Graphite**, an *allotrope* of carbon, has a two-dimensional structure.

Example

Graphite

Graphite is made up of *layers* of covalently bonded carbon atoms arranged in hexagonal rings. The layers are held together by weak intermolecular forces. Electrons which are free to move, called **delocalised electrons**, are found *between* the layers.

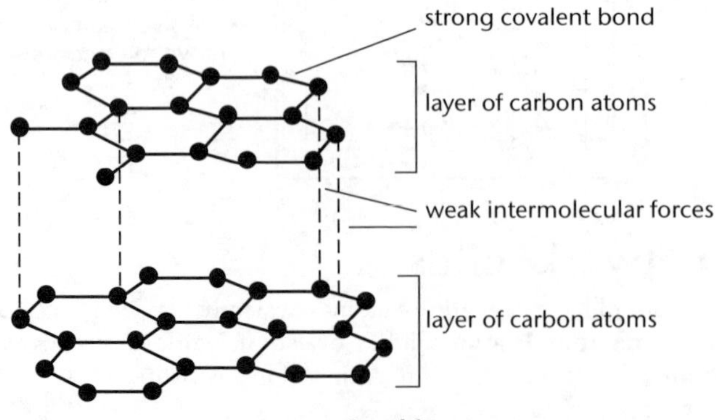

Graphite

The structure of graphite gives it the following properties:

- Electrical conductivity – it contains mobile delocalised electrons. (Graphite is the only covalent substance that conducts electricity.)
- A high melting point (3675 °C) – melting requires breaking the strong covalent bonds within the layers.
- A soft and greasy texture – the graphite layers can move easily over each other. This makes graphite an excellent 'dry' lubricant and is the substance found in 'lead' pencils.

Three-dimensional networks

All of the atoms in a 3-D network are held together by strong covalent bonds.

Three-dimensional extended covalent substances have the following properties:

- High melting points – melting requires breaking a 3-D array of strong covalent bonds.
- Low electrical conductivities – no mobile valence electrons or ions present.
- Strong, rigid structures – atoms are held fast in the 3-D crystals.
- Insoluble in water – large crystal size and non-polarity prevent extended covalent substances from dissolving.

Examples include **diamond**, SiO_2, and silicon carbide, SiC.

Example

Diamond

Diamond, an allotrope of carbon, is a 3-D network of carbon atoms. Each carbon atom is joined to four others by covalent bonds, forming a very strong, rigid structure:

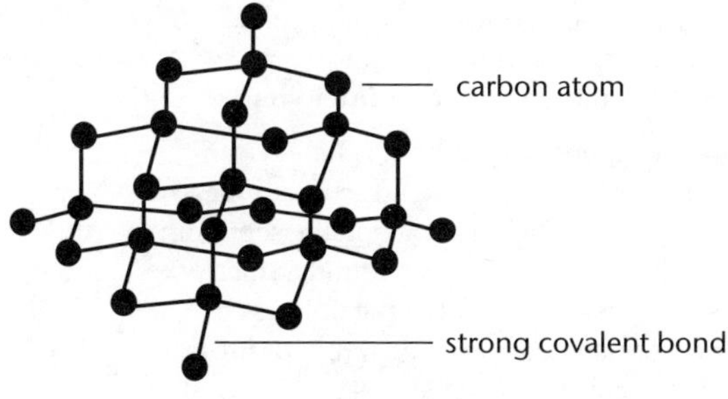

Diamond

Example

Quartz or silica

Silicon dioxide, or **silica,** is a three-dimensional covalent network solid. It has a ratio of one silicon atom to two oxygen atoms, so the empirical formula is SiO_2. Each silicon atom is bonded to *four* oxygen atoms to form a strong and rigid lattice with very high melting and boiling points.

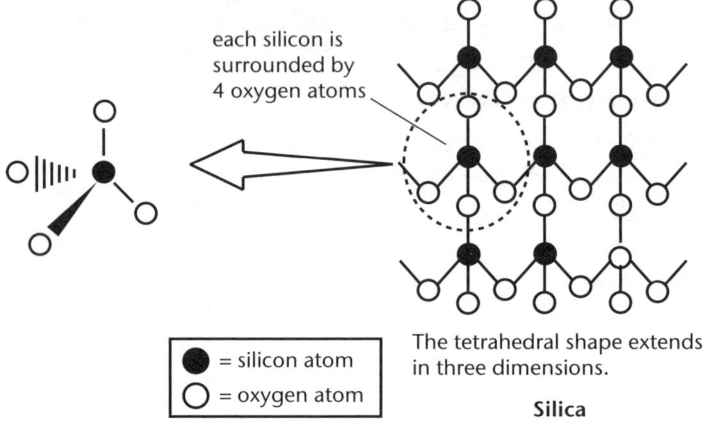

Silica

Silicon dioxide is the primary constituent of sand. It requires very high temperatures to melt sand to form glass.

Buckminsterfullerenes

In 1985, a new group of allotropic forms of carbon, called fullerenes, was discovered. The first fullerene, known as **buckminsterfullerene**, is commonly called a **buckyball**.

Buckminsterfullerenes were originally made in machines costing thousands of dollars, but since then it has been discovered they are present in natural products such as lignite, peat and the soot from an ordinary candle flame. Traces have also been found in meteorites and the interstellar dust that makes up most of the universe. Buckyballs act as superconductors at very low temperatures if metal atoms are added to them (the metal atoms exist 'within' the ball, inside the polygon framework).

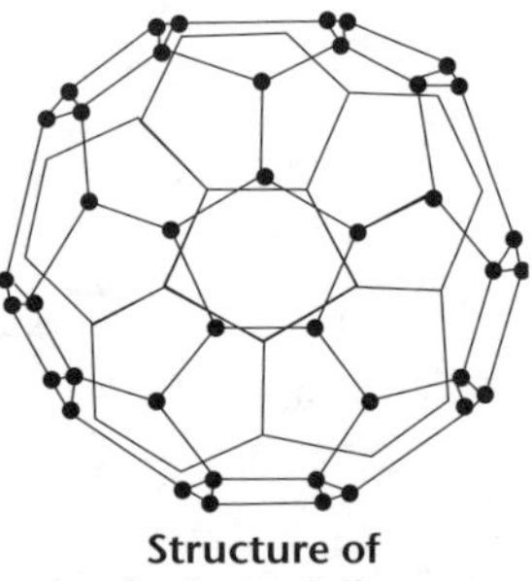

Structure of buckminsterfullerene

Molecular solids

Solids made of separate molecules packed together are known as **discrete molecular substances**.

There are two forces present in discrete molecular substances.

- The forces *within* the molecules – **intramolecular** forces – are the covalent bonds linking atoms in individual molecules.
- The forces *between* the molecules – **intermolecular** forces – link individual molecules to each other.

Intramolecular forces are very strong.

Intermolecular forces are much weaker than intramolecular forces.

Intermolecular forces determine the physical properties of discrete molecular substances.

Example

Solid carbon dioxide

Solid carbon dioxide (dry ice) is a solid at room temperature and **sublimes** (turns directly into a gas) because the forces *between* the molecules (ie the intermolecular forces) are very weak.

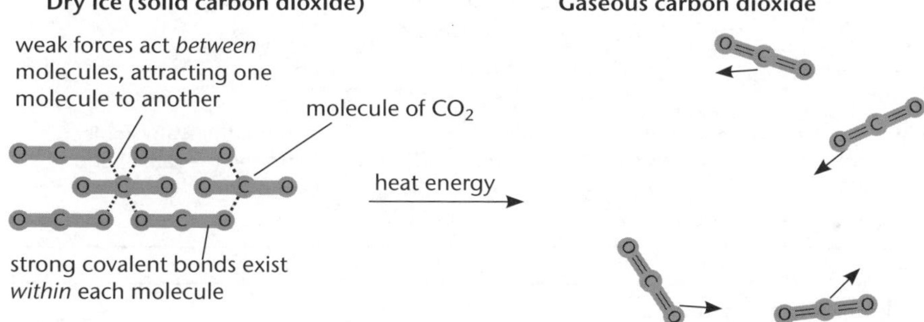

Regular arrangement of molecules held together by weak forces

Irregular arrangement of molecules, widely spaced, no forces acting between molecules

When the substance sublimes, the intramolecular forces (ie covalent bonds) remain intact. The molecules do not break down – they simply separate.

The number (and thus overall strength) of intermolecular forces increases with increasing molecular mass. Thus the boiling points (bp) of the **hydrocarbons** increases as the length of the carbon chain increases:

bp methane CH_4 < bp ethane CH_3CH_3 < bp propane $CH_3CH_2CH_3$ < bp butane $CH_3CH_2CH_2CH_3$

Few molecular substances are solids at room temperature because most have low melting and boiling points.

Solubility of molecular substances

Polar substances		Non-polar substances	
Methanol	$CH_3OH(\ell)$	Iodine	$I_2(s)$
Ethanol	$C_2H_5OH(\ell)$	Carbon dioxide	$CO_2(g)$
Water	$H_2O(\ell)$	Sulfur	$S_8(s)$
Ammonia	$NH_3(g)$	Chlorine	$Cl_2(g)$
Sulfur dioxide	$SO_2(g)$	Methane	$CH_4(g)$
		Octane	$C_8H_{18}(\ell)$

The ability of molecular substances to dissolve in water depends on their **polarity**.

- Polar molecular substances are generally soluble in water and other polar solvents.
- Non-polar molecular substances are generally insoluble in water, but will dissolve in non-polar solvents.

Solubility depends on the presence of the attractive forces between the molecular substance and solvent molecules.

Example

Solubility of ammonia

Ammonia, NH_3, is a polar molecule and is soluble in water (a polar solvent).

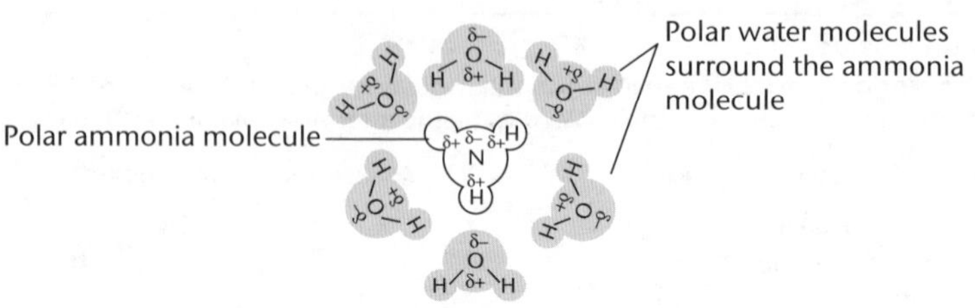

The negative end of the water molecule attracts the hydrogen (positive) end of the ammonia molecule

The positive end of the water molecule attracts the nitrogen (negative) end of the ammonia molecule

The polar ammonia molecules mix freely with water molecules.

Example

Insolubility of iodine

Iodine, I_2, is a non-polar molecule and is *insoluble* in water.

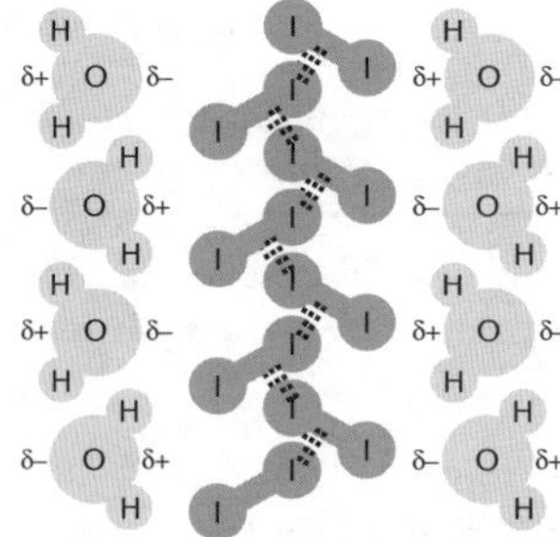

The forces of attraction between the non-polar iodine molecules and the polar water molecules are extremely weak. These forces are not strong enough to overcome the intermolecular forces between the iodine molecules or between the water molecules.

Since the water molecule is polar, the ability of polar substances to dissolve in water is sometimes described as *like dissolves like*. The expression also holds true for non-polar substances, as these are usually soluble in non-polar solvents (eg tar, a non-polar substance, can be removed from car paint by rubbing it with butter, which is also a non-polar substance).

Electrical conductivity of molecular substances

Electricity is a flow of charged particles. These charged particles will be either:

- Free-moving delocalised electrons (as in metals).
- Mobile ions (as in molten or dissolved ionic substances).

Since molecular substances consist of molecules and not charged particles, they show no electrical conductivity.

Unit 11.2 Activity 5B: Structure of solids

1. Choose from the list below properties common to ionic solids at room temperature.

A Solids with low melting points.
B Solids with fairly high melting points.
C Liquids with low melting points.
D Liquids with fairly high melting points.
E Solids that conduct electricity.
F Crystalline solids that fracture easily, retaining their smooth faces.

2. From the key list **i**.–**v**. below, select the answers to questions **a**.–**d**:

i. Diamond **ii**. Graphite **iii**. Iodine
iv. Sodium chloride **v**. Silicon dioxide

a. Which solid conducts electricity?
b. Which solid has the lowest melting point?
c. Which substance conducts electricity when molten?
d. Which substances are covalent network solids?

3. Classify each of the following substances as being molecular, ionic or metallic. Justify your answers, using the physical properties of the substances.

a. Silver, Ag **b**. Oxygen, O_2 **c**. Ethanol
d. Potassium iodide **e**. Copper(II) oxide **f**. Zinc
g. A non-conducting yellow solid which melted at just over 100 °C, forming a non-conducting liquid.
h. A white non-conducting solid that melted at 750 °C, forming a clear conducting liquid.

4. Use your knowledge of solids and their properties to copy and complete the table below. Iodine is used to show the type of answer required.

Substance	Type of solid	Particles present in the solid	Conductivity	Explanation for conductivity/ non-conductivity
Iodine	Discrete molecular	Molecules	Non-conducting	No mobile ions or delocalised electrons
Sodium bromide				
Graphite				
Aluminium				
Carbon dioxide				
Silica				

5. Explain the following:

a. Silicon dioxide has a high melting point; carbon dioxide has a low melting point.
b. Solid sodium conducts electricity; solid iodine does not.
c. Potassium chloride conducts electricity when molten; but not when solid.
d. Silver is malleable; sodium chloride is brittle.
e. Graphite leaves a black mark when rubbed on paper; diamond cuts paper.
f. Ice has a low melting point; magnesium oxide has a high melting point.

6. State the attractive forces that must be overcome when:

a. An aluminium bottle top is torn. **b**. Petrol vaporises.
c. Ice melts.

7. Complete the summary diagram below, by selecting answers for boxes **a**. to **h**. from the statement list given.

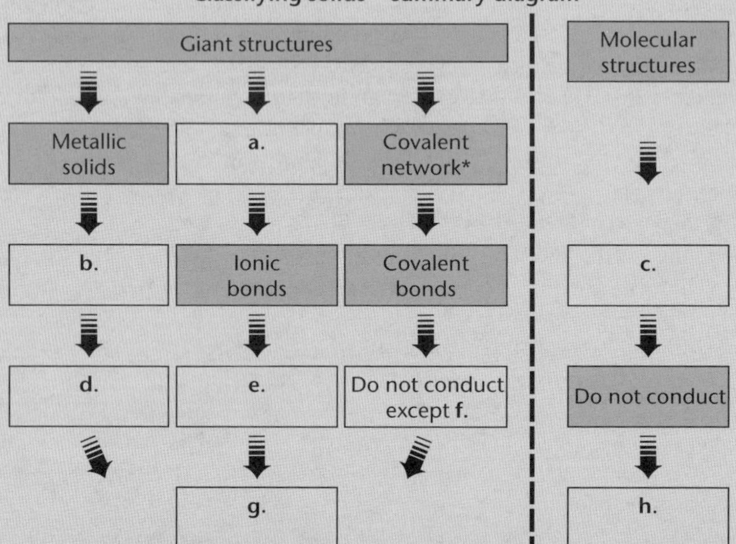

* Very few of these, eg diamond, graphite, polyethene

Statement list:

- High melting and boiling points.
- Conduct in solid and molten state.
- Covalent bonds within molecules – weak intermolecular forces between molecules.
- Low melting and boiling points.
- Metallic bonds.
- Conduct when molten or dissolved.
- Graphite.
- Ionic solids.

8. Following are four substances:

   i. Phosphorus, P_4 **ii**. Silicon dioxide, SiO_2
   iii. Magnesium chloride **iv**. Zinc, Zn

   For each of the substances:

   a. State the type of particle (atoms, ions or molecules) found in the solid substance.
   b. Specify the attractive force that is broken when the solid substance melts.
   c. Describe the attractive force existing between the particles of the solid as weak or strong.

9. Following are four substances:

   i. Aluminium, Al **ii**. Graphite, C
   iii. Hydrogen sulphide, H_2S **iv**. Calcium oxide, CaO

   For each of the substances:

   a. State the type of particle (atoms, ions or molecules) found in each solid substance.

b. Specify the attractive force existing between the particles of each solid substance.

c. Describe the relative melting point of each substance as either high or low.

10. Explain, in terms of structure and bonding within each solid, why the solid has the property described.

a. A piece of lead, Pb, can be easily reshaped without breaking into smaller pieces.

b. Solid zinc chloride, $ZnCl_2$, is a poor conductor of electricity. However, when dissolved in water, the solution is a good electrical conductor.

c. Oxygen, O_2, has a low melting point (–219 °C).

d. Diamond is one of the hardest substances known.

e. Copper, Cu, is a good conductor of electricity and is used for electrical wires.

11. For each of the uses of different crystalline solids below, discuss the property identified by relating the property to the structure and bonding within the solid.

a. Sulfur, S_8, does not conduct electricity.

b. The melting point of iron (1535 °C), is very high.

c. Silicon carbide, SiC, has a high melting point and is one of the hardest substances known.

d. Crystals of 'salt', NaCl, are hard but brittle and easily shattered.

12. Explain in terms of structure and bonding within the solid, the following properties of dry ice, $CO_2(s)$.

a. Dry ice readily sublimes.

b. Dry ice is more soluble in cyclohexane than it is in water.

c. Dry ice does not conduct electricity.

13. Iodine crystals sublime when gently heated, but crystals of potassium iodide don't melt until temperatures reach 681 °C. Iodine crystals are soluble in cyclohexane, while potassium iodide crystals are more soluble in water.

Discuss these properties of iodine and potassium iodide, in terms of the structure and bonding within each solid.

14. Carbon dioxide, CO_2, and silicon dioxide, SiO_2, are oxides of elements found in Group 14 of the periodic table. CO_2 sublimes at –78 °C, while the melting point of SiO_2 is 1700 °C.

- Account for the difference in melting point of the two oxides.
- Discuss similarities and differences in the conductivity and hardness of the two compounds.

15. Diamond and graphite are allotropes of carbon. This means that although they are both only made of carbon atoms, the atoms are arranged differently. This results in different physical forms of the same element. Diamond is used to make the teeth for saws which cut marble, while graphite is used as a solid lubricant and to make electrodes.

Discuss the structure and bonding within diamond and graphite, and relate these to the uses described above.

Unit 11.2 Chemical and Metallic Bonding

Topic 6: Periodic trends

Continuing the examination of atomic structure and bonding, we now look at periodic trends, which is the term used to describe specific patterns in the periodic table that illustrate different aspects of a certain element. In particular, we look at:

- Trends in the reactivity of metals.
- Trends in the reactivity of non-metals.
- Relationship between the trends in reactivity and the electron configuration of the element.
- Comparison of the properties of ionic compounds with covalent compounds.

In this Topic we only include the first 20 elements of the periodic table. A summary of the modern periodic table is given on p. 105.

Introduction

Two major types of element are recognised in the periodic table – metals and non-metals. Non-metals include those elements (eg helium, neon and argon) that are inert (unreactive), as well as those that are reactive.

A section of the periodic table is shown below. It contains:

- The symbols of the first 20 elements.
- Each element's atomic number (above the symbol).
- The electron configuration of an atom of the element (below the symbol).
- A dotted black-and-white line separating the metals from the non-metals.

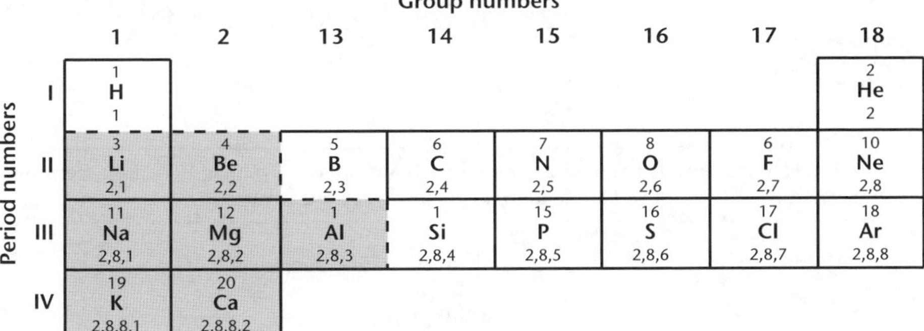

Group numbers

Period numbers	1	2	13	14	15	16	17	18
I	1 H 1							2 He 2
II	3 Li 2,1	4 Be 2,2	5 B 2,3	6 C 2,4	7 N 2,5	8 O 2,6	6 F 2,7	10 Ne 2,8
III	11 Na 2,8,1	12 Mg 2,8,2	1 Al 2,8,3	1 Si 2,8,4	15 P 2,8,5	16 S 2,8,6	17 Cl 2,8,7	18 Ar 2,8,8
IV	19 K 2,8,8,1	20 Ca 2,8,8,2						

First 20 elements in the periodic table, separating metals and non-metals

The horizontal rows in the periodic table are called **periods**. The vertical columns in the periodic table are called **groups**.

Trends in the reactivity of metals

The reactivity of a metal is related to how much energy is needed to remove electrons from the outer energy level of an atom of the metal to form positive ions. *Note*: Metals do *not* form covalent bonds.

If an atom has only one electron in its outer energy level then this electron requires little energy to be removed from the attraction of the nucleus.

Example

Lithium, sodium and potassium metals are very reactive.

Trends across a period

In a period, the number of electrons in the outer energy level of an atom of an element increases steadily and regularly as the atomic number increases.

Example

The aluminium atom has three electrons in its outer energy level. In period III, the reactivity of sodium is greater than the reactivity of magnesium, which is greater than the reactivity of aluminium. The decreasing reactivity from Na → Al occurs because the number of electrons to be removed from the atom of the metal is increasing in the series Na → Al, and the nuclear charge is increasing.

Trends down a group

In group 1, as the single electron in the outer energy level gets further away from the nucleus, less energy is required to remove it.

Example

Potassium is more reactive than sodium, which is more reactive than lithium.

The group 2 metals have two electrons in their outer energy level. Because of the greater nuclear charge, more energy is needed to remove two electrons away from the positive nucleus than is needed to remove one electron. Thus group 2 metals are not as reactive as group 1 metals.

The same trend in reactivity occurs in group 2 as in group 1.

Example

Calcium is more reactive than magnesium, which is more reactive than beryllium, because the electrons to be removed are getting further away from the nucleus.

The trends in the reactivity of metals are summarised in the diagram below.

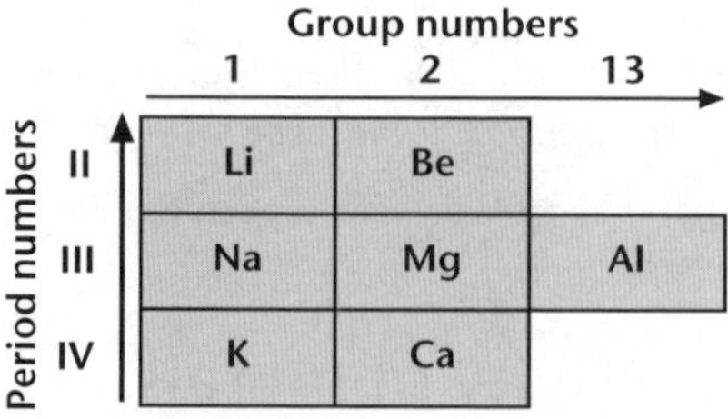

*The arrows point in the direction of **decreasing** reactivity.*

Reactivity trends for metals in the periodic table

Trends in the reactivity of non-metals

Non-metals form negative ions and also form covalent bonds. The trends in the reactivity of the non-metals can best be seen by concentrating on their formation of ions.

The group 17 non-metals have atoms with seven electrons in their outer energy levels, and only one more electron is needed to produce a stable octet. Fluorine is the most reactive of the non-metals because its atom has a small size and the nucleus of the atom can attract an electron strongly. The chlorine atom, which is larger than the fluorine atom because of an extra energy level of electrons, attracts an electron less strongly.

The group 16 non-metals need two electrons to form an octet. Once one electron has been attracted, it tends to oppose the attraction of the second electron. Oxygen atoms attract electrons more strongly than do sulfur atoms because the nucleus of the oxygen atom is closer to the outside of the atom and can attract electrons more strongly than the nucleus of a sulfur atom.

The trends in the reactivity of non-metals are summarised in the diagram below.

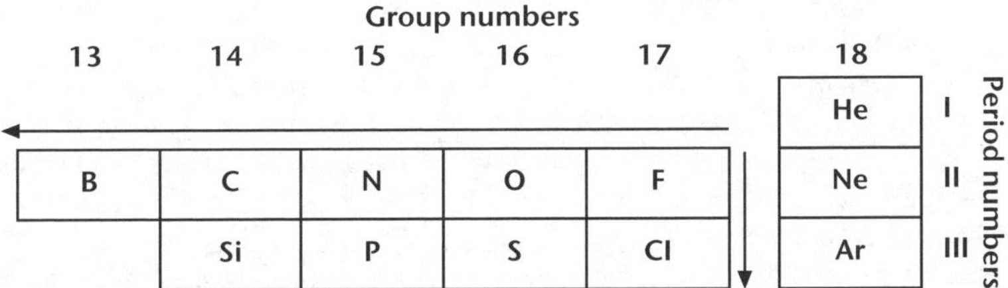

*The arrows point in the direction of **decreasing** reactivity. The inert gases (ie He, Ne and Ar) have been separated from the other non-metals to emphasise that they are unreactive.*

Reactivity trends for non-metals in the periodic table

Of the first twenty elements in the periodic table, the most vigorous reaction occurs between potassium and fluorine – the element whose atom loses an electron most readily (potassium) combining with the element whose atom gains an electron most readily (fluorine).

Unit 11.2 Activity 6A: Trends in reactivity of metals and non-metals

1. a. State the two main types of element that are recognised.

b. In the section of the periodic table below, three types of element are represented by three types of shading. Identify each type of element.

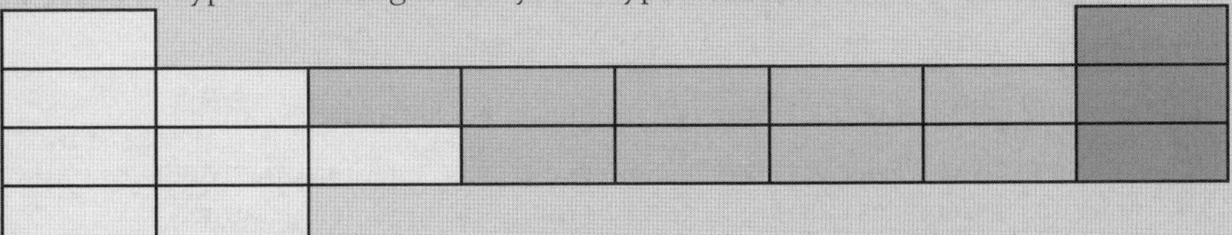

c. State the names given to:

i. The horizontal rows in the table.

ii. The vertical columns in the table.

2. Identify what type of element (metal; non-metal; inert gas) has the electron configuration of:

a. 2

b. 2, 2.

c. 2, 8, 7.

d. 2, 8, 8.

e. 2, 8, 8, 1.

3. Select the appropriate words from the word list below (in alphabetical order) to complete the following account of the reactivity of elements. The number of letters in each missing word is indicated by the spaces provided.

Elements whose atoms have _ _ _ _ _ electrons in their outer energy levels show no tendency to react. These elements are known as the _ _ _ _ _ _ _ _ _ _. If an element has an atom with one, two or three electrons in its outer energy level then these electrons can be _ _ _ _ _ _ _ to an atom of an element that has _ _ _ _, _ _ _ or _ _ _ _ _ electrons in its outer energy level. In this way, _ _ _ _ are formed. Atoms that donate electrons form _ _ _ _ _ _ _ _ ions and negative ions are formed by atoms _ _ _ _ _ _ _ _ _ electrons. _ _ _ _ _ _ form positive ions and _ _ _ - _ _ _ _ _ _ form negative ions.
The reactivity of a metal is related to the _ _ _ _ with which electron(s) in the atoms of the element are donated. Sodium is more reactive than magnesium, which is more reactive than _ _ _ _ _ _ _ _ _. There is _ _ _ electron in the outer energy level of an atom of sodium, two electrons in the outer energy level of an atom of _ _ _ _ _ _ _ _ _ and three electrons in the outer energy level of an atom of _ _ _ _ _ _ _ _ _. The _ _ _ _ _ _ needed to remove one electron from an atom is less than that needed to remove two, which is less than that needed to remove three.
Similarly, one electron is attracted more _ _ _ _ _ _ _ _ to an atom of a non-metal than are two or three electrons. Hence, fluorine is more reactive than _ _ _ _ _ _, which is more reactive than _ _ _ _ _ _ _ _.

Word list: accepting, aluminium, aluminium, donated, ease, eight, energy, five, inert gases, ions, magnesium, metals, nitrogen, non-metals, one, oxygen, positive, seven, six, strongly

4. Arrange the elements in each of the following sets in decreasing order of reactivity.

a. Aluminium, magnesium, sodium.

b. Carbon, helium, neon, nitrogen.

Comparison of compounds with ionic and covalent bonding

Molecules contain covalent bonds. Properties of molecular substances include:

- Molecules have weak forces *between* the molecules. These weak attractive forces result in the molecular substances existing as gases or liquids at room temperature.
- Many molecular substances are not soluble in water. Water molecules have a polarity, due to uneven distribution of charge in their molecule.

  The solubility process involves: separating molecules in the molecular substance (low energy required); separating water molecules (higher energy required due to the polarity); repayment of this energy debt by forces forming between the molecules and the water particles (low energy released). The repayment energy is not sufficient to allow dissolving to occur.

- In the liquid state or in solution (if the molecules dissolve in water), molecular substances carry no charge and therefore the liquid or the solution does not conduct electricity.

In contrast, the charged nature of ions means that:

- The particles attract each other strongly and form the solid state at room temperature.
- The ions attract the polar water molecules, which enables the solid to dissolve in water.
- Solutions of ionic substances, and molten ionic solids, conduct electricity (because they are composed of charged particles).

The properties of covalent (molecular) substances and ionic substances are compared below using two compounds – methane (covalent bonding) and sodium chloride (ionic bonding).

Property	Sodium	Methane
Common name(s)	Salt, brine (in solution), saline solution	Natural gas
Formula	NaCl	CH_4
Type of bonding	Ionic	Covalent
Particles present	*Sodium and chloride in a 1 : 1 ratio*	*A molecule of methane*
3-D diagram	*Sodium chloride structure*	*A methane molecule*
Physical state at room temperature	White crystalline solid	Colourless gas
Melting point	Very high (800 °C); due to strong attraction between ions	Very low (–182 °C); due to weak attraction between the molecules
Solubility in water	High; due to ions being attracted to water molecules	Low; due to the molecules having weak attraction for the water molecules
Electrical conductivity	Good both in the molten state and in aqueous solution – the ion, when free to move, can carry charge	Very poor; molecules carry no charge

Comparison of properties of compounds with ionic and covalent bonding

Unit 11.2 Activity 6B: Comparison of covalently bonded and ionically bonded compounds

1. State which of the compounds **i.** – **vi.** have atoms bonded by:
   - **a.** Electron sharing (covalency).
   - **b.** Electron transfer (ionic bonding).
     - **i.** Nitrogen oxide.
     - **ii.** Chlorine oxide.
     - **iii.** Calcium oxide.
     - **iv.** Phosphorus hydride.
     - **v.** Aluminium sulfide.
     - **vi.** Potassium fluoride.
2. State which of the formulae **i.** – **viii.** represent compounds containing atoms that are:
   - **a.** Covalently bonded.
   - **b.** Ionically bonded.
     - **i.** Na_2O.
     - **ii.** N_2O.
     - **iii.** F_2O.
     - **iv.** SiO_2.
     - **v.** K_2O.
     - **vi.** Na_3N.
     - **vii.** ClF.
     - **viii.** MgF_2.
3. Identify the type of bonding:
   - **i.** Covalent
   - **ii.** Ionic

   in each of the following compounds. Explain why you have chosen the type of bonding.
   - **a.** Colourless liquid; boiling point below 100 °C; liquid does not conduct electricity.
   - **b.** White solid; melts easily; does not dissolve in water; molten material does not conduct electricity.
   - **c.** White solid; melting point 2072 °C; does not dissolve in water; molten material conducts electricity.
   - **d.** Red-brown liquid; boiling point below 59 °C; slightly soluble in water; liquid does not conduct electricity but aqueous solution does.
4. For each of the substances **a.** – **d.**, predict the physical properties **i.** – **iii.** as high or low:

	i. Melting point	**ii.** Solubility in water	**iii.** Electrical conductivity of liquid form of substance
a. Nitrogen trichloride			
b. Carbon disulfide			
c. Sodium sulfide			
d. Potassium fluoride			

Unit 11.2 Chemical and Metallic Bonding

Topic 7: Periodic trends (extension)

In Topic 6 we included the first 20 elements only of the periodic table. Topic 7 is a revision and an extension of the previous Topic, giving you a fuller and more complete understanding of periodic trends and the nature of structure and binding in different substances.

Periodic table

In 1869, a Russian scientist, Dmitri Mendeléev, published a periodic table which contained all of the elements known at that time. He arranged them, with a few exceptions, in order of increasing atomic mass and showed that elements with similar **chemical properties** occurred at regular intervals in the same group or vertical column of the periodic table.

Modern periodic table

Atomic number	1 H 1.0	Molar mass / g mol^{-1}

1	2	3	4	5	6	7	8	9	10	11	12	13	14	15	16	17	18
																	2 He 4.0
3 Li 6.9	4 Be 9.0											5 B 10.8	6 C 12.0	7 N 14.0	8 O 16.0	9 F 19.0	10 Ne 20.2
11 Na 23.0	12 Mg 24.3											13 Al 27.0	14 Si 28.1	15 P 31.0	16 S 32.1	17 Cl 35.5	18 Ar 40.0
19 K 39.1	20 Ca 40.1	21 Sc 45.0	22 Ti 47.9	23 V 50.9	24 Cr 52.0	25 Mn 54.9	26 Fe 55.9	27 Co 58.9	28 Ni 58.7	29 Cu 63.6	30 Zn 65.4	31 Ga 69.7	32 Ge 72.6	33 As 74.9	34 Se 79.0	35 Br 79.9	36 Kr 83.8
37 Rb 85.5	38 Sr 87.6	39 Y 88.9	40 Zr 91.2	41 Nb 92.9	42 Mo 95.9	43 Tc 98.9	44 Ru 101	45 Rh 103	46 Pd 106	47 Ag 108	48 Cd 112	49 In 115	50 Sn 119	51 Sb 122	52 Te 128	53 I 127	54 Xe 131
55 Cs 133	56 Ba 137	71 Lu 175	72 Hf 179	73 Ta 181	74 W 184	75 Re 186	76 Os 190	77 Ir 192	78 Pt 195	79 Au 197	80 Hg 201	81 Tl 204	82 Pb 207	83 Bi 209	84 Po 210	85 At 210	86 Rn 222
87 Fr 223	88 Ra 226	103 Lr 262	104 Rf 261	105 Db 262	106 Sg 263	107 Bh 264	108 Hs 265	109 Mt 268	110 Ds 281	111 Rg 280							

Lanthanide Series	57 La 139	58 Ce 140	59 Pr 141	60 Nd 144	61 Pm 147	62 Sm 150	63 Eu 152	64 Gd 157	65 Tb 159	66 Dy 163	67 Ho 165	68 Er 167	69 Tm 169	70 Yb 173
Actinide Series	89 Ac 227	90 Th 232	91 Pa 231	92 U 238	93 Np 237	94 Pu 239	95 Am 241	96 Cm 244	97 Bk 249	98 Cf 251	99 Es 252	100 Fm 257	101 Md 258	102 No 259

The names and symbols of elements 104–111 have only very recently been finalised.

> Element 110, darmstadtium, named after the place it was first synthesised, was officially recognised only in 2003.
>
> In 2004, element 111, discovered in early 2003, was given the name roentgenium and the symbol Rg.

Elements 101–111 are called *transfermium* elements because of their very short existences.

Mendeléev's order of elements was in doubt in a few cases, so Mendeléev decided to give priority to the properties of the elements and thus the group they were in, rather than their atomic masses. His foresight has stood the test of time.

Example

Groups in the periodic table

The elements beryllium, magnesium and calcium are all found in group 2 of the periodic table. They have similar chemical properties which are different from the properties of elements in other groups.

In group 17, fluorine, chlorine and bromine all share similar properties.

After the periodic table was developed, the electron arrangements and the atomic numbers of the elements were revealed. These showed that the similarity of the chemical properties of the elements within a group arises because they have the same number of valence electrons. This supported Mendeléev's decision to give emphasis to properties over atomic mass.

Example

Groups and valence electrons

Magnesium, electron configuration Mg (2, 8, 2), and calcium, Ca (2, 8, 8, 2), are both in group 2 and both have 2 valence electrons.

Periods in the periodic table

Silicon, Si (2, 8, 4), and chlorine, Cl (2, 8, 7), are in the *third* period because they both have their valence electrons in the *third* energy level.

* The stepped line on the RHS of the periodic table shows where elements change from metals (on the left) to non-metals (on the right). The elements close to the line – such as germanium, Ge, arsenic, As and antimony, Sb – are known as **metalloids** or **semi-metals**, because they have properties similar to those of both metals and non-metals.

When Mendeléev developed his table there were some elements that had not yet been discovered. He left spaces for these in his table and predicted the existence and properties of the elements which later filled these gaps.

Example

Unknown elements

Mendeléev predicted the existence of germanium, which he called *eka-silicon*, since it would come just below silicon in the table. When germanium was discovered, it had properties close to those Mendeléev had predicted.

As well as individual elements missing from his periodic table, Mendeléev completely left out group 18, because no one had considered the existence of these (virtually) inert elements.

The modern periodic table contains all the known elements with atomic numbers between 1 and 111. There is space for heavier elements being identified in the future.

The first 20 elements are found in groups 1, 2 and 13 to 18 across the table in periods 1 to 4:

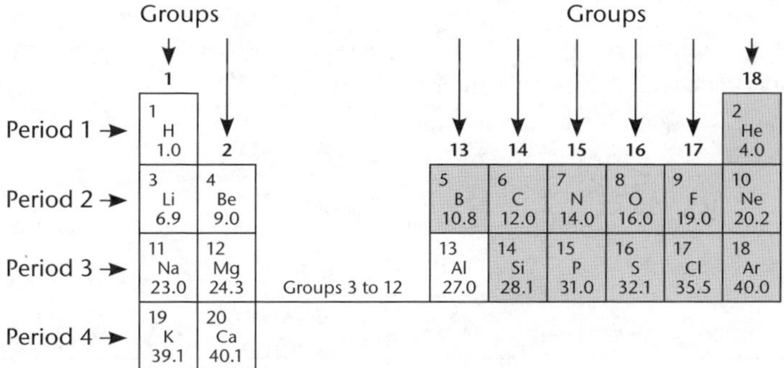

Metallic elements of groups 1, 2 and 13, on the left of the periodic table, *lose* electrons in chemical reactions to form positive ions (cations).

Example

Aluminium

Aluminium is a metal in group 13:

$Al \rightarrow Al^{3+} + 3e^-$

Elements in groups 14 to 18 on the right of the periodic table are non-metallic. The atoms can *gain* electrons to form negative ions (anions), by reacting with atoms of metals, or they can share electrons with atoms of other non-metals to form covalent bonds.

Example

Oxygen

Oxygen, in group 16, has atoms that need two extra electrons to achieve a full **valence level.**

Oxygen can *gain* electrons to form oxide ions, O^{2-}, eg with magnesium to form magnesium oxide, MgO

$\cdot\ddot{\underset{\cdot\cdot}{O}}\cdot + 2e^- \rightarrow [:\ddot{\underset{\cdot\cdot}{O}}:]^{2-}$

Oxygen can *share* electrons, eg with hydrogen to form water, H_2O

Helium and other elements in group 18 are virtually unreactive, because their atoms have a full valence level. They are called the noble or **inert gases**.

The 10 elements after calcium, elements 21 to 30, occur in groups 3 to 12 (in the centre of the table) and are called **transition metals**.

Unit 11.2 Activity 7A: Features of the periodic table

1. Key List:

A. H **B.** He **C.** C **D.** Na **E.** Cl

Answer the following questions by choosing from the key list above:

a. Which atom has four valence electrons?

b. Which two atoms have one valence electron?

c. Which atom has the second energy level as its valence level?

d. Which atom has seven valence electrons?

e. Which two atoms belong in the first period?

2. Sketch an outline of the periodic table.
   a. On it, in the appropriate places, put the symbols for the first 20 elements.
   b. Add to this the elements bromine, iodine and krypton.
   c. Circle the element which is in:
      i. The third period, group 2.
      ii. The second period, group 14.
3. Hydrogen appears in various positions on different periodic tables.
   Justify each of the following:
   a. Hydrogen sometimes appears above group 1.
   b. Hydrogen sometimes appears above group 17.

Properties of compounds

When compounds form from elements, the properties of the compounds formed depend upon the type of bonding, which in turn relates to the position of the elements on the periodic table.

The type of bonding between the elements in a compound can be predicted from the:

- Electron arrangement of the elements – elements with fewer than four electrons in their valence shell tend to form ionic compounds.
- Melting point and conductivity of the compound in its solid and liquid state – ionic compounds are usually solids with high melting points and conduct electricity when molten.
- Acid–**base** character of the compound.

  Acidic compounds generally involve covalent bonding. Acidic compounds dissolve in water producing **hydronium ions** or, if insoluble, react with a base.

  Basic compounds involve ionic bonding. Basic compounds dissolve in water producing hydroxide ions or, if insoluble, react with an **acid**.

Example

Compounds of sodium and sulfur

Sodium, a group 1 metal, is found on the *left-hand* side of the periodic table. When sodium forms compounds with non-metals, the compounds are ionic and usually basic or neutral.

Sulfur, a non-metal in group 16, is found on the *right-hand* side of the periodic table.When sulfur forms compounds with other non-metals, the compounds are covalent and sometimes acidic.

The elements of group 17 – the halogens

'Halogen' means 'salt-maker', which refers to the high reactivity of the elements found in group 17. The halogens exist as diatomic molecules.

Halogen	Formula	State at room temperature	Description	Melting point (°C)	Boiling point (°C)	Molecular mass (g mol^{-1})
Fluorine	F_2	Gas	Pale yellow	–220	–188	38
Chlorine	Cl_2	Gas	Yellow-green	–101	–35	71
Bromine	Br_2	Liquid	Red-brown	–7	59	160
Iodine	I_2	Solid	Purple-black	114	183	254

All halogen molecules are diatomic, with a single covalent bond between the atoms.

- Halogens have seven electrons in the outer energy level of their atoms. They bond covalently to form a diatomic molecule with a single covalent bond. One halogen atom shares an outer energy level electron with an outer energy level electron from another atom: a stable octet of electrons in the outer energy level is achieved for both atoms.
- Halogen molecules are non-polar – atoms forming the molecule are identical and the electrons forming the bond are equally shared between the two atoms. Weak forces exist between the molecules.
- Intermolecular forces increase in strength as the number of electrons in the molecule increases (ie as the mass increases). As a result, melting points and boiling points increase down the group (reflected in the change of state from gas to liquid to solid). At room temperature, fluorine and chlorine are gases, bromine is a liquid, iodine is a solid.

Unit 11.2 Activity 7B: Periodic trends for the halogens

1. Why are all halogen molecules diatomic?

2. If a compound can be formed between iodine and chlorine, predict the:

a. Type of bonding between the atoms.

b. Molecular formula.

c. Appearance.

d. **i.** Melting point. **ii.** Boiling point.

3. The element astatine, At, the halogen with the highest molar mass, is placed under iodine on the periodic table.

Explain why:

a. Its molecular formula is At_2 .

b. It is a grey solid with a melting point of approximately 300 °C.

Halides of the third period

The third period of the periodic table goes from sodium to argon.

Halides of the third period are described as 'the metal salts of the group 17 elements' (eg sodium fluoride, NaF; sodium chloride, NaCl; sodium bromide, NaBr; and sodium iodide, NaI).

The chemical and physical properties of the metal halides change across the period.

Chlorides of the third period

Group	1	2	13	14	15	16	17	18
Element	Na	Mg	Al	Si	P	S	Cl	Ar
Formula(e) of chloride(s)	NaCl	$MgCl_2$	$AlCl_3$	$SiCl_4$	PCl_3, PCl_5	SCl_2	Cl_2	–
Bonding	Ionic	Ionic	Ionic (< 180 °C); covalent (> 180 °C)	Polar covalent			Pure covalent	–
Acid–base behaviour	Neutral	Neutral	Acidic	Acidic	Acidic Acidic	Acidic	Acidic	–
Melting point (°C)	801	712	Sublimes at 180	–70	–91, 148 (sublimes)	–80	–101	–
Appearance	White solid	White solid	White solid	Colourless liquid	Colourless liquid, White solid	Dark red liquid	Pale green gas	–
Conductivity	Conducts in aqueous solution and in molten state		Non-conductor					

Sodium and magnesium chlorides

Sodium and magnesium form the white ionic solids sodium chloride, NaCl, and magnesium chloride, $MgCl_2$, which:

- Have moderately high melting points (because of their ionic bonding).
- Do not conduct electricity as solids but will conduct when molten or in aqueous solutions.
- Are soluble in water, forming neutral solutions.

Aluminium chloride

Aluminium chloride, $AlCl_3$, is an ionic solid which sublimes at moderately low temperatures. If heated under pressure, it melts to give a liquid. During the change from solid to liquid, the structure changes and some covalently bonded aluminium chloride *dimers* are produced.

Dimers of liquid aluminium chloride

Cl Cl Cl
Al Al
Cl Cl Cl

Dimers consist of two aluminium chloride **monomer** molecules joined together.

Silicon tetrachloride

Silicon tetrachloride, $SiCl_4$, is a colourless liquid formed when silicon is heated in chlorine gas. It has a non-polar molecular structure with a low melting point, and does not conduct electricity since it contains no mobile ions or free-moving electrons.

Chlorides of phosphorus

Phosphorus forms two covalent compounds with chlorine.

- Phosphorus(III) chloride, PCl_3, is a non-conducting liquid with a low melting point – PCl_3 is a covalently bonded molecule with only weak attractive forces between the molecules.
- Phosphorus(V) chloride, PCl_5, is a solid at room temperature with a fairly low melting point.

Sulfur dichloride

Chlorine reacts with molten sulfur in the presence of a $FeCl_3$ catalyst to form the poisonous and foul-smelling gas sulfur dichloride, SCl_2:

$$Cl_2(g) + S(\ell) \xrightarrow{FeCl_3} SCl_2(g)$$

Sulfur dichloride has a very low melting point and no electrical conductivity, consistent with its molecular structure:

S
Cl Cl

Chlorine

The physical properties of chlorine are:

- Yellow-green gas at room temperature.
- Pungent smell and corrosive if inhaled – chlorine gas was used to kill people in trench warfare in World War I.
- More dense than air.

Trends of the third-period chlorides

Moving across the third period from sodium to chlorine, the following trends are apparent:

- A change from ionic bonding (metal chlorides) to covalent bonding (non-metal chlorides).
- A change from high melting point (ionic compounds) to low melting point (molecular compounds).

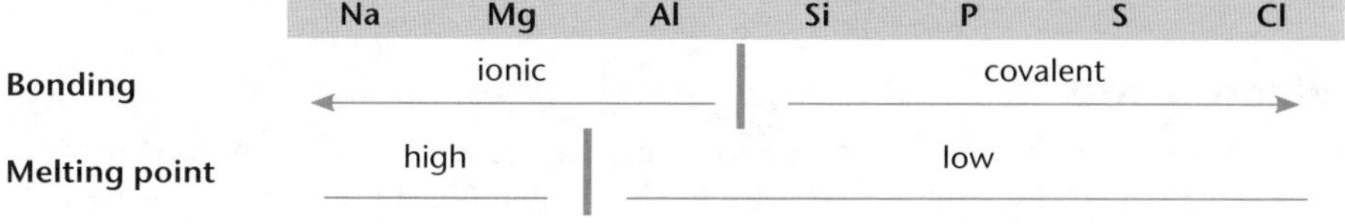

Oxides of the third period

Group	1	2	13	14	15	16	17	18
Element	Na	Mg	Al	Si	P	S	Cl	Ar
Formula(e) of oxide(s)	Na_2O	MgO	Al_2O_3	SiO_2	P_4O_6 P_4O_{10}	SO_2 SO_3	Cl_2O ClO_2 Cl_2O_7	–
Bonding	Ionic			Polar covalent				–
Structure	Ionic lattice			Extended network	Molecular			–
Melting point (°C)	1275 (sublimes)	2800	2045	1700	24 585	–73 17	–20 –60 –92	–
Appearance	White solid	White solid	White solid	White solid	White solid	Colourless gas White crystals	Cl_2O, ClO_2 are pale yellow gases; Cl_2O_7 is a colourless liquid	–
Conductivity	In aqueous solution and when molten	When molten	When molten	Non-conductors				

Oxides of sodium, magnesium and aluminium

Sodium, magnesium and aluminium form white ionic oxides, consistent with their metallic nature. Their atoms lose valence electrons to oxygen atoms and form positive ions.

These oxides are solids at room temperature, have high melting points and do not conduct electricity in the solid form. However, if ions become mobile (eg in molten or aqueous solution), the oxides show good conductivity.

Sodium oxide dissolves readily in water. Magnesium and aluminium oxides are insoluble in water.

Silicon oxide

Silicon (group 14) is the first non-metallic element in the third period. The oxide SiO_2 exists as a network covalent structure called **silica** or **silicon dioxide**. It is the basis of quartz and sand.

Silica has a three-dimensional network structure (like diamond), in which silicon atoms achieve a stable octet by using all four valence electrons in single covalent bonds with oxygen atoms.

The properties of silica are:

- High melting point due to strong 3-D covalent bonding.
- Does not conduct heat or electricity.
- Insoluble in water.

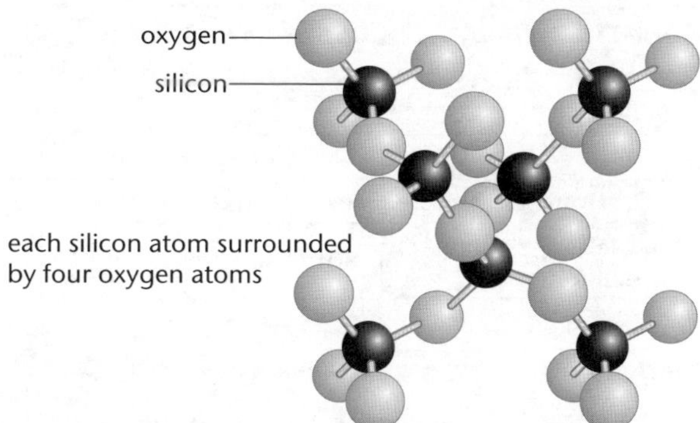

Oxides of phosphorus and sulfur

Phosphorus (group 15) is a non-metal which can form two different covalent oxides, phosphorus(III) oxide, P_4O_6, and phosphorus(V) oxide, P_4O_{10}, both of which can exist as several different structures.

Sulfur, a non-metal from group 16, forms two covalent oxides, sulfur dioxide, SO_2, and sulfur trioxide, SO_3.

The low melting points of the phosphorus and sulfur oxides are consistent with the molecular nature of the solids.

The oxides of phosphorus and sulfur tend to react with water to form **strong acids**.

Oxides of chlorine

Chlorine, a non-metal from group 17, forms several different oxides, all of which are discrete molecular substances which are mostly gases at room temperature. Most chlorine oxides react with water to form strong acids.

Argon

Argon is a group-18 element with a complete valence electron shell. It does not form any oxides.

Trends of the third-period oxides

Moving across the third period from sodium to chlorine, the following trends are apparent:

- A change from the high melting points of the ionic and extended covalent oxides to the low melting points of the molecular oxides.
- A change from the ionic bonding of the metals with oxygen to the covalent bonding of the non-metals with oxygen.

Unit 11.2 Activity 7C: Chlorides and oxides of the third period

1. Using the data table below for the chlorides of the third period, answer the questions that follow.

Element	Na	Mg	Al	Si	P	S	Cl	Ar
Formula of chloride	$NaCl$	$MgCl_2$	$AlCl_3$	$SiCl_4$	PCl_3	SCl_2	Cl_2	–
Melting point of chloride (°C)	801	712	193, 180 (Sublimes)	–68	–91	–80	–101	–

   a. Discuss two trends that can be determined from the table.
   b. Explain why the melting point of sodium chloride is much higher than that of phosphorus trichloride.
   c. Use the information in the table to predict the bonding in magnesium chloride and sulfur dichloride. Justify your answer.

2. a. Copy and complete the follow table:

Group							
Element							Ar
Oxides	Na_2O	MgO	Al_2O_3	SiO_2	P_4O_6 P_4O_{10}	SO_2 SO_3	ClO_2 Cl_2O	–
Bonding							

   b. i. Which oxide(s) exists as a gas at room temperature?
      ii. Which oxide(s) exists as a giant covalent lattice?

3. Melting points of the chlorides of selected third-row elements are given in the table below. Describe the type of bonding that must be broken to melt each substance. Give a reason for your answers

Name of substance	Formula	Melting point (°C)
Sodium chloride	$NaCl$	801
Magnesium chloride	$MgCl_2$	712
Silicon chloride	$SiCl_4$	–68
Sulfur dichloride	SCl_2	–80

4. The following table shows the melting and boiling points of oxides of some elements of the third row of the periodic table.

	Sodium oxide Na_2O	Magnesium oxide MgO	Silicon dioxide SiO_2	Sulfur dioxide SO_2
Melting point / °C	1132	2852	1713	–75
Boiling point / °C	1275	3600	2230	–10

   a. Describe the *trend* in melting and boiling points of oxides across the third row of the periodic table by referring to the data in the table above.
   b. Discuss reasons for the differences in melting points of all *four* oxides, shown in the table above, by referring to the particles and forces between the particles in the solids.

Unit 11.3 Types of Chemical Reactions
Topic 1: The language of chemistry

We are able to survive in this world because of the chemical reactions that are going on all around us, all the time. In Grades 9 and 10 you would have learnt the concept of writing chemical formulae and chemical equations. In this Unit, you learn to identify various chemical reactions and write balanced chemical equations; and you learn to classify chemical reactions into different types of chemical reactions. Now we look at:

- Formulae of compounds and valency.
- Writing equations in words and symbols.
- The importance of using accurate language to describe substances and their formulae.

Note that exothermic and endothermic reactions (Syllabus p. 15) are covered in Unit 11.4 Topic 2 (p. 197).

Formulae of compounds

A compound must contain at least two elements. For every compound, a **formula** can be written showing:

- The simplest unit of the compound that can be identified.
- The types of atom that are present.
- The number of each atom present.

Valency

Valency of an element is *either*:

- The number of single covalent bonds formed by one atom of the element. *Note*: A double covalent bond is the equivalent of two single covalent bonds. A triple covalent bond is the equivalent of three single covalent bonds.

or:

- The charge on the ion formed by one atom of the element.

The valencies of some elements whose atoms form covalent bonds (ie non-metals) are in the table:

Element and symbol	No. of electrons in outer energy level of atom	No. of electrons atom gains to become stable	Valency of element	Symbol used to indicate valency
Hydrogen, H	1	1	1	H—
Carbon, C	4	4	4	—C— (with bonds above and below) or =C=
Nitrogen, N	5	3	3	—N— (with bond below) or N≡
Oxygen, O	6	2	2	—O— or O=
Phosphorus, P	5	3	3	—P— (with bond below)
Chlorine, Cl	7	1	1	Cl—

Valencies of non-metals

Formulae for covalent compounds

Formulae for covalent molecules can be written from the symbols of the elements and the valency of the elements.

Example

Water has a molecule composed of two hydrogen atoms, each with a valency of one, and an oxygen atom which has a valency of two. From the table on the previous page, the formula of water is given by:

H– + **–O–** + **–H** to produce **H–O–H.**

This can be shortened to **H_2O.**

Formulae for ionic compounds

Ions of some metals and non-metals are shown in the table.

Element and symbol	No. of electrons in outer energy level of atom	No. of electrons atom gains/loses to become stable	Electrical charge on ion	Symbol used to indicate ion
Hydrogen, H	1	1 loss	1+	H^+
Lithium, Li	1	1 loss	1+	Li^+
Sodium	1	1 loss	1+	Na^+
Magnesium, Mg	2	2 loss	2+	Mg^{2+}
Aluminium, Al	3	3 loss	3+	Al^{3+}
Potassium, K	1	1 loss	1+	K^+
Calcium, Ca	2	2 loss	2+	Ca^{2+}
Fluorine, F	7	1 gain	1–	F^-
Chlorine, Cl	7	1 gain	1–	Cl^-
Oxygen, O	6	2 gain	2–	O^{2-}
Sulfur, S	6	2 gain	2–	S^{2-}
Nitrogen, N	5	3 gain	3–	N^{3-}
Phosphorus, P	5	3 gain	3–	P^{3-}

Ions for selected metals and non-metals

Ions are formed by the gain or loss of electrons, so they are electrically charged. Each ion attracts ions of opposite charge, eg positive ions attract negative ions. In an ionic compound, the ions are in a regular arrangement, called a **lattice**, where each positive ion is surrounded by negative ions and vice versa, but the overall ratio of positive to negative ions is simple, eg 1 : 1, 1 : 2, 2 : 1.

The formula of an ionic compound represents the simplest ratio of ions required to produce a neutral particle. The table of ions above can be used to write the formula of an ionic compound.

Example

Formulae of magnesium compounds

The formula of magnesium oxide contains one ion of each charge because the ions have the same charge:

$Mg^{2+} + O^{2-} \rightarrow MgO$

For magnesium chloride, one ion of 2+ charge balances two ions of 1– charge:

$Mg^{2+} + 2Cl^{-} \rightarrow MgCl_2$

For magnesium nitride, three ions, each of 2+ charge, balance two ions of 3– charge:

$3Mg^{2+} + 2N^{3-} \rightarrow Mg_3N_2$

Some compounds contain three or more elements. These molecules are termed **polyatomic**, meaning they contain many atoms. Some groups of atoms are not stable unless they carry a charge. These particles are called **polyatomic ions**.

Some common polyatomic ions are in the table.

Name of polyatomic ion	Formula of polyatomic ion
Ammonium ion	NH_4^+
Carbonate ion	CO_3^{2-}
Sulfate ion	SO_4^{2-}
Nitrate ion	NO_3^-
Hydrogen carbonate ion	HCO_3^-
Hydroxide ion	OH^-

Names and formulae of polyatomic ions

Example

Ionic compounds containing polyatomic ions

Ammonium sulfide contains two ions, each of 1+ charge, balancing one ion of 2– charge, ie,

$2NH_4^+ + S^{2-} \rightarrow (NH_4)_2S$

Aluminium sulfate contains two ions, each of 3+ charge, balancing three ions of 2– charge, ie,

$2Al^{3+} + 3SO_4^{2-} \rightarrow Al_2(SO_4)_3$

The subscripts after the brackets, eg the '2' in $(NH_4)_2$ and the '3' in $(SO_4)_3$, are multipliers for all the particles inside the bracket.

Example

$(NH_4)_2$ represents 2 × N atoms, and 8 × H atoms.

Unit 11.3 Activity 1A: Valency, formulae of compounds

1. State which of the following formulae represent compounds:
   - **a.** O_3.
   - **b.** SO_3.
   - **c.** Al_2S_3.
   - **d.** C_{60}.
   - **e.** C_6H_6.
   - **f.** $C_6H_{12}O_6$.
2. The valencies of some elements are:
   Phosphorus (P) = 3; oxygen (O) = 2; fluorine (F) = 1.
   Write the formula of:
   - **a.** Phosphorus oxide.
   - **b.** Phosphorus fluoride.
   - **c.** Oxygen fluoride.
3. Three elements you may not be familiar with are astatine (At), selenium (Se) and bismuth (Bi). Their valencies can be represented in the following manner:
   At– ; –Se– ; —Bi—. (with a single bond line drawn down from Bi)
   Write the formula of:
   - **a.** Selenium astatide.
   - **b.** Bismuth astatide.
   - **c.** Bismuth selenide.
4. If the bonding in selenium astatide, bismuth astatide and bismuth selenide is covalent, state how many electrons there are in the outer energy level of:
   - **a.** The astatine atom.
   - **b.** The selenium atom.
   - **c.** The bismuth atom.
5. The atoms of six elements are listed as A, B, C, X, Y, Z (these are not true symbols). Write the symbols for the ions formed by atoms of these elements given that the number of electrons in the outer energy level of the atom is (A one), (B two), (C three), (X five), (Y six), (Z seven), respectively.
6. Given the formulae for the following ions:
   Sodium Na^+; magnesium Mg^{2+}; aluminium Al^{3+}; chloride Cl^-; sulfide S^{2-}; nitride N^{3-}.
   Write the formula of:
   - **a.** Sodium chloride.
   - **b.** Magnesium sulfide.
   - **c.** Aluminium nitride.
7. You have just discovered an element at school and you have called it skulium (Sk). Because you are very clever, you have also discovered that the element has an atom with four electrons in its outer energy level. Write the formula of:
   - **a.** Skulium oxide.
   - **b.** Skulium chloride.
   - **c.** Skulium nitride.

8. Write the formula of:

a. The lithium ion, given the formula of lithium oxide is Li_2O.

b. The calcium ion, given the formula of calcium oxide is CaO.

c. The scandium ion, given the formula of scandium oxide is Sc_2O_3.

9. Given the formulae for the following ions:

Sodium Na^+, magnesium Mg^{2+}, aluminium Al^{3+}, ethanoate CH_3COO^-, oxalate $(COO)_2^{2-}$, phosphate PO_4^{3-}.

Write the formula of:

a. Sodium phosphate.

b. Magnesium ethanoate.

c. Aluminium oxalate.

10. You have just discovered the following elements at school and you have called them footium (Ft), hockium (Hk), squashium (Sq) and vollium (Vo). Because you are very clever, you have also discovered that the elements have atoms with the following number of electrons in their outer energy levels:

Ft (1); Hk (2); Sq (6); Vo (7).

Write:

a. The ion symbols for each element.

b. The formula for the compound formed between:

i. Footium and squashium.

ii. Hockium and vollium.

11. Write the formula of:

a. **i.** Sodium carbonate.

ii. Aluminium nitrate.

iii. Ammonium chloride.

iv. Calcium hydrogen carbonate.

v. Ammonium sulfate.

b. For each formula you have written in **a.**, write the number and type of each atom present.

Chemical equations

Chemical equations may be written in either word or symbol form.

Word equations

A **word equation** has the following features:

- It identifies the reagents and products of a reaction by name.
- It is of value as a preliminary to a symbol equation.

Word equations state the reaction in words.

Example

Word equations

lithium + oxygen → lithium oxide

calcium + sulfur → calcium sulfide

aluminium + chlorine → aluminium chloride

sodium + water → sodium hydroxide + hydrogen

magnesium + water → magnesium hydroxide + hydrogen

magnesium + steam → magnesium oxide + hydrogen

iron + steam → iron oxide + hydrogen

zinc + sulfuric acid → zinc sulfate + hydrogen

aluminium + hydrochloric acid → aluminium chloride + hydrogen

Symbol equations

When a chemical equation is written in symbol form:

- Each **reagent** and product is represented by a formula which accurately describes the smallest particle of that reagent or product that can take part in the reaction. The formula of a solid element is given as a single atom.
- The equation must balance, ie the number of atoms (which may be present in molecules or in the form of an ion) on both sides of the equation must be the same.
- The equation is **balanced** by placing whole numbers in front of the reagent or product particles. Formulae of individual reagents and products must not be changed.
- The states of the reagents and products are usually shown.

Example

The following symbols are used:

- (s) for solid – eg Cu(s).
- (ℓ) for liquid – eg $H_2O(\ell)$.
- (g) for gas – eg $O_2(g)$.
- (aq) for an aqueous solution – eg NaCl(aq).

In order to write equations in symbol form, you must know:

- The symbols for atoms of the elements.
- The valencies of the elements.
- The formula of the molecule, the polyatomic molecules or the polyatomic ions that are involved in the reactions.

Symbols

The symbols for the elements involved in the word equations above are (in alphabetical order):

Al, **Ca**, **Cl**, **Fe**, **H**, **Li**, **Mg**, **Na**, **O**, **S**, **Zn**.

If two letters are used for a symbol, the first letter is in upper case and the second letter is in lower case. The second letter in the symbol is usually the second letter in the name (but can be the third letter).

Example

Iron (Fe) and sodium (Na) have symbols based upon their Latin names, *ferrum* and *natrium*, respectively.

Example

Writing symbol equations

Word equation: calcium + sulfur → calcium sulfide

Symbol equation: $Ca(s) + S(s) \rightarrow CaS(s)$

The calcium atom forms a Ca^{2+} ion, and the sulfur atom forms a S^{2-} ion. Since the charges are of the same size and of opposite sign, the particles combine in a 1 : 1 ratio.

All the materials are in the solid state, (s), at room temperature.

Word equation: lithium + oxygen → lithium oxide

Symbol equation: $4Li(s) + O_2(g) \rightarrow 2Li_2O(s)$

The lithium atom forms a Li^+ ion; oxygen exists in the gaseous state as an O_2 molecule and becomes an O^{2-} ion in the reaction. Since the charges are in the ratio of $1(Li^+) : 2(O^{2-})$ and of opposite sign, the particles combine in a 2 : 1 ratio – hence the formula, Li_2O. To balance the equation, it is necessary to have four lithium atoms and one oxygen molecule to form two 'units' of lithium oxide.

Word equation: aluminium + chlorine → aluminium chloride

Symbol equation: $2Al(s) + 3Cl_2(g) \rightarrow 2AlCl_3(s)$

The aluminium atom forms an Al^{3+} ion; chlorine exists in the gaseous state as a Cl_2 molecule and becomes a Cl^- ion in the reaction. Since the charges are in the ratio of $3(Al^{3+}) : 1(Cl^-)$ and of opposite sign, the particles combine in a 1 : 3 ratio – hence the formula $AlCl_3$. The equation is balanced with two aluminium atoms and three chlorine molecules to form two aluminium chloride 'units'.

Word equation: zinc + sulfuric acid → zinc sulfate + hydrogen

Symbol equation: $Zn(s) + H_2SO_4(aq) \rightarrow ZnSO_4(aq) + H_2(g)$

The zinc atom forms a Zn^{2+} ion. Sulfuric acid is a polyatomic molecule which exists in aqueous solution as hydrogen ions, $2H^+(aq)$ and a sulfate ion, $SO_4^{2-}(aq)$. The hydrogen ions become a hydrogen molecule and the sulfate ion pairs with the zinc ion to form zinc sulfate. Since the charges are in the ratio of $1(Zn^{2+}) : 1(SO_4^{2-})$ and of opposite sign, the particles combine in a 1 : 1 ratio – hence the formula $ZnSO_4$. The equation is balanced with one particle of each reagent and product.

Unit 11.3 Activity 1B: Writing equations

1. Write the word equations for the reactions between:
   a. Lithium and oxygen.
   b. Magnesium and water.
   c. Iron and steam.
   d. Zinc and sulfuric acid.
   e. Aluminium and hydrochloric acid.
2. Complete the following word equations:
   a. __________ + oxygen → sodium oxide.
   b. Calcium + __________ → calcium sulfide.
   c. __________ + __________ → aluminium chloride.
   d. __________ + water → sodium hydroxide + hydrogen.
   e. __________ + steam → magnesium oxide + __________.
   f. __________ + sulfuric acid → magnesium sulfate + hydrogen.
   g. __________ + __________ → zinc chloride + hydrogen.
3. Balance the following equations (do not change any formulae).
   a. $Li(s) + S(s) \rightarrow Li_2S(s)$
   b. $Na(s) + Cl_2(g) \rightarrow NaCl(s)$
   c. $Al(s) + O_2(g) \rightarrow Al_2O_3(s)$
   d. $Ca(s) + H_2O(l) \rightarrow Ca(OH)_2(aq) + H_2(g)$
   e. $Mg(s) + HCl(aq) \rightarrow MgCl_2(aq) + H_2(g)$
4. Write the equations, in symbol form, for the reaction between:
   a. Magnesium and oxygen.
   b. Sodium and oxygen.
   c. Sodium and water.
   d. Magnesium and water.
   e. Zinc and steam.
   f. Magnesium and sulfuric acid.
   g. Zinc and hydrochloric acid.
   h. Aluminium and sulfuric acid.

The language of chemistry

The language used in chemistry needs to be accurate and as precise as possible.

Example

Choose the most precise of the following four statements.

Statement 1: Calcium added to water reacts to produce calcium hydroxide and hydrogen.

Statement 2: Calcium metal added to water reacts to produce calcium hydroxide solution and hydrogen gas.

Statement 3: Ca added to H_2O produces $Ca(OH)_2$ and H_2.

Statement 4: Ca added to H_2O produces CaOH and H.

Statements 1 and 2 both contain 'good' chemical language. Statement 2 is better than statement 1 because it provides more information (notice that water, because of its familiarity, is not described as a liquid).

Statement 3 is an attempt to use chemical shorthand but introduces confusion because *particles* are being used to represent *matter*. Statement 4 is an attempt to use chemical shorthand but introduces confusion because (i) two formulae (CaOH and H) are incorrect, and (ii) particles are being used to represent matter (eg using 'Ca' instead of 'calcium metal').

In summary:

- When writing *word* equations for reactions, you must use the *names* of the chemicals.
- An equation using the correct symbols for the reagents and products has a greater value than a word equation because the composition of the chemicals is indicated.

In the teaching and the learning of chemistry, there is confusion between *substances* and *particles*.

Example

Statement 5: Water is a colourless liquid.
Statement 6: H_2O is a colourless liquid.
Statement 7: Water contains two atoms of hydrogen and one atom of oxygen.
Statement 8: A water molecule contains two atoms of hydrogen and one atom of oxygen.

Statements 5 and 8 are accurate.
Statements 6 and 7 are misleading. The *substance* water is a colourless liquid, which is made up of molecules, H_2O. Each molecule contains two atoms of hydrogen and one atom of oxygen, ie it is the molecules (the *particles*) of water, not the substance (water), that contain two atoms of hydrogen and one atom of oxygen.

Example

The following three sentences contain information concerning a common chemical that we are all familiar with as white crystals.

The sugar in a cup of tea or coffee is $C_{12}H_{22}O_{11}$. The chemical name for the sugar is sucrose. Sucrose contains 12 carbons, 22 hydrogens and 11 oxygens.

Below is the same information, written using 'good' chemical language.

The chemical name for the sugar in a cup of tea or coffee is sucrose. The sucrose molecule has a formula $C_{12}H_{22}O_{11}$. The molecule of sucrose contains 12 carbon atoms, 22 hydrogen atoms and 11 oxygen atoms.

Unit 11.3 Activity 1C: The language of chemistry

1. Identify which of the following statements are 'good' chemical language (there are three).

a. Steam is $H_2O(g)$.
b. The particles leaving a jug of boiling water can be represented as $H_2O(g)$.
c. Stainless steel is composed of iron plus other elements present to prevent the iron from rusting.

 d. Glucose, $C_6H_{12}O_6$, is a sugar that contains hydrogens, oxygens and carbons.
 e. When you are taking exercise, you need more O_2 in your blood than when you are sitting still.
 f. The particles in sea water that give the water the taste when you go swimming in the sea can be represented as NaCl(aq).

2. Using 'good' chemical language, rewrite the three statements in question **1.** that are 'bad' chemical language.
3. In the following account of some of the properties of a chemical named ammonia, three of the sentences use good chemical language and three use bad chemical language. Rewrite the account in good chemical language.
   Ammonia is a gaseous chemical with a strong smell. The molecule of ammonia has a formula $NH_3(g)$. There are three times as many hydrogens in ammonia than there are nitrogens. $NH_3(g)$ is very soluble in water. In water, the particles present can be represented as $NH_3(aq)$. The solution smells of ammonia because some of the $NH_3(g)$ can escape from the solution easily.
4. Write the symbols for:
   a. An atom of hydrogen.
   b. A molecule of hydrogen.
   c. Two atoms of oxygen.
   d. Two molecules of oxygen.
   e. A molecule of water.
   f. A molecule of steam.
5. $K_2SO_4.Al_2(SO_4)_3.24H_2O$ is the formula of a substance that is commonly called 'alum'. Write a statement in words that describes the composition of alum.

Unit 11.3 Types of Chemical Reactions

Topic 2: Oxidation-reduction reactions

This Topic describes chemical reactions by considering:

- Definitions of oxidation and reduction.
- Oxidation and reduction reactions involving oxygen or hydrogen transfer.
- Oxidation and reduction reactions involving electron transfer.
- Word and symbol equations illustrating oxidation and reduction reactions.
- The terms oxidant and reductant.

Introduction

At the end of the 18th century, Lavoisier convinced the world that burning involved a reaction of a combustible material with oxygen in the air. (Previously, the **phlogiston theory** had explained burning as a process in which a combustible material released phlogiston (flame, soot, light, etc) to the air.)

Lavoisier's new theory led to the realisation that many reactions involved the addition of oxygen to a substance – hence the term **oxidation**. There were also many reactions in which the reverse process occurred, ie oxygen was removed from, or reduced in quantity, in a substance – hence the term **reduction**.

Later studies showed that oxidation could involve the removal of hydrogen from a substance and that reduction could involve the addition of hydrogen to a substance.

Today, it is understood that many oxidation and reduction reactions do not involve hydrogen and oxygen, but that they *always* involve the *transfer* of *electrons* between the particles involved, ie:

- The particles of substances that are oxidised *lose* electrons.
- The particles of substances that are reduced *gain* electrons.

In summary,

Oxidation is:	Reduction is:
• *Addition* of *oxygen* to an element or compound. • *Removal* of *hydrogen* from a compound. • *Removal* of *electrons* from an atom or ion.	• *Removal* of *oxygen* from a compound. • *Addition* of *hydrogen* to an element or a compound. • *Addition* of *electrons* to an atom or ion.

Oxidation and reduction

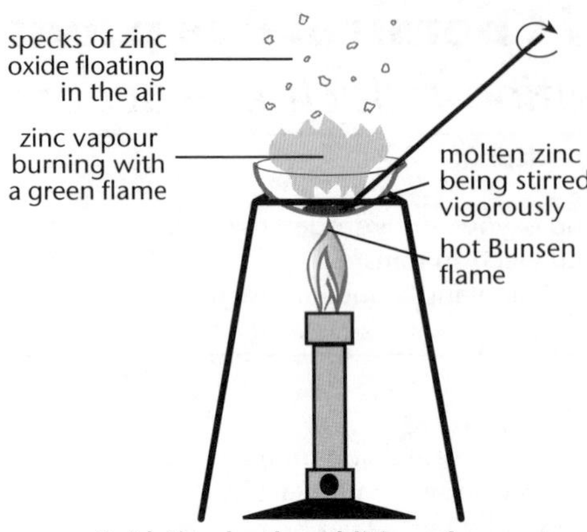

Oxidation by the addition of oxygen – *zinc has been oxidised to zinc oxide by the addition of oxygen (to be performed by a teacher)*

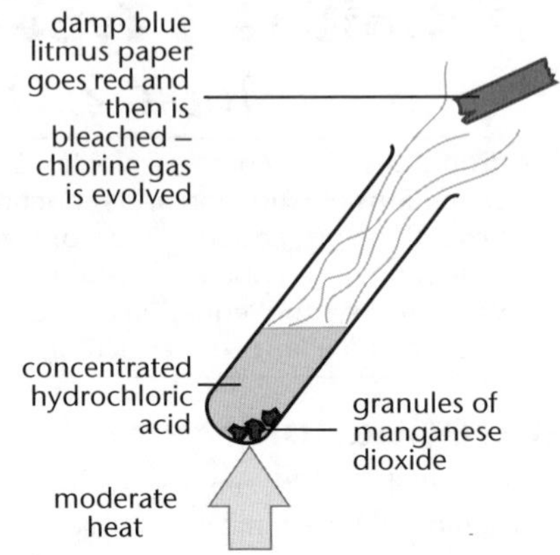

Oxidation by the removal of hydrogen – *hydrochloric acid has been oxidised to chlorine gas by the removal of hydrogen*

Oxidation as the addition of oxygen or the removal of hydrogen

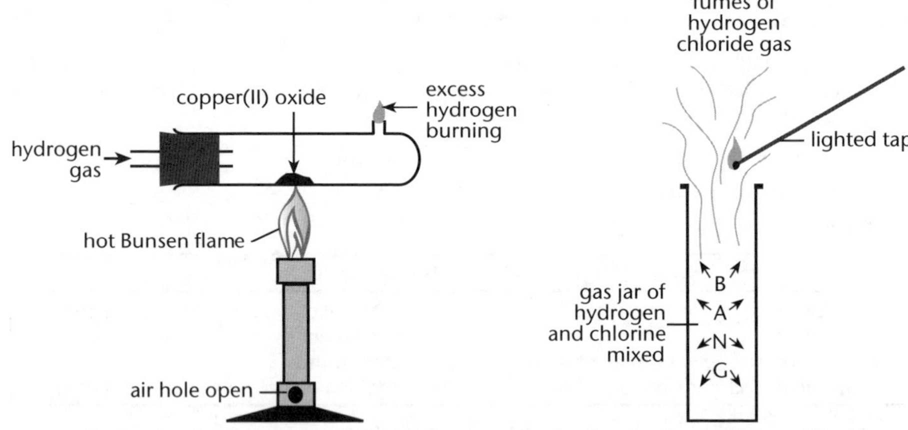

Reduction by the removal of oxygen – *copper(II) oxide has been reduced to copper by removal of oxygen*

Reduction by the addition of hydrogen – *chlorine gas has been reduced to hydrogen chloride gas (to be performed in a fume cupboard or under a fume hood)*

Reduction as the removal of oxygen or the addition of hydrogen

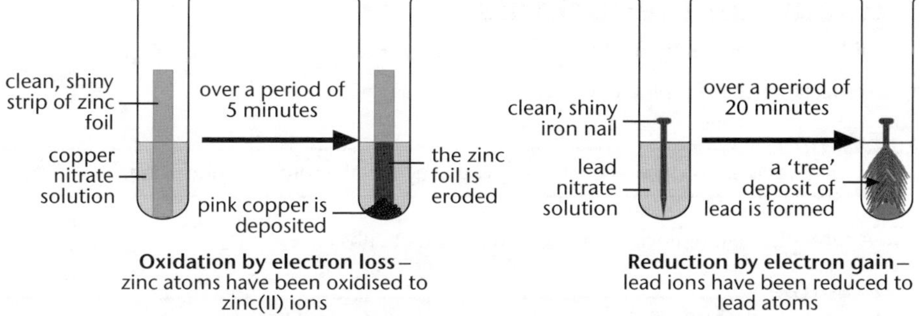

Oxidation and reduction involving the transfer of electrons

Unit 11.3 Activity 2A: Oxidation and reduction

1. **a.** Define oxidation in terms of the transfer of oxygen.
b. Define reduction in terms of the transfer of hydrogen.
c. Define oxidation in terms of electron transfer.
d. Define reduction in terms of electron transfer.

2. In each of the reactions described below, identify which substance has been oxidised and explain why the substance has been oxidised:
a. Copper oxide mixed with charcoal (a form of carbon) and the mixture heated to produce carbon dioxide and copper.
b. Steam reacting with red-hot iron to form iron oxide and hydrogen.
c. Magnesium burning in air to form magnesium oxide.

3. In each of the reactions **a.** – **c.** described below, identify which substance has been reduced and explain why the substance has been reduced:
a. Hydrogen mixed with vegetable oil and the mixture heated to produce margarine as the only product.
b. In the Saturn rocket, the fuel is hydrogen which combines with oxygen to form steam.
c. The removal of sulfur from petroleum by its conversion to hydrogen sulfide gas.

4. In each of the equations, identify which particle has been oxidised and explain why the particle has been oxidised:
a. $Mg(s) + Cu^{2+}(aq) \rightarrow Mg^{2+}(aq) + Cu(s)$
b. $2Fe^{3+}(aq) + 2I^{-}(aq) \rightarrow 2Fe^{2+}(aq) + I_2(s)$

5. In each of the equations, identify which particle has been reduced and explain why the particle has been reduced:
a. $Cl_2(aq) + 2I^{-}(aq) \rightarrow 2Cl^{-}(aq) + I_2(s)$
b. $Zn(s) + 2Fe^{3+}(aq) \rightarrow Zn^{2+}(aq) + 2Fe^{2+}(aq)$

Oxidants and reductants

An **oxidant** (**oxidising agent**) can be identified as:

- A substance that supplies oxygen to another substance.

Example

Oxygen gas is an oxidant when it reacts with zinc metal to produce zinc oxide.

- A substance that removes hydrogen from another substance.

Example

Manganese dioxide is an oxidant when it reacts with hydrochloric acid to produce chlorine gas.

- A substance whose particles will remove electrons from the particles of another substance.

Example

The copper ions in copper nitrate solution are an oxidant when they remove electrons from zinc atoms in zinc metal.

A **reductant** (**reducing agent**) can be identified as:

- A substance that removes oxygen from another substance.

Example

Hydrogen gas is a reductant when it reacts with copper(II) oxide to produce metallic copper.

- A substance that adds hydrogen to another substance.

Example

Hydrogen gas is a reductant when it reacts with chlorine gas to produce hydrogen chloride gas.

- A substance whose particles will add electrons to the particles of another substance.

Example

The atoms in iron metal are a reductant when they add electrons to lead ions in lead nitrate solution.

Oxidation and reduction *always* occur together – they are *complementary* processes. (It is not always clear that both processes are occurring.)

Example

When hydrogen gas and oxygen gas combine to form water:

- the hydrogen is oxidised by oxygen (the oxidant) *and*
- the oxygen is reduced by hydrogen (the reductant).

Example

When magnesium metal and oxygen gas combine to form magnesium oxide:

- the magnesium is oxidised by oxygen (the oxidant) *and*
- the oxygen is reduced by magnesium (the reductant).

At a particle level, magnesium atoms donate electrons to oxygen atoms to form the oxide ion.

Redox reactions

The term **redox reaction** refers to any oxidation-reduction reaction. When redox reactions involve electron transfer, the following **mnemonics** / statements are worth remembering:

- **LEO** – **L**oss of **E**lectron is **O**xidation.
- **GER** – **G**ain of **E**lectron is **R**eduction.
- A *reductant* is a good electron donor – it is *oxidised* during a redox reaction.
- An *oxidant* is a good electron acceptor – it is *reduced* during a redox reaction.

Reducing agent	Symbol	Product formed from reducing agent (by electron loss)	Species that can be reduced (by electron gain)	Observations during the reaction and the full equation
Iodine ion	$I^-(aq)$	Iodine molecule, $I_2(aq)$; $2I^-(aq) \rightarrow I_2(aq) + 2e^-$	Iron(III) ions, $Fe^{3+}(aq)$; $Fe^{3+}(aq) + e^- \rightarrow Fe^{2+}(aq)$	Solution goes brown due to formation of iodine $2Fe^{3+}(aq) + 2I^-(aq) \rightarrow 2Fe^{2+}(aq) + I_2(aq)$
Iron(II) ion	$Fe^{2+}(aq)$	Iron(III) ions, $Fe^{3+}(aq)$; $Fe^{2+}(aq) \rightarrow Fe^{3+}(aq) + e^-$	Chlorine gas, $Cl_2(g)$; $Cl_2(g) + 2e^- \rightarrow 2Cl^-(aq)$	Green solution goes yellow-brown, smell of chlorine goes $Cl_2(g) + 2Fe^{2+}(aq) \rightarrow 2Cl^-(aq) + 2Fe^{3+}(aq)$
Magnesium atom	$Mg(s)$	Magnesium ions, $Mg^{2+}(aq)$; $Mg(s) \rightarrow Mg^{2+}(aq) + 2e^-$	Copper ions, $Cu^{2+}(aq)$; $Cu^{2+}(aq) + 2e^- \rightarrow Cu(s)$	Blue solution goes colourless; pink deposit is formed $Cu^{2+}(aq) + Mg(s) \rightarrow Cu(s) + Mg^{2+}(aq)$
Zinc atom	$Zn(s)$	Zinc ions, $Zn^{2+}(aq)$; $Zn(s) \rightarrow Zn^{2+}(aq) + 2e^-$	Hydrogen ions, $H^+(aq)$; $2H^+(aq) + 2e^- \rightarrow H_2(g)$	Zinc metal dissolves; colourless gas evolved $Zn(s) + 2H^+(aq) \rightarrow Zn^{2+}(aq) + H_2(g)$

Reductants as electron donors

Oxidising agent	Symbol	Product formed from oxidising agent (by electron gain)	Species that can be oxidised (by electron loss)	Observations during the reaction and the full equation
Chlorine gas	$Cl_2(g)$	Chloride ion, $Cl^-(aq)$; $Cl_2(g) + 2e^- \rightarrow 2Cl^-(aq)$	Iodide ions, $I^-(aq)$; $2I^-(aq) \rightarrow I_2(aq) + 2e^-$	Solution goes brown due to formation of iodine $Cl_2(g) + 2I^-(aq) \rightarrow 2Cl^-(aq) + I_2(aq)$
Chlorine gas	$Cl_2(g)$	Chloride ion, $Cl^-(aq)$; $Cl_2(g) + 2e^- \rightarrow 2Cl^-(aq)$	Iron(II) ions, $Fe^{2+}(aq)$; $Fe^{2+}(aq) \rightarrow Fe^{3+}(aq) + e^-$	Green solution goes yellow-brown $Cl_2(g) + 2Fe^{2+}(aq) \rightarrow 2Cl^-(aq) + 2Fe^{3+}(aq)$
Copper(II) ions in solution	$Cu^{2+}(aq)$	Copper atoms, Cu(s); $Cu^{2+}(aq) + 2e^- \rightarrow Cu(s)$	Zinc atoms, Zn(s); $Zn(s) \rightarrow Zn^{2+}(aq) + 2e^-$	Blue solution goes colourless; pink deposit is formed $Cu^{2+}(aq) + Zn(s) \rightarrow Cu(s) + Zn^{2+}(aq)$
Lead(II) ions in solution	$Pb^{2+}(aq)$	Lead atoms, Pb(s); $Pb^{2+}(aq) + 2e^- \rightarrow Pb(s)$	Iron atoms, Fe(s); $Fe(s) \rightarrow Fe^{2+}(aq) + 2e^-$	Solution stays colourless; dark-grey shiny crystals formed $Pb^{2+}(aq) + Fe(s) \rightarrow Pb(s) + Fe^{2+}(aq)$
Iron(III) ions in solution	$Fe^{3+}(aq)$	Iron(II) ions, $Fe^{2+}(aq)$ $Fe^{3+}(aq) + e^- \rightarrow Fe^{2+}(aq)$	Iodide ions, $I^-(aq)$; $2I^-(aq) \rightarrow I_2(aq) + 2e^-$	Solution changes from yellow to deep red-brown; $2Fe^{3+}(aq) + 2I^-(aq) \rightarrow 2Fe^{2+}(aq) + I_2(aq)$

Oxidants as electron acceptors

In a redox reaction involving electron transfer, the equation for the reaction can be represented as two **half-equations**. The two half-equations for a redox reaction added together form the **overall equation**.

Example

Half-equations

Magnesium metal combines violently with sulfur when the mixture is heated. The reaction can be represented by the overall equation:

$Mg(s) + S(s) \rightarrow MgS(s)$.

The two half-equations are:

$Mg(s) \rightarrow Mg^{2+}(s) + 2e^-$ (oxidation), and

$S(s) + 2e^- \rightarrow S^{2-}(s)$ (reduction).

When these two half-equations are added together, the electrons cancel each other out and the overall equation is obtained.

Unit 11.3 Activity 2B: Oxidants and reductants

1. In each of the following reactions **a.** – **c.**, identify:

- **i.** The oxidant (oxidising agent).
- **ii.** The reductant (reducing agent).
- **iii.** The product of oxidation.
- **iv.** The product of reduction.

a. Copper oxide mixed with charcoal and the mixture heated to produce carbon dioxide and copper.

b. Steam reacting with red-hot iron to form iron oxide and hydrogen.

c. Magnesium burning in air to form magnesium oxide. *Hint*: Only one product means the product has two identities.

2. The thermit reaction is used to produce white-hot iron for welding in remote situations. The equation for the reaction is:

$2Al(s) + Fe_2O_3(s) \rightarrow Al_2O_3(s) + 2Fe(s)$.

Identify in the reaction:

a. The reductant (reducing agent).

b. The product of oxidation.

3. In the following equations, identify the:

a. Oxidising agent for $C(s) + O_2(g) \rightarrow CO_2(g)$.

b. Product of oxidation for $4H_2O_2(aq) + PbS(s) \rightarrow PbSO_4(s) + 4H_2O(\ell)$.

c. Product of reduction for $H_2(g) + I_2(g) \rightarrow 2HI(g)$.

d. Reducing agent for $Mg(s) + CuO(s) \rightarrow MgO(s) + Cu(s)$.

4. Write the full statements represented by the mnemonics:

a. L E O.

b. G E R.

5. Identify the following substances as either oxidants (oxidising agents) or reductants (reducing agents) and write an appropriate half-equation that illustrates the agent's action.

Example

Question: Iodide ion, I^-.

Answer: Reductant (electron donor), $2I^- \rightarrow I_2 + 2e^-$.

a. Chlorine molecule, Cl_2.

b. Iron(II) ion, Fe^{2+}.

c. Iron(III) ion, Fe^{3+}.

d. Copper(II) ion, Cu^{2+}.

e. Magnesium atom, Mg.

f. Oxygen molecule, O_2.

6. The oxidation and reduction processes can be identified in an equation by:

oxidation by electron loss

$$Mg(s) + 2Ag^{+}(aq) \longrightarrow Mg^{2+}(aq) + 2Ag(s)$$

reduction by electron gain

Use this example to indicate in the following equations where oxidation and reduction have occurred.

a. $Fe(s) + Cu^{2+}(aq) \rightarrow Fe^{2+}(aq) + Cu(s)$.

b. $2Fe^{3+}(aq) + 2I^{-}(aq) \rightarrow 2Fe^{2+}(aq) + I_2(aq)$.

c. $2I^{-}(aq) + Cl_2(aq) \rightarrow I_2(aq) + 2Cl^{-}(aq)$.

d. $Cl_2(aq) + 2Fe^{2+}(aq) \rightarrow 2Fe^{3+}(aq) + 2Cl^{-}(aq)$.

Unit 11.3 Types of Chemical Reactions

Topic 3: Precipitation reactions

This Topic describes chemical reactions by considering:

- Soluble and insoluble substances.
- Preparing an insoluble substance.
- Identification of positive and negative ions.

Soluble and insoluble substances

Water is often regarded as the universal **solvent** – meaning that it will dissolve any substance. This is not true. Substances can be divided into three categories, based on their solubility in water:

- Substances that are **soluble** in water – ie 1 g or more of the substance will dissolve in 100 mL of water at 25 °C.

Example

4.3 g of sodium sulfate will dissolve in 100 mL of water.

- Substances that are **sparingly soluble** in water – ie less than 1 g, but more than 0.1 g, of the substance will dissolve in 100 mL of water at 25 °C.

Example

0.63 g of calcium sulfate will dissolve in 100 mL of water.

- Substances that are **insoluble** in water – ie less than 0.1 g of the substance will dissolve in 100 mL of water at 25 °C.

Example

0.00022 g of barium sulfate will dissolve in 100 mL of water.

The following solubility rules apply:

- Nitrates are all soluble.
- Chlorides are all soluble, except silver chloride and lead chloride.
- Sulfates are all soluble, except lead sulfate, calcium sulfate and barium sulfate.
- Carbonates are all insoluble, except those of group I (sodium carbonate and potassium carbonate), and ammonium carbonate.
- Hydroxides are all insoluble, except those of group I (sodium hydroxide and potassium hydroxide), and ammonium hydroxide.

Precipitates

A **precipitate** (abbreviated to **ppt**) is an insoluble substance formed when two soluble substances in solutions are mixed.

Example

Precipitate of a metal compound

A precipitate of a metal compound will always form from the reaction of a metal nitrate with a sodium compound – the other product of this reaction is sodium nitrate. The equation written in word form is:

copper nitrate + sodium hydroxide form copper hydroxide (the precipitate) + sodium nitrate.

Spectator ions

Every ionic compound in solution contains two independent ions. In all precipitation reactions, there are two **spectator ions** – ions that need to be present in a reactant but do not take part in the reaction.

Example

Spectator ions

The symbol equation for the formation of the precipitate copper hydroxide is:

$Cu(NO_3)_2(aq) + 2NaOH(aq) \rightarrow 2NaNO_3(aq) + Cu(OH)_2(s)$

With spectator ions removed, the equation becomes:

$Cu^{2+}(aq) + 2OH^-(aq) \rightarrow Cu(OH)_2(s)$

The table lists some insoluble compounds and their formation. In the table, Solution A:

- Contains the metal component of the precipitate.
- Is always the metal nitrate solution (all nitrates are soluble).

Solution B:

- Contains the non-metal component (as an **acid radical**) of the precipitate.
- Is always the sodium salt solution (all sodium compounds are soluble).

The equations have been written in three different forms:

- The first four are *word* equations.
- The next six are *symbol* equations, using *molecular formulae*.
- The last seven *ionic* equations do not include the spectator ions (ie sodium ion and the nitrate ion).

		Precipitate		
Solution A	Solution B	Name and formula	Appearance	Equation
Silver nitrate	Sodium chloride	Silver chloride AgCl	White solid	Silver nitrate + sodium chloride form silver chloride + sodium nitrate
Lead nitrate	Sodium chloride	Lead chloride $PbCl_2$	White solid	Lead nitrate + sodium chloride form lead chloride + sodium nitrate
Barium nitrate	Sodium sulfate	Barium sulfate $BaSO_4$	White solid	Barium nitrate + sodium sulfate form barium sulfate + sodium nitrate
Calcium nitrate	Sodium sulfate	Calcium* sulfate $CaSO_4$	White solid	Calcium nitrate + sodium sulfate form calcium sulfate + sodium nitrate
Lead nitrate	Sodium sulfate	Lead sulfate $PbSO_4$	White solid	$Pb(NO_3)_2(aq) + Na_2SO_4(aq) \rightarrow PbSO_4(s) + 2NaNO_3(aq)$

Solution A	Solution B	Precipitate		
		Name and formula	Appearance	Equation
Copper(II)[#] nitrate	Sodium hydroxide	Copper(II) hydroxide $Cu(OH)_2$	Blue solid	$2NaOH(aq) + Cu(NO_3)_2(aq) \rightarrow 2NaNO_3(aq) + Cu(OH)_2(s)$
Iron(II) sulfate	Sodium hydroxide	Iron(II) hydroxide $Fe(OH)_2$	Green solid	$2NaOH(aq) + FeSO_4(aq) \rightarrow Na_2SO_4(aq) + Fe(OH)_2(s)$
Iron(III) nitrate	Sodium hydroxide	Iron(III) hydroxide $Fe(OH)_3$	Red-brown solid	$3NaOH(aq) + Fe(NO_3)_3(aq) \rightarrow 3NaNO_3(aq) + Fe(OH)_3(s)$
Zinc nitrate	Sodium hydroxide	Zinc hydroxide $Zn(OH)_2$	White solid	$2NaOH(aq) + Zn(NO_3)_2(aq) \rightarrow 2NaNO_3(aq) + Zn(OH)_2(s)$
Aluminium nitrate	Sodium hydroxide	Aluminium hydroxide $Al(OH)_3$	White solid	$3NaOH(aq) + Al(NO_3)_3(aq) \rightarrow 3NaNO_3(aq) + Al(OH)_3(s)$
Calcium nitrate	Sodium hydroxide	Calcium* hydroxide $Ca(OH)_2$	White solid	$2OH^-(aq) + Ca^{2+}(aq) \rightarrow Ca(OH)_2(s)$
Magnesium nitrate	Sodium hydroxide	Magnesium hydroxide $Mg(OH)_2$	White solid	$2OH^-(aq) + Mg^{2+}(aq) \rightarrow Mg(OH)_2(s)$
Copper(II) nitrate	Sodium carbonate	Copper(II) carbonate $CuCO_3$	Green-blue solid	$CO_3^{2-}(aq) + Cu^{2+}(aq) \rightarrow CuCO_3(s)$
Iron(II) sulfate	Sodium carbonate	Iron(II) carbonate $FeCO_3$	Grey solid	$CO_3^{2-}(aq) + Fe^{2+}(aq) \rightarrow FeCO_3(s)$
Zinc nitrate	Sodium carbonate	Zinc carbonate $ZnCO_3$	White solid	$CO_3^{2-}(aq) + Zn^{2+}(aq) \rightarrow ZnCO_3(s)$
Calcium nitrate	Sodium carbonate	Calcium carbonate $CaCO_3$	White solid	$CO_3^{2-}(aq) + Ca^{2+}(aq) \rightarrow CaCO_3(s)$
Magnesium nitrate	Sodium carbonate	Magnesium carbonate $MgCO_3$	White solid	$CO_3^{2-}(aq) + Mg^{2+}(aq) \rightarrow MgCO_3(s)$

[#]Roman numeral indicates charge on positive ion.
*Calcium sulfate and calcium hydroxide are sparingly soluble in water. Solutions A and B will need to be concentrated to ensure that both these compounds precipitate.

Some insoluble compounds and their formation

Preparation of an insoluble compound

Insoluble compounds can be prepared by a precipitation reaction. The precipitate can be separated from solution by **filtration**, then washed with distilled water to remove the unwanted solution and air dried – even gentle heating can decompose some compounds.

Example

Preparation of a pure sample of insoluble copper hydroxide

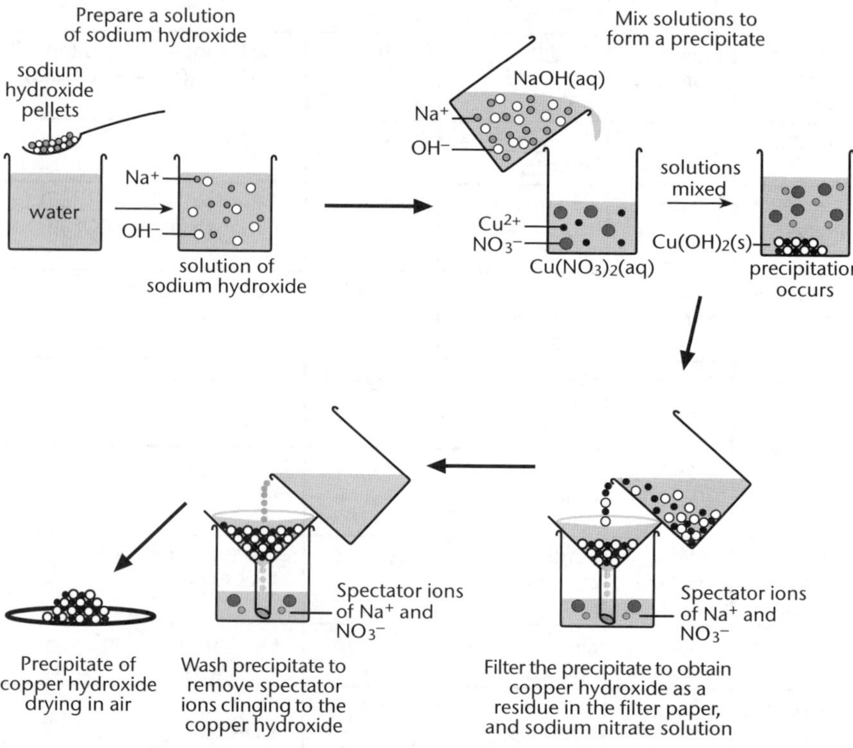

Formation and purification of an insoluble compound

The equation for this reaction may be written in three ways.

- In *word form* as: sodium hydroxide solution + copper nitrate solution forms copper hydroxide precipitate + sodium nitrate solution
- Using *molecular formulae* as: $2NaOH(aq) + Cu(NO_3)_2(aq) \rightarrow Cu(OH)_2(s) + 2NaNO_3(aq)$

Note: (aq) indicates an aqueous (water) solution and (s) indicates an insoluble solid, ie a precipitate.

- Using *ion formulae* with *spectator ions absent* as: $2OH^-(aq) + Cu^{2+}(aq) \rightarrow Cu(OH)_2(s)$

Unit 11.3 Activity 3A: Soluble and insoluble substances, preparation of an insoluble compound

1. The solubilities of three everyday compounds are listed. Decide for each compound whether it is soluble, sparingly soluble or insoluble in water.

a. Lead chromate (commonly called 'chrome yellow') is used as a yellow pigment in oil paint. Solubility = 0.000017 g per 100 g of water.

b. Sodium chloride is a necessary article of diet for human beings. Solubility = 31.7 g per 100 g of water.

c. Plaster of Paris is a form of calcium sulfate that, when mixed with water, produces a 'porridge' that will set to a hard cast for stabilising arms or legs, and allow broken bones to set. Solubility = 0.16 g per 100 g of water.

2. a. Name three soluble metal nitrates.

b. Name one insoluble metal chloride.

c. Name two insoluble metal carbonates.

d. Name two soluble metal hydroxides.

3. Explain the meaning of the term 'precipitate'.

4. Complete the following table by providing information for **a.** to **h.**

Solution A	Solution B	Precipitate		
		Name and formula	Appearance	Equation in word form
silver nitrate	sodium chloride	silver chloride AgCl	**a.**	**b.**
barium nitrate	sodium sulfate	**c.**	white solid	**d.**
e.	**f.**	iron(III) hydroxide $Fe(OH)_3$	**g.**	**h.**

5. Complete the following equations:

a. Copper(II) nitrate + sodium hydroxide → __________ + __________.

b. __________ + __________ → calcium hydroxide + sodium nitrate.

c. __________ + __________ → zinc carbonate + __________.

d. $Mg(NO_3)_2(aq) + 2NaOH(aq) \rightarrow$ __________ + __________.

e. __________ + __________ → $FeCO_3(s) + Na_2SO_4(aq)$.

f. $Ag^+(aq) + Cl^-(aq) \rightarrow$ __________.

g. __________ + __________ → $BaSO_4(s)$.

6. Describe an experiment, using words or diagrams, you could perform to produce a pure, dry sample of lead chloride. Name the reagents and describe the procedure to make lead chloride and how it could be purified. Provide an equation for the reaction.

Identification of positive (metal) ions

Metal (positive) ions – ie $\mathbf{Ag^{+}}$, $\mathbf{Cu^{2+}}$, $\mathbf{Pb^{2+}}$, $\mathbf{Fe^{2+}}$, $\mathbf{Fe^{3+}}$, $\mathbf{Zn^{2+}}$, $\mathbf{Al^{3+}}$, $\mathbf{Ca^{2+}}$, $\mathbf{Mg^{2+}}$, $\mathbf{NH_4^{+}}$ – can be detected in solution by finding a **reagent** that is specific for the ion. Analysis for positive ions relies on the formation of a precipitate, which is why the solubility rules are important.

Separation of two ions, both of which form a precipitate with a reagent, can be achieved in either of two ways:

- Adding an excess of the reagent.

Example

Zinc, lead and aluminium hydroxides will dissolve in excess sodium hydroxide solution.

- Adding an additional reagent in which only one of the precipitates will dissolve.

Example

Zinc, aluminium and lead hydroxides are soluble in sodium hydroxide solution but only zinc hydroxide will dissolve in ammonia solution.

The **analysis** flowchart below shows how to identify, in a solution, the positive ions listed.

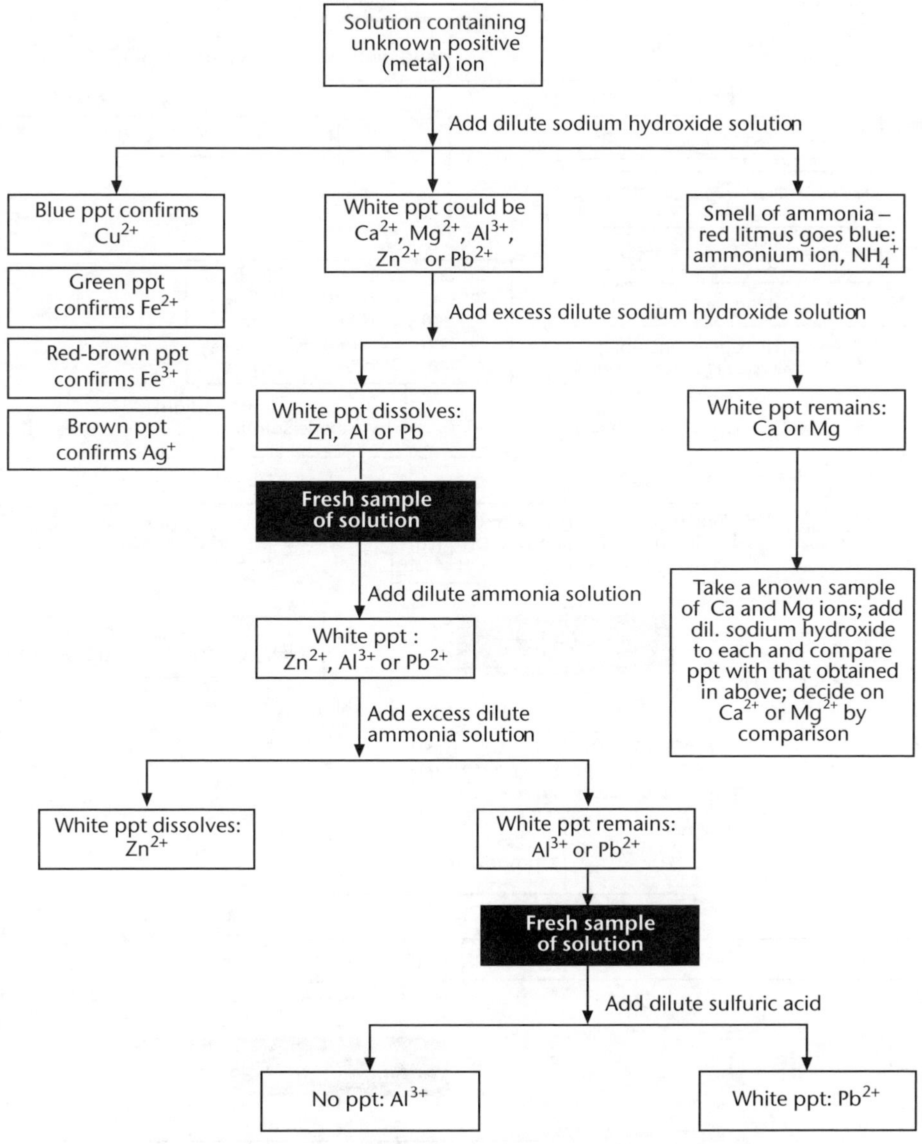

Analysis flowchart for identification of metal (positive) ions

Equations for the formation of precipitates, and for the dissolving of precipitates in excess reagent, are shown in the flowchart on the next page – spectator ions have been omitted from the equations.

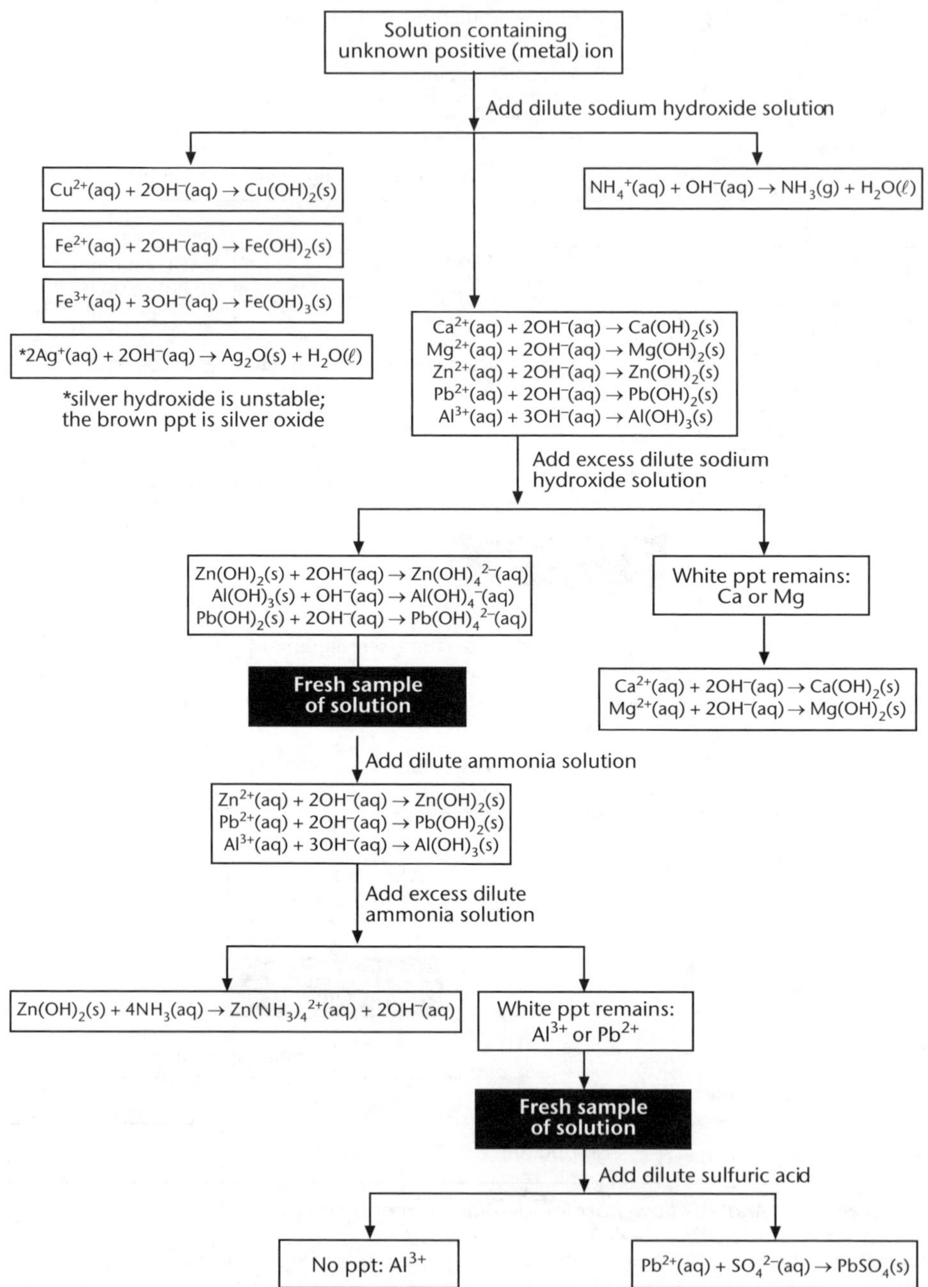

Equations for the analysis flowchart for identification of metal (positive) ions

Unit 11.3 Activity 3B: Identifying positive (metal) ions

1. Name:

- **a.** Three metal compounds that are white solids at room temperature.
- **b.** One metal compound that is a green solid at room temperature.
- **c.** One metal compound that is a blue solid at room temperature.
- **d.** One metal compound that is a brown solid at room temperature.

2.
- **a.** Name three metal nitrate solutions, which would produce a white precipitate when mixed with sodium hydroxide solution.
- **b.** Write the formula of each of the metal nitrates used in **a.**
- **c.** Write the formula of each of the three precipitates formed in the reactions indicated in **a.**
- **d.** Indicate which precipitates formed in **a.** would dissolve in excess sodium hydroxide solution.

3. Write the formula of the:

- **a.** Sodium ion.
- **b.** Iron(II) ion.
- **c.** Aluminium ion.

4. Describe all the observed effects of:

- **a.** Adding sodium hydroxide solution to magnesium nitrate solution.
- **b.** Adding sodium hydroxide solution to iron(II) sulfate solution.
- **c.** Adding silver nitrate solution to sodium chloride solution and then adding ammonia solution until in excess.

5. Complete the equations below, indicating the state of the substance missing, eg (aq), (ℓ), (s) or (g).

- **a.** $Ag^+(aq) + Cl^-(aq) \rightarrow$ ____________(_).
- **b.** $Al^{3+}(aq) +$ __________(_) $\rightarrow Al(OH)_3(s)$.
- **c.** ________(_) + ________(_) $\rightarrow BaSO_4(s)$.
- **d.** $Ag_2O(s) +$ ___$NH_3(aq) + H_2O(\ell) \rightarrow 2Ag(NH_3)_2^+(aq) + 2OH^-(aq)$.

6. Identify the metal present in the compound from the following description of experiments. Write word, molecular or ionic equations for all the reactions occurring. *Note*: Ammonia solution contains free ammonia, $NH_3(aq)$, and also ammonium hydroxide solution, $NH_4OH(aq)$ – ie both $NH_4^+(aq)$ and $OH^-(aq)$ are present in ammonia solution.

A white solid was dissolved in water and the solution was divided into two parts:

- *Part I had dilute sodium hydroxide solution added to it and a white precipitate was formed, which dissolved in excess sodium hydroxide solution.*
- *Part II had dilute ammonia solution added to it and a white precipitate was formed, which dissolved in excess of ammonia solution.*

Identification of negative (non-metal) ions

Non-metal (negative) ions (ie $\mathbf{OH^-}$, $\mathbf{Cl^-}$, $\mathbf{I^-}$, $\mathbf{SO_4^{2-}}$, $\mathbf{CO_3^{2-}}$) can be detected in solution by finding a reagent that is specific for the ion. Analysis for negative ions relies on the formation of a precipitate, which is why the solubility rules are important.

Identification of two ions, both of which form a precipitate with a reagent, can be achieved by adding an additional reagent, in which only one of the precipitates will dissolve.

Example

In non-metal (negative) ion analysis, silver chloride and silver oxide are soluble in ammonia solution but silver iodide is not.

The analysis flowchart below shows how to identify, in a solid or solution, the negative ions listed.

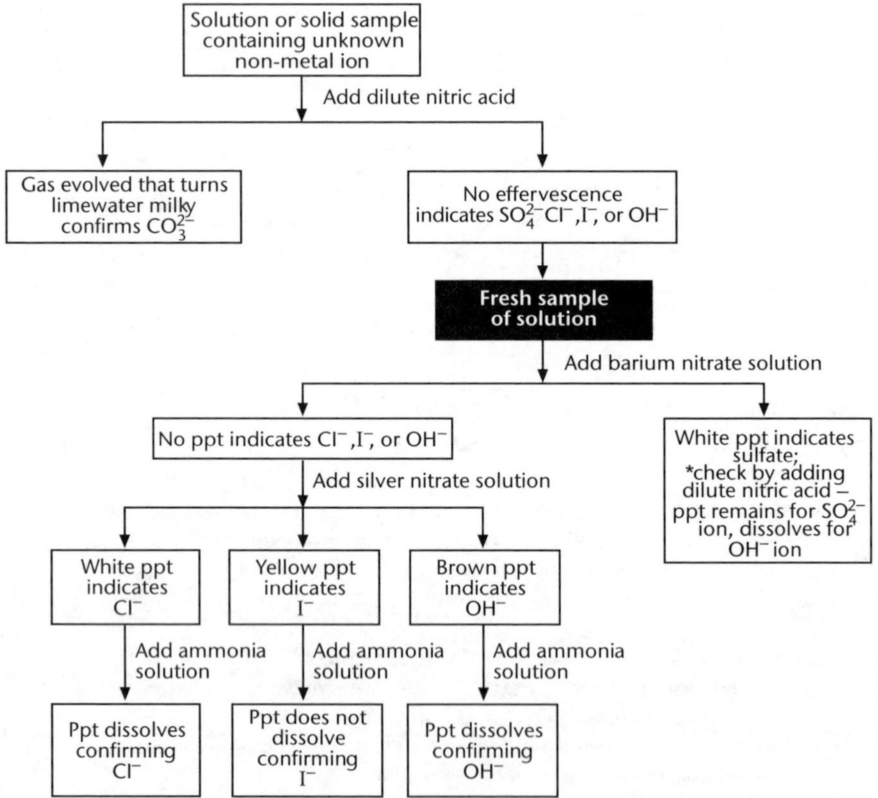

* Barium hydroxide may precipitate when barium nitrate solution is added to a hydroxide solution It depends upon the concentration of the solution used.

Effervescence is the release of gas during a reaction involving a liquid – the gas causes the liquid to froth (effervesce).

Analysis flowchart for identification of non-metal (negative) ions

Equations for the formation of precipitates, and the dissolving of precipitates in excess reagent, are shown in the flowchart below – spectator ions have been omitted from the equations.

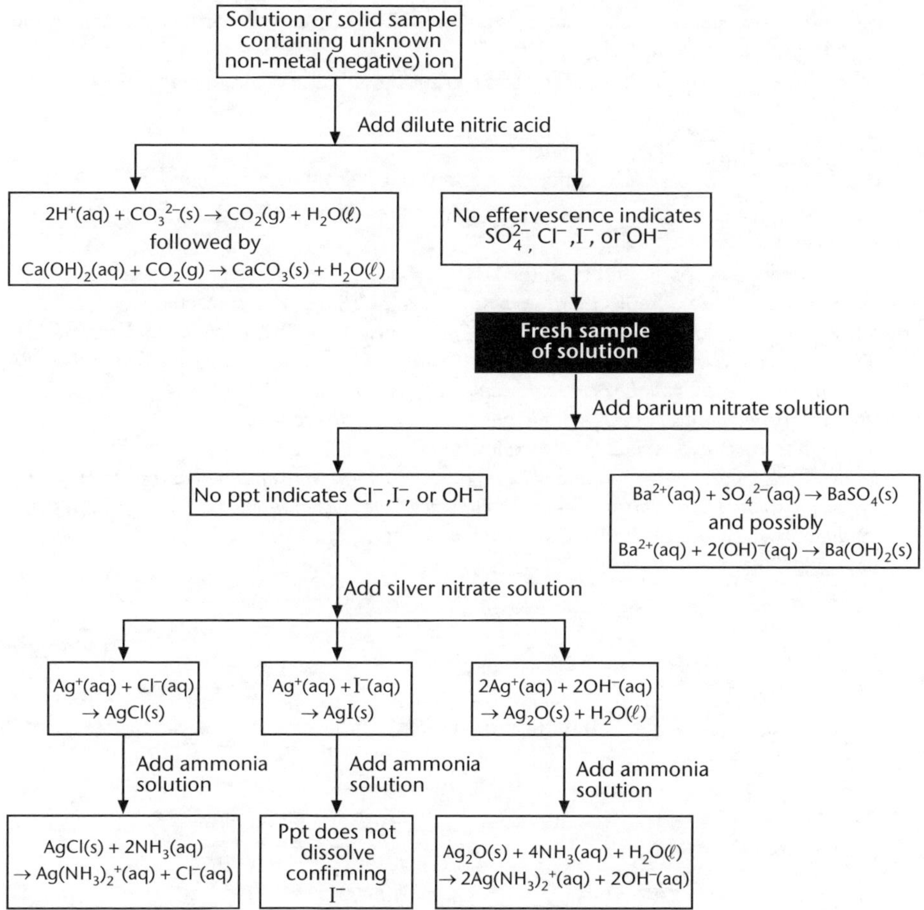

Equations for the analysis flowchart for identification of non-metal (negative) ions

Unit 11.3 Activity 3C: Identifying negative (non-metal) ions

1. Name:
   - **a.** A white solid that reacts with dilute nitric acid to release a colourless gas.
   - **b.** A solution containing a metal compound that will react with barium nitrate solution to produce a white precipitate.
   - **c.** A solution that could be added to silver nitrate solution to produce:
     - **i.** A white precipitate.
     - **ii.** A brown precipitate.
   - **d.** Two solutions that, when mixed, would produce a yellow precipitate.
2. Describe the observations you would expect to make when adding dilute nitric acid to green copper carbonate solid.
3. Identify the metal compound from the following description of experiments. Write word equations (***A*** grade) or molecular/ionic equations (**M** grade) for the reactions occurring.
   A white solid was added to dilute nitric acid. Immediately, effervescence (bubbling) occurred, producing a gas that turned limewater milky. Sodium hydroxide solution was added to the resulting solution until all the acid had been neutralised and the solution tested slightly ***alkaline*** *(using litmus paper). A moderate amount of white precipitate formed, which did not dissolve in excess sodium hydroxide solution.*
   Known samples of magnesium and calcium nitrate solution were obtained and sodium hydroxide solution was added to each. White precipitates were formed in both cases, but the magnesium hydroxide precipitated in greater quantity.
4. For each of the white solids below, draw two flowcharts for the identification of the solid – one flowchart that would identify the positive ion and one that would identify the negative ion.
   - **a.** Aluminium chloride.
   - **b.** Zinc iodide.
5. Complete the equations **a.** – **h.** indicating the state of the substance missing, eg (aq), (ℓ), (s) or (g).
   - **a.** Sodium sulfate + barium nitrate → barium sulfate + __________.
   - **b.** Sodium hydroxide + zinc nitrate → __________+ sodium nitrate.
   - **c.** $2AgNO_3(aq) + 2NaOH(aq) \rightarrow$ ______ + $H_2O(\ell) + 2NaNO_3(aq)$.
   - **d.** $Ca(OH)_2(aq) + CO_2(g) \rightarrow$ __________ + $H_2O(\ell)$.
   - **e.** $Ag^+(aq) + I^-(aq) \rightarrow$ __________(_).
   - **f.** $2H^+(aq) + CO_3^{2-}(s) \rightarrow H_2O(\ell)$ + ______(_).
   - **g.** $2Ag^+(aq) + 2OH^-(aq) \rightarrow$ _______(_) + _______(_).
   - **h.** _______(_) + _______(_) → $Ba(OH)_2(s)$.

Unit 11.3 Types of Chemical Reactions

Topic 4: Thermal decomposition reactions

This Topic describes chemical reactions by considering:

- Stable and inert substances.
- Thermal decomposition of metal hydroxides, carbonates and hydrogen carbonates.

Stable and unstable substances

Stable in chemistry means unreactive in the conditions stated:

- The elements helium, neon and argon are unreactive both to any chemical reagents and to any changes in physical conditions – these elements are **inert**.
- The elements krypton and xenon are *not inert* (fluorides and oxides of krypton and xenon have been prepared) but they are *stable* in almost all chemical conditions.
- Gold is valued because it is *stable* in air, ie is unaffected by atmospheric gases and retains its yellow **lustre** permanently. Gold is *not inert* – it reacts with 'aqua regia' (a mixture of 3 parts concentrated hydrochloric acid and 1 part concentrated nitric acid).

Thermal decomposition

When two elements combine to form a compound, the product requires heat energy for the elements to be re-formed, ie the compound is more stable than the elements at room temperature.

Many compounds undergo **thermal decomposition** – the compound breaks up into compounds of lower mass, or into its elements, when it is heated. The thermal stability of a metal compound is directly related to the chemical activity of the metal.

Example

Sodium compounds are the most thermally stable; gold compounds do not exist because they are thermally unstable.

Metal carbonates and metal hydrogen carbonates

The conditions required for the thermal decomposition of metal carbonates and metal **hydrogen carbonates (bicarbonates)** depend on the chemical activity of the metal.

Example

Sodium and potassium carbonates cannot be decomposed by heating – they are thermally stable.

Highly reactive metals form stable compounds that require high energy to undergo thermal decomposition.

Example

Calcium carbonate

Calcium is high in the **activity series** for metals. The high reactivity of calcium metal results in a stable compound and thus high energy is needed to decompose calcium carbonate.

The equation for the thermal decomposition reaction is:
$CaCO_3(s) \rightarrow CaO(s) + CO_2(g)$.

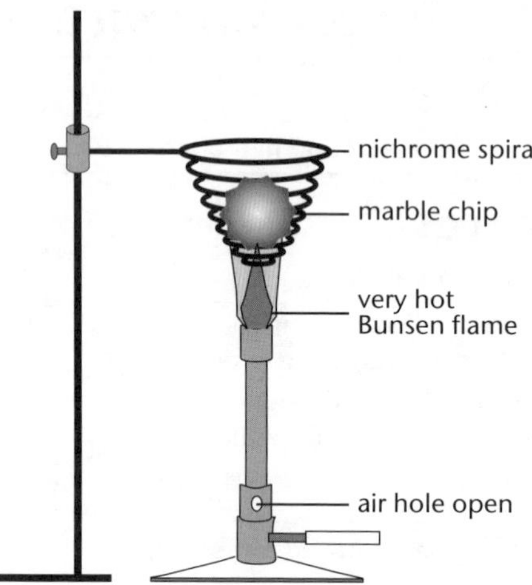

Marble is a form of calcium carbonate, $CaCO_3$.
On very strong heating, marble decomposes into calcium oxide and carbon dioxide.
The calcium oxide will remain in the spiral and the carbon dioxide will escape to the atmosphere.

Thermal decomposition of calcium carbonate

Less reactive metals form compounds that require lower energy to undergo thermal decomposition.

Example

Copper carbonate

Copper is low in the activity series of metals and so, copper carbonate is thermally decomposed with low energy.
The equation for the thermal decomposition reaction is:
$CuCO_3(s) \rightarrow CuO(s) + CO_2(g)$.

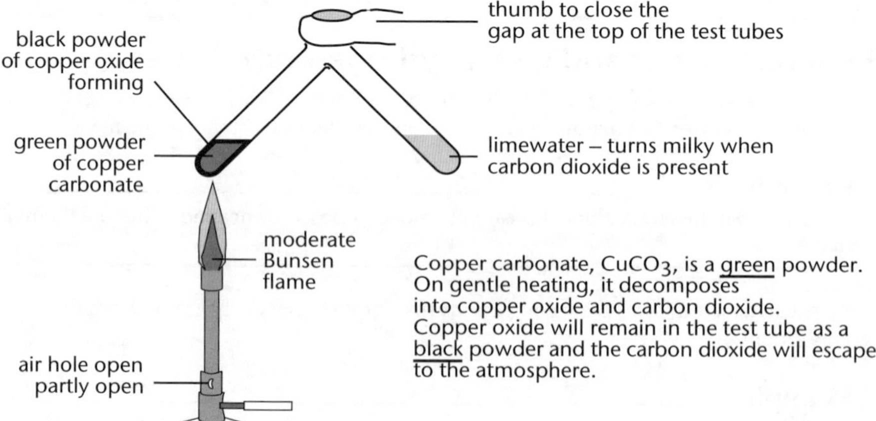

Copper carbonate, $CuCO_3$, is a green powder.
On gentle heating, it decomposes into copper oxide and carbon dioxide.
Copper oxide will remain in the test tube as a black powder and the carbon dioxide will escape to the atmosphere.

Thermal decomposition of copper carbonate

Example

Sodium hydrogen carbonate

Sodium hydrogen carbonate contains the hydrogen carbonate ion, HCO_3^- (sometimes called the bicarbonate ion). The equation for the thermal decomposition reaction is:

$2NaHCO_3(s) \rightarrow Na_2CO_3(s) + H_2O(g) + CO_2(g)$.

Sodium hydrogen carbonate is used in baking (as **baking soda**).

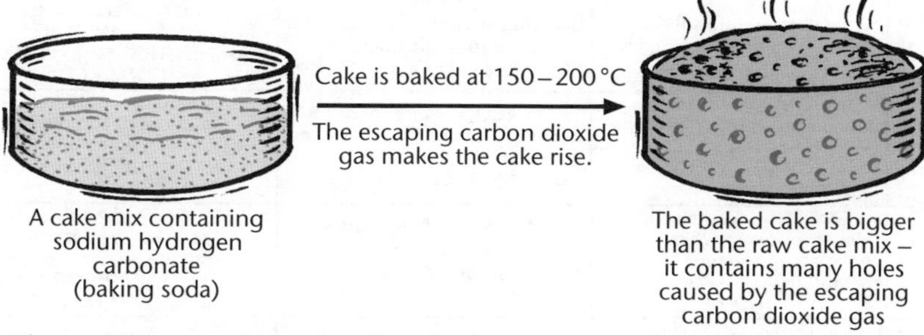

Thermal decomposition of sodium hydrogen carbonate – making a cake rise

Metal hydroxides

Metal hydroxides also undergo thermal decomposition – the products are metal oxides and steam. The quantity of heat energy needed for the thermal decomposition of a metal hydroxide depends upon the chemical activity of the metal.

The thermal stability of the metal hydroxides, carbonates and hydrogen carbonates is summarised in the table below.

Metal	Formula – appearance at room temperature; Chemical change on heating; Equation for thermal decomposition		
	Hydroxide	Carbonate	Hydrogen carbonate
Sodium	$NaOH$ – white solid; none	Na_2CO_3 – white solid; none	$NaHCO_3$ – white solid; decomposes to white carbonate, steam and carbon dioxide; $2NaHCO_3(s) \rightarrow Na_2CO_3(s) + H_2O(g) + CO_2(g)$
Lithium	$LiOH$ – white solid; none	Li_2CO_3 – white solid; decomposes to white oxide and carbon dioxide; $Li_2CO_3(s) \rightarrow Li_2O(s) + CO_2(g)$	$LiHCO_3$ – only stable in solution; decomposes to carbonate, water and carbon dioxide; $2LiHCO_3(aq) \rightarrow Li_2CO_3(aq) + H_2O(\ell) + CO_2(g)$
Calcium	$Ca(OH)_2$ – white solid; decomposes to white oxide and steam; $Ca(OH)_2(s) \rightarrow CaO(s) + H_2O(g)$	$CaCO_3$ – white solid; decomposes to white oxide and carbon dioxide; $CaCO_3(s) \rightarrow CaO(s) + CO_2(g)$	$Ca(HCO_3)_2$ – only stable in solution; decomposes to carbonate, water and carbon dioxide; $Ca(HCO_3)_2(aq) \rightarrow CaCO_3(s) + H_2O(\ell) + CO_2(g)$
Magnesium	$Mg(OH)_2$ – white solid; decomposes to white oxide and steam; $Mg(OH)_2(s) \rightarrow MgO(s) + H_2O(g)$	$MgCO_3$ – white solid; decomposes to white oxide and carbon dioxide; $MgCO_3(s) \rightarrow MgO(s) + CO_2(g)$	$Mg(HCO_3)_2$ – only stable in solution; decomposes to carbonate, water and carbon dioxide; $Mg(HCO_3)_2(aq) \rightarrow MgCO_3(s) + H_2O(\ell) + CO_2(g)$
Aluminium	$Al(OH)_3$ – white solid; decomposes to white oxide and steam; $2Al(OH)_3(s) \rightarrow Al_2O_3(s) + 3H_2O(g)$	*DNE	DNE
Zinc	$Zn(OH)_2$ – white solid; decomposes to yellow (hot)/ white (cold) oxide and steam; $Zn(OH)_2(s) \rightarrow ZnO(s) + H_2O(g)$	$ZnCO_3$ – white solid; decomposes to yellow (hot)/ white (cold) oxide and carbon dioxide; $ZnCO_3(s) \rightarrow ZnO(s) + CO_2(g)$	DNE
Iron	$Fe(OH)_2$ – green solid; decomposes to black oxide and steam; $Fe(OH)_2(s) \rightarrow FeO(s) + H_2O(g)$. Oxide changes to rust-red iron(III) oxide in contact with air; $4FeO(s) + O_2(g) \rightarrow 2Fe_2O_3(s)$	$FeCO_3$ – black solid; decomposes to black oxide and carbon dioxide; $FeCO_3(s) \rightarrow FeO(s) + CO_2(g)$. Oxide changes to rust-red iron(III) oxide in contact with air; $4FeO(s) + O_2(g) \rightarrow 2Fe_2O_3(s)$	DNE
Iron	$Fe(OH)_3$ – rust-red solid; decomposes to rust-red oxide and steam; $2Fe(OH)_3(s) \rightarrow Fe_2O_3(s) + 3H_2O(g)$	DNE	DNE
Lead	$Pb(OH)_2$ – white solid; decomposes to pink (hot)/ yellow (cold) oxide and steam; $Pb(OH)_2(s) \rightarrow PbO(s) + H_2O(g)$	$PbCO_3$ – white solid; decomposes to pink (hot)/ yellow (cold) oxide and carbon dioxide; $PbCO_3(s) \rightarrow PbO(s) + CO_2(g)$	DNE
Copper	$Cu(OH)_2$ – blue-green solid; decomposes to black oxide and steam; $Cu(OH)_2(s) \rightarrow CuO(s) + H_2O(g)$	$CuCO_3$ – blue-green solid; decomposes to black oxide and carbon dioxide; $CuCO_3(s) \rightarrow CuO(s) + CO_2(g)$	DNE
Silver	DNE	Ag_2CO_3 – yellow solid; decomposes to silver metal, carbon dioxide and oxygen; $2Ag_2CO_3(s) \rightarrow 4Ag(s) + 2CO_2(g) + O_2(g)$	DNE
Gold	DNE	DNE	DNE

*DNE = does not exist

The metals in the table are arranged in order of decreasing chemical activity – sodium is the most reactive and gold is the least reactive.

Thermal stability of metal hydroxides, carbonates and hydrogen carbonates

Unit 11.3 Activity 4A: Thermal decomposition

1. Explain the meaning of the terms:
  a. Stable.
  b. Thermal decomposition.
  c. Inert.

2. Name and give the formula of:
  a. A metal carbonate.
  b. A metal hydrogen carbonate.

3. Describe the colour changes that occur when copper carbonate is heated.

4. Describe how you would test a gas jar of gas to show that the gas was carbon dioxide.

5. Explain why calcium carbonate requires a greater temperature to decompose than copper carbonate.

6. Write word equations for the thermal decomposition of:
  a. Copper carbonate.
  b. Sodium hydrogen carbonate.

7. Write a molecular equation for the thermal decomposition of calcium carbonate.

8. Write the formula for:
  a. The carbonate ion.
  b. The hydrogen carbonate ion.

9. Name a white metal carbonate that when heated:
  a. Does not appear to change.
  b. Turns yellow when hot.
  c. Forms a pink solid.

10. Complete the following equations. The substance on the left-hand side of the equation has been heated in a test tube using a Bunsen burner.
  a. Sodium hydrogen carbonate → sodium carbonate + ________ + ________.
  b. ________ ________ → lithium oxide + carbon dioxide.
  c. $ZnCO_3(s) \rightarrow$ ________ + ________.
  d. ____________ $\rightarrow CaCO_3(s) + H_2O(\ell) + CO_2(g)$.

Unit 11.3 Types of Chemical Reactions

Topic 5: Combustion reactions

In Topic 5 we examine combustion reactions, in particular:

- General, spontaneous and non-spontaneous combustion.
- Combustion of organic materials.
- Combustion of non-organic materials.
- Applications of combustion reactions.

General combustion reactions

When a substance burns, it is a combustion reaction. Combustion requires a combustible material and an **oxidant** (usually oxygen). The reaction produces heat and light. For example, dry grass and twigs are combustible materials that burn in air to produce heat and light.

Combustion reactions can be classified according to their ignition source (spontaneous and non-spontaneous) or their completeness (complete and incomplete).

Non-spontaneous combustion reactions require an external ignition source. Spontaneous combustion reactions do not require an external heat source.

Some combustion reactions do not involve oxygen. For example, hydrogen gas (H_2) burns in chlorine gas $Cl_2(g)$ to form hydrogen chloride gas (HCl) with the production of heat and light:

$$H_2(g) + Cl_2(g) \rightarrow 2HCl(g)$$

Spontaneous combustion

Spontaneous combustion occurs without an external ignition source. A pyrophoric substance is one that will self-combust in air at room temperature – it has a relatively low ignition temperature. The process starts when oxidation or **fermentation** occurs and the pyrophoric substance begins to release heat. The heat cannot escape and the temperature of the material rises above its ignition point. If a strong oxidant such as oxygen is present, then combustion will occur.

The element sodium (Na) is a pyrophoric substance that undergoes a kind of spontaneous (and potentially very violent) explosion when exposed to oxygen, water or moisture in the air.

Large-scale spontaneous combustion

Spontaneous combustion can occur on a large scale and affect large populations and areas. Large amounts of material can spontaneously combust if not properly managed or kept under the right conditions.

There are many examples of spontaneous combustion. Bacterial fermentation in haystacks, and piles of compost and cow manure produce heat, which can result in spontaneous combustion. If coal is stored without adequate ventilation, reaction with oxygen can cause it to heat up and self-ignite. Also, in hot weather, areas of vegetation can spontaneously combust, although the mechanism for this is not yet understood.

Non-spontaneous combustion

Non-spontaneous combustion requires an external ignition source, such as a flame or an electrical fault. Non-spontaneous combustion can occur in organic substances (that is, substances made of carbon, hydrogen and oxygen), such as grass, wood and natural gas, and non-organic substances, such as the non-wood parts of buildings and cars. In PNG, many bush fires are non-spontaneous, usually having a human cause.

Complete combustion reactions

Complete combustion occurs in the presence of excess oxygen to yield carbon dioxide (CO_2) and water. The result is a colourless gas, heat, light and no black smoke. Combustion reactions are almost always exothermic (that is, they give off heat and light).

Wood is an organic substance that burns in the presence of O_2 to produce **charcoal**, CO_2 and a lot of heat and light. The general expression for the combustion for wood is:

$$\text{Wood} + \text{oxygen} \rightarrow \text{burning mass} + \text{heat} + CO_2$$

A general equation for the combustion of a **hydrocarbon** in oxygen can be expressed. Suppose x represents the number of carbon and y represents the number of hydrogen atoms making up the hydrocarbon. Then, for complete combustion the amount of oxygen required per mole of hydrocarbon is $x + \frac{y}{4}$ and the amounts of carbon dioxide and water produced are x and $\frac{y}{2}$ respectively.

$$C_xH_y + (x + \frac{y}{4})O_2 \rightarrow xCO_2 + \frac{y}{2}\,H_2O + \text{heat}$$

The following examples serve to illustrate the principle.

The **alkanes** are a group of saturated hydrocarbons. The combustion equation of the alkane methane (CH_4) is:

$$CH_4(g) + 2O_2(g) \rightarrow CO_2(g) + 2H_2O(l) + \text{heat}$$

$$\text{Methane} + \text{oxygen} \rightarrow \text{carbon dioxide} + \text{water} + \text{heat}$$

The combustion equation of the alkane propane (C_3H_8) is:

$$C_3H_8(g) + 5O_2(g) \rightarrow 3CO_2(g) + 4H_2O(l) + \text{heat}$$

$$\text{Propane} + \text{oxygen} \rightarrow \text{carbon dioxide} + \text{water} + \text{heat}$$

Ethene (C_2H_4) belongs to the class of **unsaturated** hydrocarbons called **alkenes**. The combustion equation of ethene is:

$$C_2H_4(g) + 3O_2(g) \rightarrow 2CO_2(g) + 2H_2O(l) + \text{heat}$$

$$\text{Ethene} + \text{oxygen} \rightarrow \text{carbon dioxide} + \text{water} + \text{heat}$$

Benzene (C_6H_6) is an example of the class of unsaturated hydrocarbons called arenes. The combustion equation of benzene is:

$$C_6H_6(l) + 7.5O_2\,(g) \rightarrow 6CO_2\,(g) + 3H_2O(l) + \text{heat}$$

$$\text{Benzene} + \text{oxygen} \rightarrow \text{carbon dioxide} + \text{water} + \text{heat}$$

Methanol (CH_3OH) is an **alcohol**. The combustion equation of methanol is:

$$CH_3OH(l) + O_2(g) \rightarrow CO_2(g) + 2H_2O(l) + \text{heat}$$

Methanol + oxygen → carbon dioxide + water + heat

Incomplete combustion of organic materials

Not all combustion reactions release carbon dioxide and water. If there is insufficient oxygen, **incomplete combustion** occurs to produce carbon monoxide (CO) and water. The flame of incomplete combustion is red-yellow due to the presence of red-hot soot.

The general equation for the incomplete combustion of a hydrocarbon in oxygen can also be expressed with *x* and *y* defined the same way as for complete combustion and *z* is the number of moles of the hydrocarbon. The equation is:

$$zC_xH_y + z(\frac{x}{2} + \frac{y}{4})O_2 \rightarrow zxCO + \frac{zy}{2} H_2O + \text{heat}$$

The following example serves to illustrate the principle.

The equation for the incomplete combustion of propane is:

$$2C_3H_8(g) + 7O_2(g) \rightarrow 2C(s) + 2CO(g) + 8H_2O(l) + \text{heat}$$

Propane + insufficient oxygen → unburnt carbon (soot) + carbon monoxide + water + heat

Combustion in air

In most industrial applications and in fires, air is the source of oxygen (O_2). When there is incomplete combustion, some of carbon is converted to carbon monoxide.

Air contains other gases; some of them may also be involved in the combustion reaction. For example, oxygen makes up 21% while nitrogen makes up 78.9%. That is, for every 100 molecules, 21 are oxygen and 78.9 are nitrogen. Thus, for every molecule of oxygen there are 3.76 molecules of nitrogen. Nitrogen does not take part in the combustion reaction, but it will react with oxygen at high temperatures to form nitrogen oxides, NO_x (usually 0.002–1%).

When air is the oxygen source for a combustion reaction, nitrogen can be added to the equation (although it does not react) to show the composition of the flue gas.

The general simple stoichiometric equation for complete combustion of hydrocarbons in air is:

$$C_xH_y + (x+ \frac{y}{4})O_2 + 3.76(x+ \frac{y}{4})N_2 \rightarrow xCO + \frac{y}{2} H_2O + 3.76(x+ \frac{y}{4})N_2 + \text{heat}$$

Hydrocarbon + air → carbon dioxide + water + nitrogen + heat

A more complete set of equations for complete combustion of hydrocarbons in air is required for each gas involved.

Example

The equations for the complete combustion of methane in air are:

$CH_4 + 2O_2 \rightarrow CO_2 + 2H_2O$

$2CH_4 + 3O_2 \rightarrow 2CO + 4H_2O$

$N_2 + O_2 \rightarrow 2NO$

$N_2 + 2O_2 \rightarrow 2NO_2$

Example

The equation for the burning of propane in air is:
$C_3H_8 + 5O_2 + 18.8N_2 \rightarrow 3CO_2 + 4H_2O + 18.8N_2$

Unit 11.3 Activity 5A: Balanced equations

1. Write a balanced equation for the complete combustion of 3 moles of propane in oxygen.
2. Write a balanced equation for the incomplete combustion of 4 moles of methane in oxygen.
3. Write a balanced equation for the complete combustion of 1 mole of propane in air.

Combustion reactions of non-organic materials

Combustion of inorganic and non-metallic materials yields heat and oxides. Any carbon in the sample is oxidised to carbon dioxide (CO_2).

Sulfur is converted to sulfur dioxide (SO_2) and heat:

$$S(s) + O_2(g) \rightarrow SO_2(g) + \text{heat}$$

Sulfur + oxygen → sulfur dioxide + heat

Hydrogen gives off water and heat:

$$2H_2(g) + O_2(g) \rightarrow 2H_2O(g) + \text{heat}$$

Hydrogen + oxygen → water + heat

Combustion of metallic materials also yields heat and oxides. Metals can be classified with combustible materials like wood and coal. Most metals and their alloys will burn in oxygen if heated to a high enough temperature in special furnaces.

The combustion equation for magnesium (Mg) is:

$$2Mg(s) + O_2(g) \rightarrow 2MgO(g) + \text{heat}$$

Magnesium + oxygen → magnesium oxide + heat

The combustion equation for iron (Fe) is:

$$Fe(s) + 0.5O_2(g) \rightarrow FeO(g) + \text{heat}$$

Iron + oxygen → iron oxide + heat

The combustion equation for aluminium (Al) is:

Aluminium + oxygen → aluminium oxide + heat

$$2Al(s) + O_2(g) \rightarrow 2AlO(g) + \text{heat}$$

The combustion of inorganic materials can be hastened through the use of an accelerator. Examples of single-component non-toxic accelerators are glass frit, niobium pentoxide and inorganic phosphate compounds. Multi-component accelerators are formed from combinations of single-component accelerators.

Explosions

Rapid combustion releases large amounts of heat and light energy, which often results in a flame. Sometimes large volumes of gas are also liberated, which can also create high pressures and loud noise; that is, an explosion.

A container of ethanol, or petrol vapour mixed with air, undergoes rapid combustion.

Applications of combustions

Combustion reactions provide the energy to perform work. Throughout history, humans have used the combustion of wood as a source of heat and light energy. Fire has also been used in warfare, to clear forested areas, in metallurgy and to produce smoke for signalling purposes. Combustion reactions have been used to power machinery and vehicles and to generate electricity.

Flash point and fire point are two important properties of **fuels**. The *flash point* of a volatile material is the lowest temperature at which it can vaporise to form an ignitable mixture in air. It is often used as a descriptive characteristic of liquid fuel, and it is also used to help characterise the fire hazards of liquids. Flash point refers to both flammable liquids and combustible liquids. The flash point is different from the autoignition temperature, which does not require an ignition source.

The *fire point* is the temperature at which the vapour continues to burn after being ignited. Both flash point and fire point are not dependent on the temperature of the ignition source, which is much higher.

Solid fuels

The combustion of solid fuels consists of three phases.

In the preheating phase, the unburnt fuel is heated up to its flash point and then fire point. This results in the production of flammable gases. During the **distillation** phase or gaseous phase, the mix of flammable gases and oxygen is ignited to produce energy in the form of heat and light. Flames are often visible. Heat transfer from the combustion reaction to the solid maintains the evolution of flammable vapours. The **charcoal** phase or **solid** phase is when the output of flammable gases is too low to maintain a flame and the charred fuel does not burn rapidly any more but just glows and then smoulders.

Liquid fuels

With a liquid fuel, it is the vapour that burns, not the liquid. A liquid will normally catch fire at temperatures above its flash point – the lowest temperature at which enough liquid can evaporate to form an ignitable mix with air.

The flash points of some fuels are ethanol 17 °C, propane –104 °C, petrol –43 °C, diesel >62 °C, jet fuel >60 °C, vegetable oil (canola) 327 °C, biodiesel >130 °C and kerosene 52 °C.

Fuel economy

As the world's demand for energy increases, it is important that we are able to measure the efficiency of various fuels. This is done by measuring the energy output per mole, per gram or per litre of fuel. The fuel with larger energy output is the better fuel.

Example

Which is a better fuel: methane or hydrogen?

By consulting a data table, we can see that 1 mole (16 g) of methane liberates 890 kJ of heat energy and 1 mole (2 g) of hydrogen produces 285 kJ of heat energy. So methane produces more energy per mole but hydrogen produces more energy per gram. In this case, hydrogen appears to be a better fuel. But we would not replace methane with hydrogen because hydrogen is very explosive and dangerous to use in its current form.

We buy fuel in mass and litres.

Combustion of fuels and climate change

Rising carbon dioxide emissions are thought to contribute to climate change and reduced quality of life on the planet. About 80% of the world's energy is produced from combustion of **fossil fuels**. The graph shows how annual CO_2 emissions from fossil fuel combustion have increased over the last two centuries.

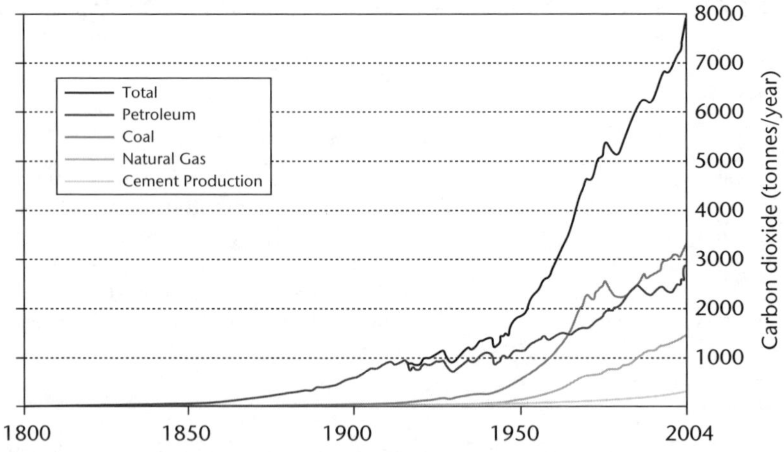

Annual carbon dioxide emissions from various fuel types during 1800–2004

Unit 11.1 Activity 5B: Combustion reactions – multiple choice

1. In a combustion reaction which of the following is *not* true?

A. The reaction is exothermic.
B. The reaction always requires oxygen.
C. Carbon dioxide is formed from combustion of hydrocarbons.
D. Water is formed from burning of organic materials.

2. In a spontaneous combustion reaction which of the following is *not* true?

A. Ignition is effected by humans.
B. Large areas of grassland can be consumed.
C. Pyrophoric substances can ignite the materials.
D. Spontaneous combustions are preventable.

3. In the complete combustion of propane, the number of moles of oxygen per mole of propane required is:

A. 1
B. 2
C. 3
D. 4
E. 5

4. In the incomplete combustion of propane, the number of moles of CO per mole of propane formed is:

A. 1
B. 2
C. 3
D. 4
E. 5

5. Which of the following is true of the combustion of metals?

A. Metals do not undergo combustion reactions.
B. Metals burn at low temperatures.
C. Metals burn at high temperatures, giving off CO_2 gas.
D. Metals burn at high temperatures, forming an oxide.

6. The cause of explosion during combustion is the:

A. kind of fuel used.
B. kind of ignition mode.
C. ignition of rapidly expanding gas.
D. igniting of the slowly expanding gases.

7. Which of the following is *not* true of natural gas?

A. It is composed mostly of methane.
B. It is the cleanest of the fuels.
C. It is not a fossil fuel.
D. It liberates fewer by-products during combustion.

Unit 11.3 Types of Chemical Reactions

Topic 6: Molecular mass calculations

This Topic describes chemical reactions by considering:

- Atomic and molecular masses.
- Calculating percentage composition of compounds.
- Empirical formulae and their calculation from data.
- Mass ratios of chemicals involved in chemical reactions.

Atomic masses

Atoms of all elements have mass, measured in **amu** (**atomic mass units**). The amu is a relative value, which is based on the carbon-12 isotope being assigned a mass of 12.0000 atomic mass units.

Example

If it were possible to weigh one carbon atom on a set of scales, using grams as the unit of mass, the reading would be 0.000 000 000 000 000 000 000 02 g or 2×10^{-23} g. Because this mass is so small, the relative scale is used.

Example

The hydrogen atom has a relative atomic mass of 1.008, which is taken as 1.0.

$\mathbf{A_r}$ values are not whole numbers due to the existence of **isotopes**.

Example

Hydrogen has three isotopes, with masses of 1.0, 2.0 and 3.0 on the atomic scale – it is the 'average' hydrogen atom that has a mass of 1.008 (1.0 to one decimal place).

Example

A_r values

H = 1.0, C = 12.0, N = 14.0, O = 16.0, Na = 23.0, S = 32.0.

Molecular masses

Relative molecular mass values, $\mathbf{M_r}$, of compounds can be calculated from **relative atomic mass** values and formulae.

Formulae

Formulae (eg NaOH, Na_2CO_3) represent the simplest recognisable unit of each compound and indicate the number of atoms of each element. (The term 'molecule' applies only to purely covalently bonded particles.)

Example

Na_2CO_3 contains 2 sodium (Na) atoms, 1 carbon (C) atom and 3 oxygen (O) atoms.
$Fe_2(SO_4)_3$ contains 2 Fe atoms, 3 S atoms and 12 (3 × 4) O atoms.

Some examples of calculating relative molecular masses, M_r, are given in the following table.

Compound	Formula	A_r values for component atoms	M_r value for molecule
Water	H_2O	H = 1.0; O = 16.0	2 × 1.0 + 1 × 16.0 = 18.0
Carbon dioxide	CO_2	C = 12.0; O = 16.0	1 × 12.0 + 2 × 16.0 = 44.0
Ammonia	NH_3	H = 1.0; N = 14.0	3 × 1.0 + 1 × 14.0 = 17.0
Sulfuric acid	H_2SO_4	H = 1.0; O = 16.0; S = 32.0	2 × 1.0 + 1 × 32.0 + 4 × 16.0 = 98.0

Relative molecular masses

Unit 11.3 Activity 6A: Calculating M_r values

The following A_r values are to be used for all calculations in Activities 6A to 6D.

A_r values: H = 1.0, C = 12.0, N = 14.0, O = 16.0, Mg = 24.3, Al = 27.0, P = 31.0, S = 32.0, Cl = 35.5, Ca = 40.1, Ti = 47.9, Fe = 55.9, Cu = 63.6, Zn = 65.4, Hg = 200.6.

1. Calculate M_r values for:
- **a.** Ethyne, C_2H_2.
- **b.** Propene, C_3H_6.
- **c.** Butane, C_4H_{10}.

2. Calculate relative molecular masses for:
- **a.** Carbon monoxide, CO.
- **b.** Sulfur trioxide, SO_3.
- **c.** Hydrazine, N_2H_4.
- **d.** Methanol, CH_3OH.
- **e.** Ethanol, C_2H_5OH.
- **f.** Ammonium chloride, NH_4Cl.
- **g.** Ammonium sulfate, $(NH_4)_2SO_4$.
- **h.** Aluminium phosphate, $AlPO_4$.
- **i.** Aluminium sulfate, $Al_2(SO_4)_3$.
- **j.** Aluminium nitrate, $Al(NO_3)_3$.

3. Calculate M_r values for:
- **a.** Sucrose, $C_{12}H_{22}O_{11}$.
- **b.** Nicotine, $C_{10}H_{14}N_2$.
- **c.** Heroin, $C_{17}H_{17}NO(C_2H_3O_2)_2$.

Calculating percentage composition

Calculations can be made of the percentage of an element, or elements, in a compound.

Example

Calculate the percentage of oxygen in water

$M_r(H_2O) = 2 \times 1.0 + 16.0 = 18.0$

16 parts out of 18 parts of water are oxygen.

Percentage oxygen = $\frac{16}{18} \times 100\% = 88.9\%$

Example

Calculate the percentage of oxygen in sodium carbonate

$M_r(Na_2CO_3) = 2 \times 23.0 + 12.0 + 3 \times 16.0 = 106.0$

(3 x 16) or 48 parts out of 106 parts of sodium carbonate are oxygen.

Percentage oxygen = $\frac{48}{106} \times 100\% = 45.3\%$

Example

Calculate the percentage of oxygen in washing soda ($Na_2CO_3.10H_2O$)

$M_r(Na_2CO_3.10H_2O) = 2 \times 23.0 + 12.0 + 3 \times 16.0 + 10 \times 18.0 = 286.0$. A 'full stop' in a formula implies addition (eg $Na_2CO_3 + 10H_2O$), and *not* multiplication (eg $Na_2CO_3 \times 10H_2O$).

48 + 160 (10 × 16) = 208 parts out of 286 parts of washing soda are oxygen.

Percentage oxygen = $\frac{208}{286} \times 100\% = 72.7\%$

Calculating yield

Calculations can be made of quantities of chemicals obtainable from, or present in, other chemicals.

Example

Calculate how much water can be made from a tonne of hydrogen

2 parts of hydrogen can make 18 parts of water.

1 tonne of hydrogen can make $\frac{18}{2} \times 1 = 9$ tonnes of water.

Example

Calculate how much carbon is present in 1 kg of carbon dioxide

12 parts of carbon are present in 44 parts of carbon dioxide.

Mass of carbon in 1 kg of carbon dioxide = $\frac{12}{44} \times 1 = 0.273$ kg = 273 g

Example

Calculate how much washing soda can be made from 273 g of carbon dioxide

44 parts of 286 parts of washing soda are carbon dioxide.

Mass of washing soda that can be made from 273 g of carbon dioxide

$$= \frac{286}{44} \times 273 = 1774.5 \text{ g} = 1.775 \text{ kg}$$

Unit 11.3 Activity 6B: Calculating composition and yields of chemicals

1. Calculate the percentage of carbon in each of the following organic compounds:

- **a.** Ethane, C_2H_6.
- **b.** Ethene, C_2H_4.
- **c.** Ethyne, C_2H_2.
- **d.** Ethanol, C_2H_5OH.
- **e.** Glucose, $C_6H_{12}O_6$.

2. Calculate the percentage of the named element in each of the following *minerals*:

- **a.** Carbon in *marble*, $CaCO_3$.
- **b.** Sulfur in *iron pyrite*, FeS_2.
- **c.** Magnesium in *dolomite*, $CaCO_3.MgCO_3$.
- **d.** Titanium in *titanomagnetite*, $Fe_3O_4.TiO_2$.
- **e.** Oxygen in *gypsum*, $CaSO_4.2H_2O$.

3. Calculate the mass of:

- **a.** Oxygen present in 100 g of hydrogen peroxide, H_2O_2.
- **b.** Oxygen that can be obtained from 100 g of mercury oxide, HgO.
- **c.** Nitrogen is present in 100 g of urea, $CO(NH_2)_2$.
- **d.** Aluminium sulfide, Al_2S_3, that could be made from 100 g of sulfur.

4. Calculate the mass of:

- **a.** Zinc that can be extracted from 1 kg of zinc blende, ZnS.
- **b.** Zinc needed to make 1 kg of calamine, $ZnCO_3$.
- **c.** Nitrogen needed to make 1 kg of ammonium nitrate, NH_4NO_3.
- **d.** Phosphorus that can be extracted from 1 tonne of rock phosphate, $Ca_3(PO_4)_2$.

Empirical formulae

An empirical formula is the simplest formula of a compound. It shows:

- The ratio of *atoms* in a molecule, or
- The ratio of *ions* in an ionic compound.

For some compounds, the empirical formula is the *same* as the molecular formula.

Example

The formulae H_2O and CO_2 represent the simplest ratio of atoms in molecules of water and carbon dioxide respectively, and also represent the composition of the molecule of each compound.

For other compounds, the empirical and molecular formulae are *different*.

Example

Hydrogen peroxide has an empirical formula of HO but the molecule has the formula H_2O_2. Ethane – an **alkane** (a compound of carbon and hydrogen) – has the empirical formula of CH_3 but the molecule has the formula C_2H_6.

Calculating empirical formulae

The empirical formula of a compound can be calculated from percentage composition of a compound.

Example

A compound contains 88.9% oxygen and 11.1% hydrogen. Calculate the empirical (simplest) formula of the compound.

The number of oxygen atoms in 88.9 amu of oxygen is 88.9 ÷ 16.0 = 5.56.

The number of hydrogen atoms in 11.1 amu of hydrogen is 11.1 ÷ 1.0 = 11.1.

Look for a simple ratio between these two numbers of atoms. This is best done by dividing the larger number by the smaller number, ie 11.1 ÷ 5.56 = 2.

Hence, there are two atoms of hydrogen to every one atom of oxygen in this molecule, leading to the empirical formula H_2O.

Example

Calculating an empirical formula

A compound contains 94.1% oxygen and 5.9% hydrogen. Calculate the empirical formula of the compound.

The calculation procedure can be shown in a table.

Element	O	H
Composition as a percentage	94.1	5.9
Divide composition by atomic mass	$\frac{94.1}{16.0} = 5.9$	$\frac{5.9}{1.0} = 5.9$
Look for the simplest ratio: divide all numbers by the smallest number	$\frac{5.9}{5.9} = 1$	$\frac{5.9}{5.9} = 1$
Empirical formula is **HO**		

Unit 11.3 Activity 6C: Empirical formulae

1. State which of the following compounds will have a different empirical formula from the molecular formula that is given. Where the empirical formula is different from the molecular formula, write the empirical formula of the compound.
   - **a.** Methane, CH_4.
   - **b.** Ethene, C_2H_4.
   - **c.** Ethyne, C_2H_2.
   - **d.** Ethanol, C_2H_5OH.

- **e.** Glucose, $C_6H_{12}O_6$.
- **f.** Ethanoic acid, CH_3COOH.

2. Calculate the empirical formula of:
- **a.** A compound of iron and sulfur that contains 63.6% iron.
- **b.** A compound of iron and sulfur that contains 46.6% iron.
- **c.** A compound of nitrogen and hydrogen that contains 87.5% nitrogen.
- **d.** A compound that contains 37.5% carbon, 50% oxygen and 12.5% hydrogen.
- **e.** A compound that contains 52.2% carbon, 34.8% oxygen and 13.0% hydrogen.
- **f.** A compound that contains 40.0% carbon, 53.3% oxygen and 6.7% hydrogen.

Calculating mass ratios

Mass ratio calculations using equations for reactions can best be explained with examples.

Example

What mass of oxygen is needed for the complete combustion of 6 g of carbon?
First obtain the equation for the reaction: $C(s) + O_2(g) \rightarrow CO_2(g)$.
This equation shows that 12 g of carbon combine with 32 g of oxygen,
$\Rightarrow$ (6 ÷ 12) × 32 g of oxygen combine with 6 g of carbon, or **16 g.**

Example

What is the percentage purity of a sample of limestone, 50 g of which produces 21 g of carbon dioxide on heating?
Equation: $CaCO_3(s) \rightarrow CaO(s) + CO_2(g)$
This equation shows that 44 g of carbon dioxide ($M_r(CO_2) = 44$) is obtained from 100 g of limestone ($M_r(CaCO_3) = 100$),
$\Rightarrow$ 22 g of carbon dioxide is obtained from 50 g of limestone,
$\Rightarrow$ percentage purity of limestone = (21 ÷ 22) × 100 = **95.5%.**

Example

What mass of chlorine can be obtained by the **electrolysis** of 100 tonnes of salt?
Equation: $2NaCl(s) \rightarrow Cl_2(g)$ at the **anode**.
$M_r(NaCl) = 23 + 35.5 = 58.5$, $M_r(Cl_2) = 2 \times 35.5 = 71$
$\Rightarrow$ 71 g of chlorine is obtained from 117 g of salt
$\Rightarrow$ 100 tonnes of salt will produce (71 ÷ 117) × 100 = **60.7** tonnes of chlorine.

Example

What mass of carbon dioxide can be made from 71.5 g of washing soda by addition of dilute hydrochloric acid?
Equation: $Na_2CO_3.10H_2O(s) + 2HCl(aq) \rightarrow 2NaCl(aq) + H_2O(\ell) + CO_2(g)$
This equation shows that 44 g of carbon dioxide can be obtained from 286 g of washing soda ($M_r(Na_2CO_3.10H_2O) = 286$),
$\Rightarrow$ (71.5 ÷ 286) × 44 g of carbon dioxide can be obtained from 71.5 g of washing soda, or **11 g.**

Unit 11.3 Activity 6D: Calculating masses of chemicals in a reaction

1. Calculate the mass of carbon dioxide that could be obtained from 100 g of sodium carbonate when reacted with hydrochloric acid. The equation for the reaction between sodium carbonate and hydrochloric acid is:

$$Na_2CO_3(s) + 2HCl(aq) \rightarrow 2NaCl(aq) + H_2O(\ell) + CO_2(g).$$

2. Calculate what mass of zinc metal would be needed to produce 10 g of hydrogen from dilute sulfuric acid. The equation for the reaction between zinc and sulfuric acid is:

$$Zn(s) + H_2SO_4(aq) \rightarrow ZnSO_4(aq) + H_2(g).$$

3. Calculate what mass of oxygen is needed for the complete combustion of 1 litre of petrol. Assume the petrol is octane (C_8H_{18}) and that 1 litre of octane weighs 700 g. The equation for the reaction between octane and oxygen is:

$$2C_8H_{18}(\ell) + 25O_2(g) \rightarrow 16CO_2(g) + 18H_2O(\ell).$$

4. Calculate what mass of copper would be required to produce 10 g of copper nitrate crystals from concentrated nitric acid. The equation for the reaction between copper and nitric acid is:

$$Cu(s) + 4HNO_3(aq) \rightarrow Cu(NO_3)_2(aq) + 2H_2O(\ell) + 2NO_2(g).$$

The crystals are formed from the evaporation of water from the copper nitrate solution:

$$Cu(NO_3)_2(aq) \rightarrow Cu(NO_3)_2.3H_2O(s) + H_2O(g).$$

5. Calculate how much baking soda (sodium hydrogen carbonate) is required to produce 10 g of carbon dioxide when the baking soda is heated. The equation for the reaction when baking soda is heated is:

$$2NaHCO_3(s) \rightarrow Na_2CO_3(s) + H_2O(g) + CO_2(g).$$

6. Calculate the quantities of ammonia gas and sulfuric acid required to produce 1 tonne of sulfate of ammonia (ammonium sulfate) fertiliser. The equation for the reaction to produce the sulfate of ammonia is:

$$2NH_3(g) + H_2SO_4(\ell) \rightarrow (NH_4)_2SO_4(s).$$

Unit 11.3 Types of Chemical Reactions
Topic 7: pH and solutions

In this Topic we look at thermochemical and equilibrium principles. The Topic deals with:

- Explaining the behaviour of strong and weak acids and bases in terms of the concentration of H_3O^+ or OH^- ions.
- Calculating pH, $[H_3O^+]$ and $[OH^-]$ for aqueous solutions.

The pH scale

The concentration of **hydrogen ions** (or **hydronium ions**) in a solution can be expressed as a **pH**. pH is a number and has no units. The pH of different solutions forms a **pH scale** and usually ranges between 0 and 14:

The pH scale

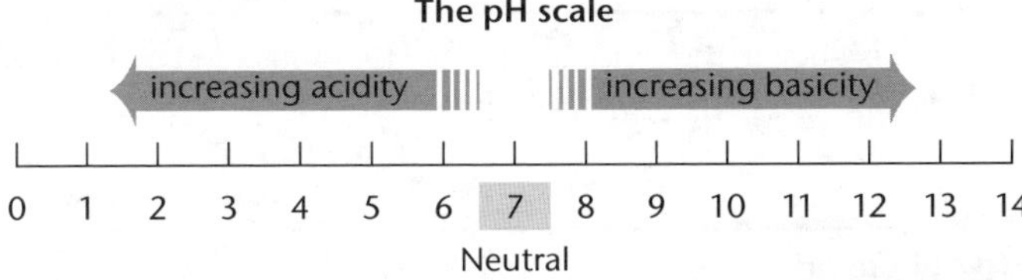

On the pH scale:

- An acidic solution has a pH number *less* than 7.
- A basic solution has a pH number *greater* than 7.
- A neutral solution has a pH number of *exactly* 7.
- A solution becomes more acidic if its pH decreases.
- A solution becomes less acidic (more basic) if its pH increases.

pH of 'everyday' substances	
Substance	**pH**
0.1 mol L^{-1} HCl	1.0
Lemon juice	2.5
Vinegar	3.0
Milk	6.8
Pure water	7.0
Blood	7.4
Window cleaner	11.0
0.1 mol L^{-1} NaOH	13.0

The pH of a solution can be determined using a pH meter or **universal indicator**. Universal indicator is a mixture of dyes and produces a range of colours corresponding to different pH values:

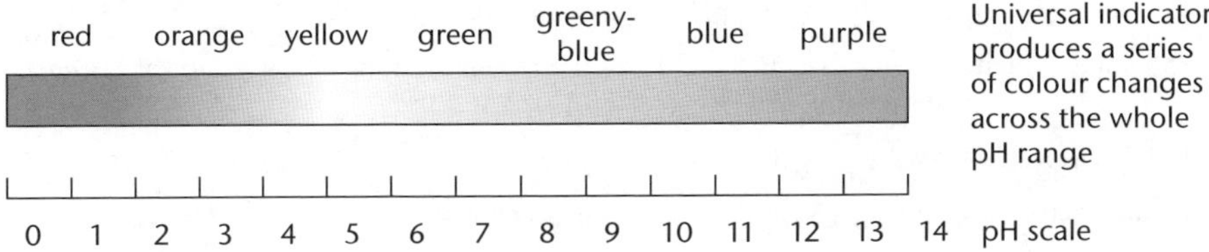

Universal indicator produces a series of colour changes across the whole pH range

Calculating pH

pH can be calculated from the concentration of hydronium (or hydrogen) ions.

Mathematically, pH is defined using logarithms:

pH = negative logarithm of the hydronium ion concentration

$$pH = -\log [H_3O^+(aq)] \qquad (\text{or } pH = -\log_{10} [H_3O^+(aq)])$$

Example

Calculating pH

Calculate the pH of a solution where $[H_3O^+] = 4.0 \times 10^{-5}$ mol L^{-1}.

$pH = -\log [H_3O^+]$

$= -\log 4.0 \times 10^{-5}$ [substituting]

$= 4.4$

If the pH of a solution is known, the concentration of H_3O^+ ions can be calculated using a rearrangement of the equation $pH = -\log [H_3O^+]$, ie:

$$[H_3O^+] = 10^{-pH}$$

Example

Finding $[H_3O^+]$ from pH

Calculate the H_3O^+ ion concentration of a solution of pH = 11.4.

$[H_3O^+] = 10^{-pH}$

$= 10^{-11.4}$ [substituting]

$= 4.0 \times 10^{-12}$ mol L^{-1}

Ionic product, K_w

In pure water, most of the particles present are H_2O molecules. However, there will also be a few ionised (or dissociated) water molecules:

$$H_2O(\ell) + H_2O(\ell) \rightleftharpoons H_3O^+(aq) + OH^-(aq)$$

$$(\text{or } H_2O(\ell) \rightleftharpoons H^+(aq) + OH^-(aq))$$

(Double-headed arrows $\rightleftharpoons$ are used since a **dynamic** equilibrium exists between water molecules and $H_3O^+(aq)$ and $OH^-(aq)$ ions.)

The dissociation is only slight. At 25 °C, only about *one* water molecule per six hundred million will be dissociated. For pure water, which is neutral:

$$[H_3O^+(aq)] = [OH^-(aq)] = 1 \times 10^{-7} \text{ mol L}^{-1}$$

Note: From $pH = -\log[H_3O^+(aq)]$, the pH of pure water is 7 ($-\log 1 \times 10^{-7} = 7$).

Experimental evidence shows that the product of the concentrations of $H_3O^+(aq)$ (or $H^+(aq)$) and $OH^-(aq)$ has a constant value of 1×10^{-14} mol^2 L^{-2} at 25 °C:

ie $[H_3O^+(aq)] \times [OH^-(aq)] = 1 \times 10^{-14}$ mol^2 L^{-2} at 25 °C

The product of the concentrations is called the **dissociation constant** or **ionic product** of water, K_w;

ie $K_w = [H_3O^+(aq)][OH^-(aq)] = 1 \times 10^{-14}$ at 25 °C

The dissociation constant, K_w, can be used to calculate the concentration of hydronium (or hydrogen) ions in a solution if the concentration of hydroxide ions is known, and vice versa.

Example

Finding $[H_3O^+]$ from K_w and $[OH^-]$

Calculate the $[H_3O^+]$ in a solution with hydroxide concentration $[OH^-] = 5 \times 10^{-5}$ mol L^{-1}.

$$K_w = [H_3O^+][OH^-] = 1 \times 10^{-14}$$

$$\therefore [H_3O^+] = \frac{K_w}{[OH^-]} \quad \text{[rearranging]}$$

$$= \frac{10^{-14}}{5 \times 10^{-5}} \quad \text{[substituting]}$$

$$= 2 \times 10^{-10} \text{ mol L}^{-1}$$

Note: Once the $[H_3O^+]$ is calculated from $K_w = [H_3O^+][OH^-]$, the pH can be found using $pH = -\log [H_3O^+]$.

The expression $K_w = [H_3O^+][OH^-]$ is constant at constant temperature. This means that if the concentration of the hydrogen ions *increases* then the concentration of the hydroxide ions must *decrease*.

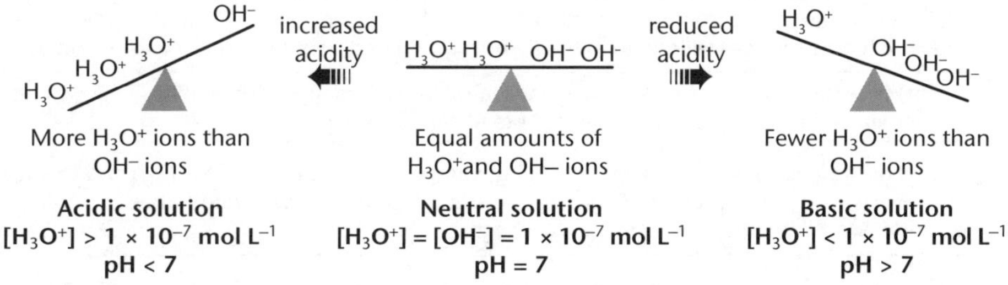

Acidic solution
$[H_3O^+] > 1 \times 10^{-7}$ mol L^{-1}
$pH < 7$

Neutral solution
$[H_3O^+] = [OH^-] = 1 \times 10^{-7}$ mol L^{-1}
$pH = 7$

Basic solution
$[H_3O^+] < 1 \times 10^{-7}$ mol L^{-1}
$pH > 7$

Neutralisation

When a basic solution is added to an acidic solution, the acidic solution becomes *less* acidic, because a process called **neutralisation** takes place. The pH of the solution increases so that it approaches or becomes a pH of 7. A solution with a pH of 7 is described as **neutral**.

Similarly, when an acidic solution is added to a basic solution, lowering the pH, the basic solution is being neutralised.

Example

Neutralising hydrochloric acid

When sodium hydroxide solution is added to hydrochloric acid solution, a neutralisation reaction occurs. The hydrogen ions, $H_3O^+(aq)$, from the acid, react with the hydroxide ions, $OH^-(aq)$, from the base. The reaction occurring is:

$$H_3O^+(aq) + OH^-(aq) \longrightarrow 2H_2O(\ell)$$

The neutralisation reaction $H_3O^+(aq) + OH^-(aq) \longrightarrow 2H_2O(\ell)$ can be shown diagramatically:

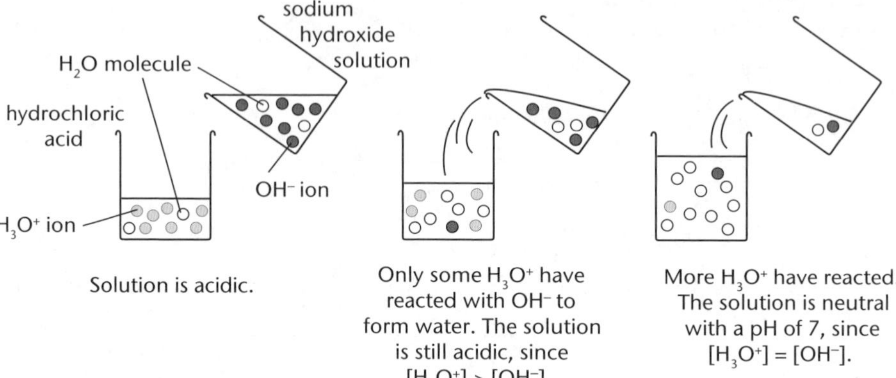

The neutralisation reaction could be monitored using an acid–base indicator. Universal indicator would produce a red colour in acid, but as hydroxide ions were added, the colour would become orange, yellow, then finally turn green as a neutral solution was obtained.

In practice, 'adding' hydroxide ions is hard to do – very small amounts of base have to be added each time – this is only really feasible using a **burette**.

The resulting neutral solution would contain water molecules, $H_2O(\ell)$, $Na^+(aq)$ and $Cl^-(aq)$ and a very few $H_3O^+(aq)$ and $OH^-(aq)$ ions. If the solution was evaporated, the salt, sodium chloride, would be isolated.

Neutralisation of an acid does *not* always produce a neutral solution.

Example

Neutralising ethanoic acid with sodium hydroxide

Ethanoic acid can be 'neutralised' by adding sufficient sodium hydroxide solution to react with all the $H_3O^+(aq)$ from the acid:

$$CH_3COOH(aq) + NaOH(aq) \rightleftharpoons CH_3COONa(aq) + H_2O(\ell)$$

The resulting solution is not neutral, because the ethanoate ions formed react with water to produce $OH^-(aq)$ ions:

$$CH_3COO^-(aq) + H_2O(\ell) \rightleftharpoons CH_3COOH(\ell) + OH^-(aq)$$

The resulting solution is *basic* (*not* neutral) due to the excess $OH^-(aq)$ ions.

Unit 11.3 Activity 7A: pH

1. Following are three solutions:

i. pH = 4 changes to pH = 8.

ii. pH = 2 changes to pH = 5.

iii. pH = 10 changes to pH = 4.

a. For each solution **i.** to **iii.**, state whether the solution is becoming more acidic or more basic.

b. In each case, does $[H_3O^+]$ get larger or smaller?

2. a. Find $[OH^-]$ when $[H_3O^+] = 10^{-8}$ mol L^{-1}, given $K_w = [H_3O^+][OH^-] = 1.00 \times 10^{-14}$.

b. Which ion is present in the greater concentration in the solution in **a**., H_3O^+ or OH^-?

c. Is the solution in **a**. acidic or basic?

3. **a**. Find $[H_3O^+]$ if $[OH^-] = 10^{-11}$ mol L^{-1}, given $K_w = [H_3O^+][OH^-] = 1.00 \times 10^{-14}$.

b. Which ion is present in the greater concentration in **a**., H_3O^+ or OH^-?

c. Is the solution in **a**. acidic or basic?

4. $K_w = 1.00 \times 10^{-14}$.

a. Find $[H_3O^+]$ if $[OH^-] = 2 \times 10^{-3}$ mol L^{-1}.

b. Find $[H_3O^+]$ if $[OH^-] = 10^{-7}$ mol L^{-1}.

5. Describe the following solutions as acidic or basic, given that $K_w = [H_3O^+][OH^-] = 1.00 \times 10^{-14}$.

a. $[H_3O^+] = 10^{-9}$ mol L^{-1} **b**. $[OH^-] = 10^{-4}$ moL l^{-1}

c. $[H_3O^+] = 2 \times 10^{-2}$ mol L^{-1} **d**. $[OH^-] = 5 \times 10^{-10}$ mol L^{-1}

6. Copy and complete the table by calculating $[OH^-]$ and pH.
One column has been done for you.

$[H_3O^+]$	10^{-1}	10^{-2}	10^{-7}	10^{-10}	10^{-14}
$[OH^-]$		10^{-12}			
pH		2			
solution		acidic			

7. Find the pH of the following solutions $[K_w = [H_3O^+] \times [OH^-] = 1.00 \times 10^{-14}]$:

a. $[H_3O^+] = 10^{-4}$ mol L^{-1} **b**. $[H_3O^+] = 10^{-3}$ mol L^{-1}

c. $[OH^-] = 10^{-2}$ mol L^{-1} **d**. $[OH^-] = 10^{-11}$ mol L^{-1}

e. $[H_3O^+] = 5 \times 10^{-9}$ mol L^{-1} **f**. $[H_3O^+] = 4 \times 10^{-6}$ mol L^{-1}

8. Find the hydrogen concentration, $[H_3O^+]$, for the following solutions:

a. Gastric juice, pH = 2.0 **b**. Ammonia solution, pH = 10

c. Cow's milk, pH = 6.4 **d**. Vinegar, pH = 2.9

e. Human blood, pH = 7.4 **f**. *Ajax* (cleaner), pH = 10.2

9. 5 mL of 0.1 mol L^{-1} nitric acid solution is diluted to 100 mL by adding 95 mL of water. Calculate:

a. $[H_3O^+]$ in the diluted solution.

b. pH of the diluted solution.

10. 10 mL of 0.01 mol L^{-1} NaOH is added to 90 mL of water to make a total volume of 100 mL. Calculate each of the following, given $K_w = [H_3O^+][OH^-] = 1.00 \times 10^{-14}$.

a. $[OH^-]$ in the 10 mL of 0.01 mol L^{-1} NaOH.

b. $[OH^-]$ in the diluted solution.

c. $[H_3O^+]$ in the diluted solution.

d. pH of the diluted solution.

11. **a**. Calculate the pH of a solution with a hydronium ion concentration, $[H_3O^+]$, of 0.0850 mol L^{-1}. State your answer to 3 **significant figures**.

b. If a solution of sodium hydrogencarbonate has a pH of 9.22, calculate the concentration of hydroxide ions, $[OH^-]$, present in the solution. State your answer to 3 significant figures.

c. If a solution of methanoic acid has a pH of 2.22, calculate $[H_3O^+]$ and $[OH^-]$.

d. A solution of sodium fluoride has $[H_3O^+]$ of 8.86×10^{-8} mol L^{-1}. Calculate the pH of this solution.

e. A glass cleaner solution has a hydroxide ion concentration of 0.00667 mol L^{-1}. Calculate the pH of this solution.

12. Copy and complete the table below to show the hydronium ion concentration, hydroxide ion concentration, and pH for the solutions **A** to **I**.

$K_w = 1.00 \times 10^{-14}$

Solution	$[H_3O^+]$ / mol L^{-1}	$[OH^-]$ / mol L^{-1}	pH
A	0.0672		
B			10.6
C		1.93×10^{-11}	
D			5.9
E	6.35×10^{-10}		
F		0.0055	
G	0.00150		
H		8.77×10^{-9}	
I			7.4

13. Account for each of the following. Include chemical equations in your discussions.

a. Slaked lime, $Ca(OH)_2$, will raise the pH of acidic soils.

b. The treatment for indigestion ('acid stomach') might involve either:

i. Drinking a solution of baking soda, $NaHCO_3$.

ii. Swallowing calcium carbonate tablets.

Strong and weak acids

Hydrochloric acid, sulfuric acid and nitric acid are **strong acids**. **Organic acids**, such as ethanoic acid, are **weak acids**.

Acids

Strong acids

Strong acids donate protons readily in aqueous solution to become *completely* ionised or dissociated.

Example – when hydrogen chloride gas reacts with water, it dissociates fully:

$HCl(g) + H_2O(\ell) \longrightarrow H_3O^+(aq) + Cl^-(aq)$

The solution formed contains mostly water molecules, hydronium ions, H_3O^+, and chloride ions, Cl^-. Very few, if any, HCl molecules will be present.

Weak acids

Weak acids donate protons only to a limited extent in aqueous solution, so are only partly ionised or dissociated.

Example – when ethanoic acid reacts with water, and an equilibrium is established, only a few acid molecules ionise:

$CH_3COOH(\ell) + H_2O(\ell) \rightleftharpoons CH_3COO^-(aq) + H_3O^+(aq)$

The solution formed contains mostly water molecules and ethanoic acid molecules, CH_3COOH. Only a few hydronium ions, H_3O^+, and ethanoate ions, CH_3COO^-, will be present, since only a small number of acid molecules dissociate.

Example

Comparing strong and weak acids

Comparing equal volumes of 0.1 mol L^{-1} hydrochloric and 0.1 mol L^{-1} ethanoic acids:

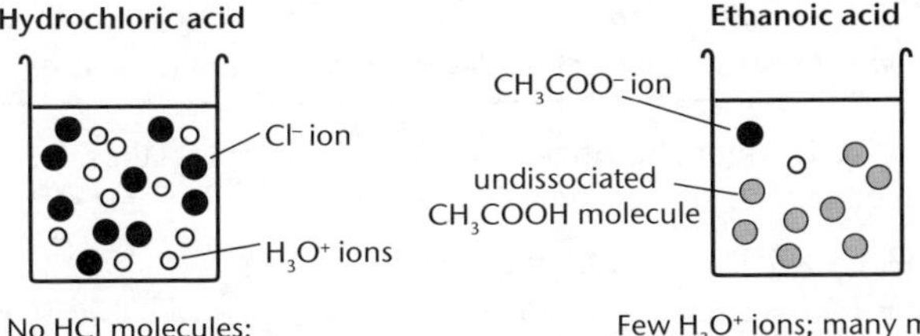

No HCl molecules; many H_3O^+ ions

Few H_3O^+ ions; many molecules of CH_3COOH which have not dissociated

If a solution of a strong acid is compared with that of a weak acid of the *same concentration*:

- The strong acid has more ions, so is a better **electrolyte** and therefore will be a better conductor of electricity.
- The strong acid will have a *greater* concentration of hydrogen ions, so has a *lower pH*.
- The strong acid solution will react faster with metals and carbonates, because of the higher concentration of H_3O^+.

If there is the same amount (ie equal moles) of stong acid and weak acid, the amount of product(s) formed for a reaction will be the same; however, the strong acid will react faster than the weak acid (ie the reaction will get to completion more quickly for the stong acid).

Example

Zinc reacting with strong and weak acids

One mole of hydrochloric acid, HCl, will consume the same mass of zinc as will one mole of ethanoic acid, CH_3COOH. The difference is that HCl will consume the zinc much more quickly (ie faster reaction rate).

In HCl solution, all the HCl molecules have dissociated to form Cl^- and H_3O^+ ions; all the H_3O^+ ions are available to react with the added zinc. In CH_3COOH solution, most of the CH_3COOH molecules have not dissociated to form CH_3COO^- and H_3O^+ ions; only a few H_3O^+ ions are therefore available to react with the added zinc. However, when an H_3O^+ ion from CH_3COOH does react, it is 'replaced' (from the equilibrium reaction of $CH_3COOH + H_2O \rightleftharpoons CH_3COO^- + H_3O^+$) with another H_3O^+ until all the CH_3COOH has been 'used up'. Since only a relatively small amount of H_3O^+ ions are present at any one time, the reaction of ethanoic acid with zinc is slow compared with the reaction of hydrochloric acid with zinc.

Strong and weak bases

Strong bases readily accept protons; **weak bases** accept protons only to a limited extent. Sodium and potassium hydroxide are strong bases. Ammonia solution (ammonium hydroxide) is a weak base.

Example

Ammonia and sodium hydroxide

- Ammonia is a weak base because only a few $OH^-(aq)$ ions form when ammonia is added to water:

  $NH_3(aq) + H_2O \rightleftharpoons NH_4^+(aq) + OH^-(aq)$

An ammonia solution contains mainly NH_3 molecules and water; very few $NH_4^+(aq)$ and $OH^-(aq)$ ions are present.

- Sodium hydroxide is a strong base because it ionises fully in water:

$$NaOH(s) \xrightarrow{H_2O} Na^+(aq) + OH^-(aq)$$

The hydroxide ions released will react readily with hydronium (or hydrogen) ions.

A solution of sodium hydroxide is more basic than a solution of ammonia of the same concentration, because the NaOH solution produces many more OH^- ions than the ammonia solution.

- One litre of 1 mol L^{-1} NaOH produces 1 mol of $OH^-(aq)$ ions.
- One litre of 1 mol L^{-1} NH_3 produces only 1×10^{-5} mol of $OH^-(aq)$ ions.

Concentrated and dilute acidic and basic solutions

Moderately **concentrated** hydrochloric acid will contain *more* $H_3O^+(aq)$ ions than the *same volume* of **dilute** hydrchloric acid.

> In highly concentrated HCl ([HCl > 5 mol L^{-1}), there are not enough water molecules present to 'allow' all the HCl molecules to dissociate; very highly concentrated acids may not contain a lot of H_3O^+ ions!

Example

Hydrochloric acid

Comparing 1 L of 1 mol L^{-1} HCl solution with 1 L of 0.1 mol L^{-1} HCl solution:

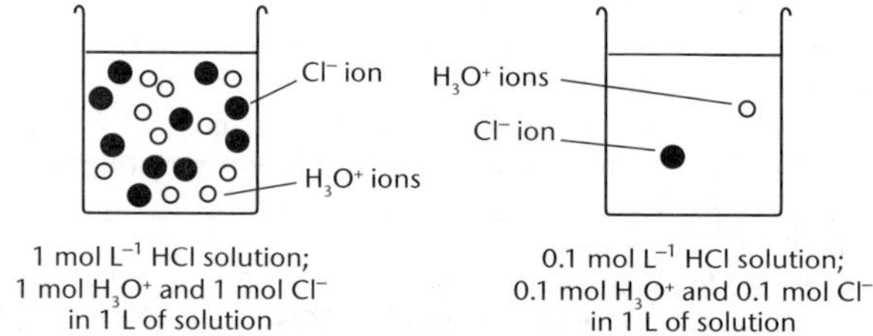

1 mol L^{-1} HCl solution; 1 mol H_3O^+ and 1 mol Cl^- in 1 L of solution

0.1 mol L^{-1} HCl solution; 0.1 mol H_3O^+ and 0.1 mol Cl^- in 1 L of solution

Note: Water molecules are not shown.

Compared with the dilute solution, the more concentrated hydrochloric acid solution will:

- Be a better electrolyte and therefore a better conductor of electricity, because there are more ions present.
- React faster with metals and other reactants because there are more frequent collisions between H_3O^+ ions and the other reactants. This is because there is a higher concentration of H_3O^+.

A similar situation occurs with basic solutions. A moderately concentrated sodium hydroxide solution will contain more OH^- ions than the same volume of dilute sodium hydroxide solution.

Concentration of an acid (or base) refers to the amount of H_3O^+ (or OH^-) ions in a given volume.

Strength of an acid (or base) refers to the ease with which a substance will undergo proton transfer.

Example

Strength of acids and pH

1 L of 1 × 10^{-1} mol L^{-1} HCl contains 1 × 10^{-1} mol of H_3O^+ ions; pH = 1, so *HCl* is a *strong* acid.

1 L of 1 × 10^{-1} mol L^{-1} CH_3COOH contains only 1 × 10^{-3} mol of H_3O^+ ions; pH = 3, so *CH_3COOH* is a *weak* acid.

The strong acid, HCl, has a lower pH (ie solution is more acidic) than a weak acid of the same concentration; pH for monoprotic acids of the same concentration depends on the *strength* of the acid.

Concentration of acids and pH

1 L of 1 × 10^{-3} mol L^{-1} HCl contains 1 × 10^{-3} mol of H_3O^+ ions; pH = 3, so HCl is a *dilute* solution of a *strong* acid.

1 L of 1 × 10^{-1} mol L^{-1} CH_3COOH contains 1 × 10^{-3} mol of H_3O^+ ions; pH = 3, so CH_3COOH is a *concentrated* solution of a *weak* acid.

The strong (but dilute) acid (the HCl) has the same pH as the weaker (but more concentrated) acid (the CH_3COOH); pH depends mostly on the *concentrations* of the acids.

Unit 11.3 Activity 7B: Strong and weak acids and bases

1. 50 mL of a 0.01 mol L^{-1} NaOH solution is added to 50 mL of a 0.01 mol L^{-1} HCl solution. $K_w = [H_3O^+][OH^-] = 1.00 \times 10^{-14}$.
   a. Calculate the pH of the original NaOH solution.
   b. Calculate the pH of the combined solution.

2. The pH of 0.01 mol L^{-1} nitric acid solution is found to be approximately 2, whereas the pH of a 0.01 mol L^{-1} ethanoic acid solution is found to be approximately 5. Explain why these two acids of the same concentration have different pH values.

3. **a.** What amount of hydrochloric acid is contained in 20 mL of 0.01 mol L^{-1} hydrochloric acid solution?
   b. What amount of sodium hydroxide is needed to completely neutralise the solution in **a**? What is the pH of the resulting solution?

4. Explain the difference between a weak acid and a dilute acid.

5. A piece of calcium carbonate is added to samples of hydrochloric acid, HCl, and ethanoic acid, CH_3COOH, both of the same concentration.
   a. Describe the expected observations and account for any differences between the reactions of the two acids.
   b. Which of the two acids would have the higher initial pH? Justify your answer.

6. The pH of a 0.10 mol L^{-1} solution of acid HX and the pH of a 0.10 mol L^{-1} solution of acid HY are measured. The pH of acid HX is 3 and the pH of acid HY is 1.
   a. Which of the two acids, HX or HY, is the stronger acid? Justify your answer in terms of the measured pH.
   b. Describe another test that could be carried out to confirm that the acid you selected in **a**. is the stronger of the two. Describe what you would do and what you would expect to observe.

7. Two acids of the same concentration, nitric acid, HNO_3, and methanoic acid, HCOOH, have pH of 1.00 and 2.37 respectively.
   a. Account for the difference in pH. Include balanced equations in your discussion.
   b. Which of the two acids, HNO_3 or HCOOH, will be the better electrolyte (conductor of electricity)? Justify your answer.

8. Account for the variation in the pH values of the acids in the following table:

Acid	Concentration (mol L^{-1})	pH
HA	0.100	1.00
HB	0.100	2.50
HC	0.00100	3.00

Unit 11.3 Types of Chemical Reactions

Topic 8: Thermochemical and equilibrium principles (extension)

In this Topic we revise and extend our understanding of thermochemical and equilibrium principles. Topic 8 deals with:

- Describing the dynamic nature of equilibrium.
- Writing equilibrium constant expressions and recognising the significance of the size of K_c.
- Predicting the effect of changes to an equilibrium system – changes are in temperature, concentration, pressure, or the addition of a catalyst.

Introduction

Some physical and chemical processes continue until one of the reactants has been used up – the reaction is said to have *gone to completion*.

Example

Water evaporating

If a beaker of water is left unsealed in a warm environment, evaporation occurs and the level of water drops until no water is left:

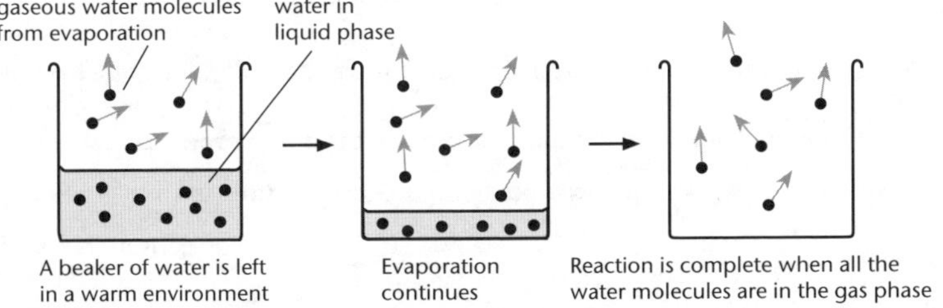

Reactions that go to completion form an **open system** – one (or all) of the products can 'escape' (ie is/are continually removed).

Example

Zinc in excess dilute acid solution

An unsealed test tube forms an open system. The reaction between zinc and excess dilute acid solution in the unsealed test tube goes to completion, ie continues until all the zinc has dissolved. Acid was in excess; some acid still remains unreacted

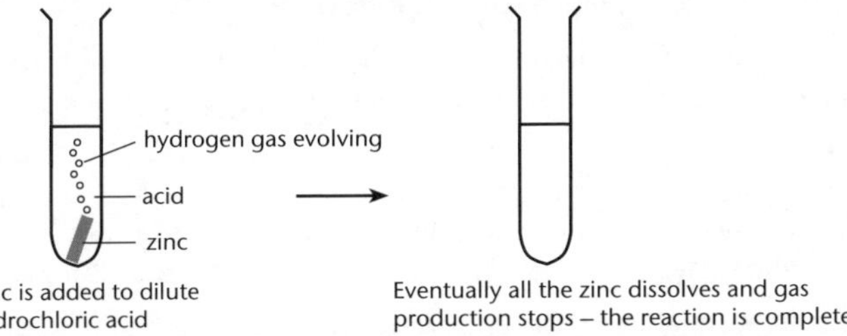

In the reactions in the two previous examples, the forward reactions continued until the reactant(s) is/are completely used up:

Reactants ⟶ Products

Some reactions, however, are **reversible reactions**, and the products may break down again to reform the reactants.

Example

Evaporation and condensation

If a beaker of water is sealed with cling film then left in a warm environment, the level of water drops slightly then remains constant.

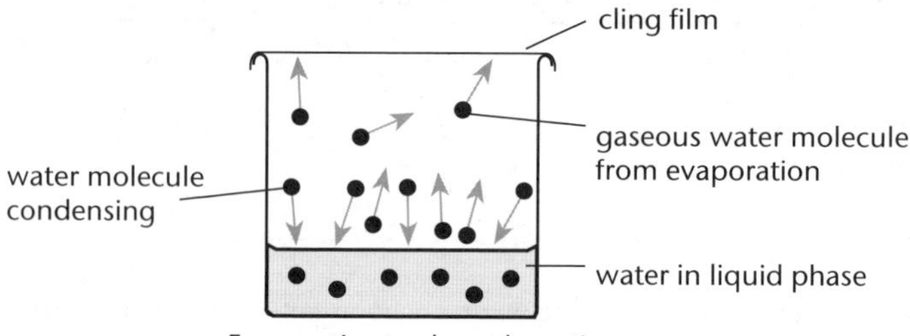

Evaporation and condensation occur at the same rates

The sealed beaker forms a **closed system**, as water molecules cannot escape from the container.

The reaction mixture contains a mixture of both reactants and products.

Evaporation is occurring, but so is the *opposing* process, **condensation**:

$$H_2O(\ell) \xrightarrow{\text{evaporation}} H_2O(g) \quad \text{and} \quad H_2O(g) \xrightarrow{\text{condensation}} H_2O(\ell)$$

This can be written as $H_2O(\ell) \rightleftharpoons H_2O(g)$

Note:
- The forward reaction is that from left to right:
  $H_2O(\ell) \longrightarrow H_2O(g)$.
- The reverse reaction is $H_2O(g) \longrightarrow H_2O(\ell)$.
- The reaction mixture is *all* the water, both $H_2O(\ell)$ and $H_2O(g)$.

Equilibrium

When reversible reactions reach the point where the rate of forward reaction equals the rate of backward reaction, the system is said to have reached **equilibrium**.

This equilibrium is said to be dynamic, indicating that the forward and backward reactions do not stop, but rather proceed at the same rates – so the amounts of reactants and products remain constant.

When the system in the Example above reaches equilibrium, each time some water evaporates the *same* amount condenses. As a consequence, the amount of water in the beaker in either liquid or vapour form stays the same.

An equilibrium system can be recognised by the presence of both reactants and products.

Physical equilibrium

The term *equilibrium* applies to both physical and chemical changes.

In physical changes, no new substances are formed.

The previous example is a physical equilibrium, as the water:

- Does not react with any other substances.
- Remains as either liquid or gaseous water molecules throughout the process.

A **saturated solution** is an example of dynamic equilibrium.

Example

Saturated sodium chloride solution

If enough solid sodium chloride is added to a beaker of water, the solution becomes saturated and no more sodium chloride *appears* to dissolve. However, the solid sodium chloride does continue to dissolve, but dissolved sodium chloride precipitates out of the solution at the same rate:

$$NaCl(s) \underset{}{\overset{H_2O(\ell)}{\rightleftharpoons}} Na^+(aq) + Cl^-(aq)$$

Chemical equilibrium

Many chemical reactions behave as a dynamic equilibrium.

Example

The iron(III) thiocyanate equilibrium

Iron(III) ions, Fe^{3+}, and thiocyanate ions, SCN^-, react to form $FeSCN^{2+}$, a bright blood-red complex ion. The reaction is:

$$Fe^{3+}(aq) + SCN^-(aq) \longrightarrow FeSCN^{2+}(aq)$$

orange colourless bright red

In the reaction:

- The red colour forms quickly, then remains constant.
- The system is closed – the ions Fe^{3+}, SCN^- and $FeSCN^{2+}$ are contained within the same aqueous solution.

The reaction has not gone to completion, because it can be shown (by a precipitation reaction involving Fe^{3+}) that there are unreacted Fe^{3+} (and thus also SCN^-) ions in the red solution. The reverse reaction is also occurring:

$$FeSCN^+(aq) \longrightarrow Fe^{3+}(aq) + SCN^-(aq)$$

The equilibrium for the reaction is written:

$$Fe^{3+}(aq) + SCN^-(aq) \rightleftharpoons FeSCN^{2+}(aq)$$

At equilibrium, the rate at which the complex ions form is equal to the rate at which the complex ions break up.

The relative concentrations of particles during an equilibrium reaction can be shown schematically. The following diagram represents the iron(III) and thiocyanate reactions and equilibrium in the previous example.

	Before mixing	Just after mixing	Equilibrium reached	Later on, reaction still at equilibrium
$Fe^{3+}(aq)$ ●	●●●●●●●●	●●●●	●●	●●
$SCN^{-}(aq)$ ○	○○○○○○○○○○○○○○○○○○○○	○○○○○○○○○○○○○○○○	○○○○○○○○○○○○○○	○○○○○○○○○○○○○○
$FeSCN^{2+}(aq)$ ●○		●○ ●○ ●○ ●○	●○ ●○ ●○ ●○ ●○ ●○	●○ ●○ ●○ ●○ ●○ ●○

Once equilibrium is established, the concentrations of all the ions do not change. However, the concentrations of the different ions are not necessarily the same.

The diagram does not show that both the forward and reverse reactions are still occurring at equilibrium. It does show that the overall concentrations are not changing.

Unit 11.3 Activity 8A: Equilibrium

1. Which of the following reactions are reversible? Give evidence to support your answers.
   - **a.** $CuS_4.5H_2O \longrightarrow CuSO_4(s) + 5H_2O(\ell)$
   - **b.** Blue litmus ⟶ Red litmus
   - **c.** $Zn(s) + Cu^{2+}(aq) \longrightarrow Zn^{2+}(aq) + Cu(s)$
   - **d.** Haemoglobin + O_2 ⟶ Oxyhaemoglobin
   - **e.** $Mg(s) + 2H^{+}(aq) \longrightarrow Mg^{2+}(aq) + H_2(g)$
   - **f.** $H^{+}(aq) + 2CrO_4^{2-}(aq) \longrightarrow Cr_2O_7^{2-}(aq) + OH^{-}(aq)$
     yellow (for $2CrO_4^{2-}$), orange (for $Cr_2O_7^{2-}$)
2. Which of the following represent dynamic equilibrium situations? Explain your anwers.
   - **a.** Children on a seesaw.
   - **b.** Throwing sand over a wall.
   - **c.** Walking up a downwards escalator but staying in the same place.

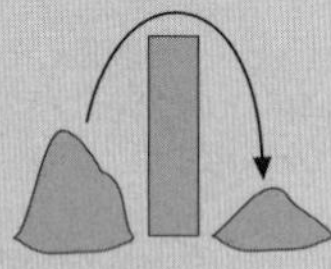

Equilibrium constant

A general equation can be written for a reversible reaction:

aA	+	bB	⇌	cC	+	dD
'a' moles of substance A	+	'b' moles of substance B		'c' moles of substance C	+	'd' moles of substance D

The ratio of 'products' (C and D) to 'reactants' (A and B), is expressed as the **equilibrium constant**,

$$K_c = \frac{[C]^c \times [D]^d}{[A]^a \times [B]^b}$$, usually written as:

$$K_c = \frac{[C]^c[D]^d}{[A]^a[B]^b}$$ where: [A] is the concentration of A in mol L^{-1} at equilibrium, [B] is the concentration of B in mol L^{-1} at equilibrium, etc.

K_c only gives information on how far a reaction proceeds. It provides no information on the rate of the reaction (ie how *fast* a reaction proceeds).

Example

Writing equilibrium constant expressions

For the reaction:

$$CO(g) + H_2O(g) \rightleftharpoons CO_2(g) + H_2(g), \quad K_c = \frac{[CO_2][H_2]}{[CO][H_2O]}$$

For the reaction:

$$N_2(g) + 3H_2(g) \rightleftharpoons 2NH_3(g), \quad K_c = \frac{[NH_3]^2}{[N_2][H_2]^3}$$

For equilibrium systems involving solids and liquids, neither the solids nor the liquids are included in the equilibrium constant expression.

Example

Equilibria involving solids

For the reaction:

$$CO_2(g) + C(s) \rightleftharpoons 2CO(g), \quad K_c = \frac{[CO]^2}{[CO_2]}$$

Equilibrium constants are specific for a given reaction and temperature.

An equilibrium constant is useful for determining *how far* a reaction has proceeded:

- If K_c is large (> 1), then, at equilibrium, there will be more products than reactants. The reaction is said 'to favour products'.
- If K_c is very large (> 10^{20}), the reaction will have gone to completion.
- If K_c is small (≤ 1), then, at equilibrium, only a small amount of reactant is used up and only a small amount of product has formed. The reaction is said 'to favour reactants'.
- If K_c is very small (< 10^{-20}), then the reaction will not occur.

Unit 11.3 Activity 8B: Equilibrium expressions

1. Write equilibrium expressions for the following:
   - **a.** $H_2(g) + I_2(g) \rightleftharpoons 2HI(g)$
   - **b.** $2NO(g) + O_2(g) \rightleftharpoons 2NO_2(g)$
   - **c.** $Fe^{3+}(aq) + SCN^-(aq) \rightleftharpoons FeSCN^{2+}(aq)$
   - **d.** $2SO_3(g) \rightleftharpoons 2SO_2(g) + O_2(g)$
   - **e.** $C(s) + H_2O(g) \rightleftharpoons CO(g) + H_2(g)$

2. Write equations from which the following equilibrium constant expressions are derived:

a. $K_c = \dfrac{[CO_2(g)]^2[H_2O(g)]^2}{[C_2H_4(g)]\ [O_2(g)]^3}$

b. $K_c = \dfrac{[CS_2(g)]\ [H_2(g)]^4}{[CH_4(g)]\ [H_2S(g)]^2}$

c. $K_c = \dfrac{[NO(g)]^2[Br_2(g)]}{[NOBr(g)]^2}$

3. Consider the system:

$N_2(g) + O_2(g) \rightleftharpoons 2NO(g)$, at 1700 °C, $K_c = 4 \times 10^{-4}$.

a. Write the equilibrium expression for the reaction.

b. Which species will be present in the greatest concentration at equilibrium?

c. What information, if any, does *K* provide about the rate at which the reaction proceeds?

4. a. $2SO_2(g) + O_2(g) \rightleftharpoons 2SO_3(g)$

i. Write the equilibrium constant expression for the above reaction.

ii. At 25 °C, the value of K_c is 1.7×10^8. Which species is present in the greater amount at equilibrium – $SO_2(g)$ or $SO_3(g)$? Justify your choice.

b. $2NOCl(g) \rightleftharpoons 2NO(g) + Cl_2(g)$

i. Write the equilibrium constant expression for the above reaction.

ii. At 25 °C, the value of K_c is 1.6×10^{-5}. Which species is present in the greater amount at equilibrium – NOCl(g) or NO(g)? Justify your choice.

5. a. $N_2O_4(g) \rightleftharpoons 2NO_2(g)$

i. Write the equilibrium constant expression for the above reaction.

ii. At 25 °C, the value of K_c is 6.1×10^{-3}. Which species is present in the greater amount at equilibrium – $N_2O_4(g)$ or $NO_2(g)$? Justify your choice.

b. $2H_2(g) + S_2(s) \rightleftharpoons 2H_2S(g)$

i. Write the equilibrium constant expression for the above reaction.

ii. At 25 °C, the value of K_c is 1.08×10^8. Which species is present in the greater amount at equilibrium – $S_2(g)$ or $H_2S(g)$? Justify your choice.

Equilibrium changes

Once a reversible reaction has reached equilibrium, it will remain at equilibrium indefinitely unless changes occur which upset the equilibrium. These changes, which arise from variations in *temperature*, *pressure* or *concentration*, result in either an increase in the forward reaction or an increase in the reverse reaction.

- An increase in the rate of the forward reaction changes more reactants into products. The equilibrium position is said to 'move to the right'.

  Reactants $\longrightarrow$ Products

- An increase in the rate of the reverse reaction increases the rate at which products break down to reactants. The equilibrium position is said to 'move to the left'.

  Reactants $\longleftarrow$ Products

When equilibrium is re-established, the rates of the forward and reverse reactions become equal again:

Reactants $\rightleftharpoons$ Products

Henri Louis Le Chatelier (1850–1936) was a French scientist who studied equilibrium systems. He established a rule which can be used to predict the effect different changes would have on a system at equilibrium. This is known as **Le Chatelier's principle**, and states:

'When a change is applied to a system at equilibrium, the system responds so that the effects of the change are minimised.'

The principle is used to predict the *direction* of change. It does not predict the extent of change.

Concentration changes

Concentration changes involve either:

- extra reactant or product being *added* to the reaction mixture

  or

- reactant or product being *remove*d from the reaction mixture.

Generally, when extra reactant or product is added to a reaction mixture, the response is to 'consume' the extra material added.

Example

Consuming added reactant

Extra $Fe^{3+}(aq)$ is added to the equilibrium:

$$Fe^{3+}(aq) + SCN^-(aq) \rightleftharpoons FeSCN^{2+}(aq)$$

The position of equilibrium will shift to the right to use up some of the added $Fe^{3+}(aq)$ ions. To do this, the forward reaction $Fe^{3+}(aq) + SCN^-(aq) \longrightarrow FeSCN^{2+}(aq)$ speeds up so more $FeSCN^{2+}(aq)$ ions form (ie products are favoured). The red colour will become more intense, so this change can easily be seen.

If product or reactant is *removed*, the equilibrium will shift to favour production of the removed species ('replacing' the material removed).

Example

Removing product

The reaction between $SO_2(g)$ and $O_2(g)$ forms an equilibrium:

$$2SO_2(g) + O_2(g) \rightleftharpoons 2SO_3(g)$$

If $SO_3(g)$ is removed, the forward reaction increases to replace the $SO_3(g)$. The equilibrium position moves to the right, so more $SO_2(g)$ and $O_2(g)$ are used up.

In industry, the process of removing product from an equilibrium as the product forms is commonly used because it promotes the forward reaction and causes *more* product to form.

Example

The Haber process

In the **Haber process**, ammonia is produced according to the reaction:

$$N_2(g) + 3H_2(g) \rightleftharpoons 2NH_3(g)$$

The ammonia formed is removed by absorbing it in water or by condensing it to form liquid ammonia. As ammonia is removed, the remaining reactants react to produce more ammonia.

For equilibrium reactions in aqueous solutions, the equilibrium position can be shifted by **precipitation**.

Example

Equilibria and precipitation

The equilibrium between orange dichromate ions, $Cr_2O_7^{2-}$, and yellow chromate ions, CrO_4^-, is:

$$\underset{\text{orange}}{Cr_2O_7^{2-}(aq) + OH^-(aq)} \rightleftharpoons \underset{\text{yellow}}{2CrO_4^{2-}(aq) + H^+(aq)}$$

Chromate ions, CrO_4^{2-}, can be removed by adding $Ba^{2+}(aq)$ ions to form a precipitate of yellow insoluble $BaCrO_4$. When CrO_4^{2-} ions are removed, the forward reaction speeds up, producing more CrO_4^{2-} ions. The extent of the reaction can be monitored by looking at the colour change as orange dichromate ions form yellow chromate ions.

Pressure changes

Pressure changes have little or no effect on equilibrium reactions *unless* there are reactants or products in the gas phase.

Particles in the gaseous phase are spread far apart and move rapidly in all directions. Gas particles in a container collide with the walls of the container, exerting a force called the **pressure** of the gas. The more collisions between gas particles and the walls of the container, the greater the gas pressure. Gas pressure can be increased by:

- Adding more gas (particles) to a closed system.
- Reducing the volume of a closed system.

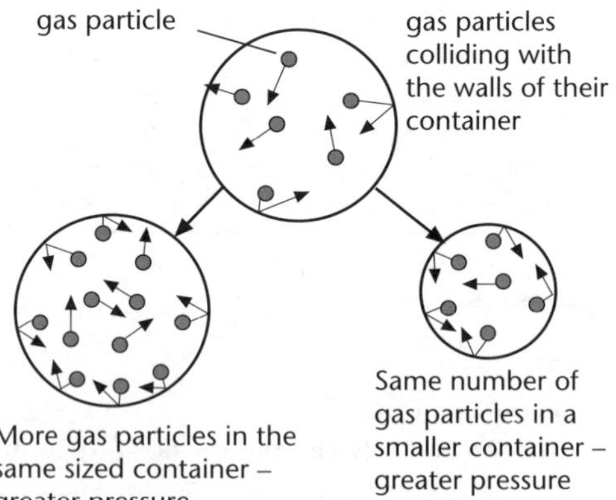

A mixture of gases at equilibrium will be affected by a change in pressure. A change in pressure is brought about by altering the volume of the container or altering the amount of one of the reactant gases.

- An *increase* in pressure will be opposed by *reducing* the number of gas particles in an equilibrium mixture.
- A *decrease* in pressure will be opposed by *increasing* the number of gas particles in a reaction mixture.

Example

Pressure changes

Consider the equilibrium of the reversible reaction:

$$2NO_2(g) \rightleftharpoons N_2O_4(g)$$

brown colourless

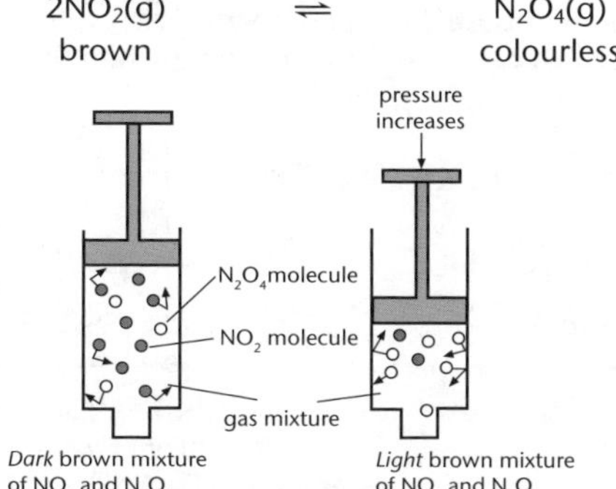

As the pressure increases, the system acts to oppose the change, by reducing the pressure by decreasing the amount (number) of gas particles present.

The forward reaction (in which *two* $NO_2(g)$ molecules combine to form *one* molecule of $N_2O_4(g)$) speeds up. This reduces the number of gas molecules, reducing the effect of the pressure increase. The shift in equilibrium can be seen by the change in colour from dark brown to light brown.

In reactions which involve the same number of moles of gas particles on both sides of the balanced equation, pressure changes have *no* effect on the equilibrium.

Example

Further pressure changes

Both the left-hand side and right-hand side of the following equation have two gas particles:

$$H_2(g) + I_2(g) \rightleftharpoons 2HI(g)$$

Pressure changes have *no* effect on the equilibrium position.

Temperature changes

A *decrease* in temperature causes a reaction to move in the **exothermic** direction. This releases heat energy to replace energy removed by lowering the temperature.

An *increase* in temperature causes a reaction to move in the **endothermic** direction. This absorbs the increased heat energy supplied to the system.

The sign of $\Delta_r H$ determines the effect of a temperature change on the equilibrium system.

Example

Temperature changes

For the equilibrium:

$$2NO_2(g) \rightleftharpoons N_2O_4(g) \qquad \Delta_r H = -54 \text{ kJ}$$

brown colourless

The forward reaction is *exothermic,* so the reverse reaction is *endothermic.* (In equilibrium reactions, the sign of $\Delta_r H$ *always* refers to the forward reaction.)

- Heating the mixture will favour the reverse, endothermic reaction, which will absorb heat to oppose the change. More NO_2 will be formed, and the mixture will become a darker brown colour as the equilibrium position moves to the left.
- Cooling the mixture will favour the exothermic (or forward) reaction, in order to produce more heat to oppose the change. More N_2O_4 will form and the mixture will become a lighter brown colour as the reaction moves to the right.

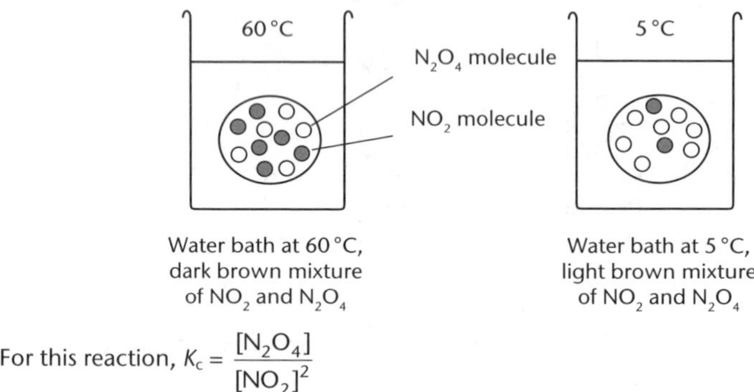

For this reaction, $K_c = \dfrac{[N_2O_4]}{[NO_2]^2}$

At higher temperatures, $[NO_2]$ is increased, so size of K_c will decrease.

Effect of a catalyst

In equilibrium reactions, catalysts speed up *both* the forward and reverse reactions – this causes equilibrium to be established more rapidly.

Since there is no overall change to the relative amounts of product or reactant present, there is *no change* in the equilibrium position and no change in the equilibrium constant K_c.

Summary of responses to equilibrium system to changes

Change in condition		Direction of shift in equilibrium position	Effect on K_c
Concentration	Increase products	In the reverse direction.	No change in K_c.
	Decrease products	In the forward direction.	No change in K_c.
	Increase reactants	In the forward direction.	No change in K_c.
	Decrease reactants	In the reverse direction.	No change in K_c.
Pressure	Increase	In the direction with the least number of moles of gas.	No change in K_c.
	Decrease	In the direction with the greater number of moles of gas.	No change in K_c.
Temperature	Increase	In the direction of the endothermic reaction.	K_c changes.
	Decrease	In the direction of the exothermic reaction.	K_c changes.
Catalyst added		No change in equilibrium position. Equilibrium reached more quickly (ie reaction rate changes).	No change in K_c.

Unit 11.3 Activity 8C: Equilibrium changes

1. For the reaction, $PCl_3(g) + Cl_2(g) \rightleftharpoons PCl_5(g)$, predict the effect on the equilibrium of:

a. **i**. Decreasing the pressure. **ii**. Adding chlorine gas.

iii. Adding a catalyst.

b. For each of **i**. – **iii**. above, explain your prediction.

2. Predict, with reasons, the effect of increasing the temperature on each of the following equilibria:

a. $N_2(g) + 3H_2(g) \rightleftharpoons 2NH_3(g)$ $\Delta_rH = -92$ kJ

b. $N_2(g) + O_2(g) \rightleftharpoons 2NO(g)$ $\Delta_rH = +176$ kJ

c. $CO(g) + 2H_2(g)) \rightleftharpoons CH_3OH(g)$ $\Delta_rH = +91$ kJ

3. The equilibrium system:

$2NO_2(g) \rightleftharpoons N_2O_4(g)$ $\Delta_rH = -54 \text{ kJ mol}^{-1}$

was in a container of fixed volume. Describe three changes in conditions which would decrease the amount of $N_2O_4(g)$. Explain your reasoning.

4. The equilibrium between nitrogen dioxide and dinitrogen tetroxide is represented by:

$2NO_2(g) \rightleftharpoons N_2O_4(g)$ $\Delta_rH = -54 \text{ kJ mol}^{-1}$

dark brown colourless

Some of the equilibrium mixture is put into a syringe, as shown in the diagram:

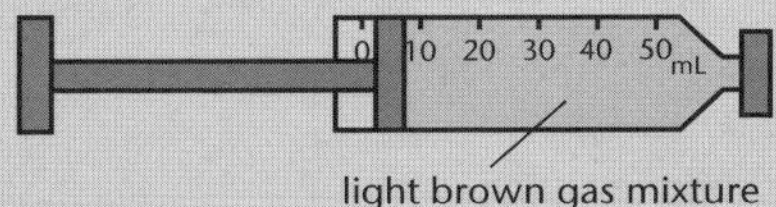

a. What would be the effect on the number of molecules of nitrogen dioxide if the plunger was pushed into the 30 mL mark on the syringe while the temperature was kept constant? Explain your answer.

b. What colour change would occur if the plunger is held in position but the syringe is placed in iced water? Explain why this colour change occurs.

5. The equation for the decomposition of ammonia, NH_3, is:

$2NH_3(g) \rightleftharpoons N_2(g) + 3H_2(g)$ $\Delta_rH = +92 \text{ kJ mol}^{-1}$.

The system has reached equilibrium. What would happen if the temperature was increased? Explain your answer.

6. The following equilibrium system is established when thiocyanate ions, SCN^-, are added to iron(III) ions, Fe^{3+}. The resulting aqueous solution is a dark red colour. The equation representing the equilibrium system and the colours of each species are:

$Fe^{3+}(aq) + SCN^-(aq) \rightleftharpoons FeSCN^{2+}(aq)$

orange colourless dark red

a. Write the equilibrium constant expression for the above reaction.

b. When iron(III) ions, Fe^{3+}, are removed from the equilibrium mixture (by adding sodium fluoride), a colour change is observed. Describe the colour change you would expect to see and explain why it occurs.

c. Explain why the reaction mixture gets darker when crystals of $FeCl_3$ are added.

7. Explain how the concentration of methanol, $CH_3OH(g)$, in the equilibrium system below will be affected by the changes described. Give reasons for your answers.

$$CO(g) + 2H_2(g) \rightleftharpoons CH_3OH(g) \quad \Delta_r H = -91 \text{ kJ mol}^{-1}$$

   a. The temperature is increased.
   b. The pressure is increased.
   c. $CO(g)$ is added to the system.

8. Explain how the concentration of hydrogen gas in the equilibrium system following will be affected by the changes described. Give reasons for your answers.

$$C(s) + H_2O(g) \rightleftharpoons CO(g) + H_2(g) \quad \Delta_r H = 131 \text{ kJ mol}^{-1}$$

   a. The temperature is increased.
   b. The pressure is decreased.
   c. $CO(g)$ is removed from the system.
   d. $C(s)$ is added to the system.

9. Explain how the concentration of chlorine gas in the equilibrium system following will be affected by the changes described. Give reasons for your answers.

$$P_4(s) + 10Cl_2(g) \rightleftharpoons 4PCl_5(g) \quad \Delta_r H = -1528 \text{ kJ mol}^{-1}$$

   a. The temperature is increased.
   b. The pressure is decreased.
   c. $PCl_5(g)$ is removed from the system.
   d. A catalyst is added to the system.

10. Soft drinks are produced by dissolving carbon dioxide gas in water to produce *soda water*. The soda water can then be flavoured. The equation for producing soda water is:

$$CO_2(g) + H_2O(\ell) \rightleftharpoons HCO_3^-(aq) + H^+(aq)$$

   a. Explain how the 'fizziness' would be affected if:
      i. Sodium hydrogencarbonate, $NaHCO_3$, was added to the soda water.
      ii. Citric acid was added to the soda water.
   b. A few drops of universal indicator were added to soda water, resulting in an orange-coloured solution.
      i. The solution was heated. The colour changed, indicating a decrease in pH. Is the reaction $CO_2 + H_2O \rightleftharpoons HCO_3^- + H^+$ exothermic or endothermic? Explain your answer.
      ii. How will heating affect the value of K_c? Explain your answer.
   c. Explain how the pH is affected when a can of soft drink is opened.

11. Describe the conditions which would give the greatest yield of sulfur trioxide in the following reaction:

$$2SO_2(g) + O_2(g) \rightleftharpoons 2SO_3(g) \quad \Delta_r H = -100 \text{ kJ mol}^{-1}$$

12. Temperature effects and yields were investigated for the reaction:

$N_2(g) + 3H_2(g) \rightleftharpoons 2NH_3(g), \quad \Delta_r H < 0.$

The results are shown in graph form.

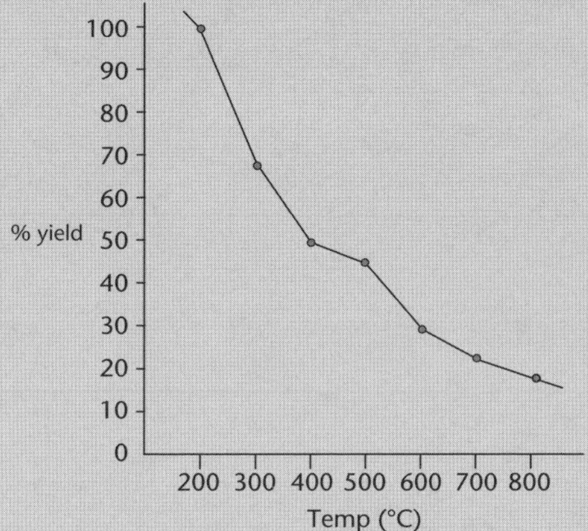

Explain why these results are consistent with those predicted by Le Chatelier's Principle.

13. The industrial preparation of ethanol, $CH_3CH_2OH(g)$, from ethene can be represented by the following equation:

$CH_2CH_2(g) + H_2O(g) \rightleftharpoons CH_3CH_2OH(g)\ \Delta_r H = -48\text{ kJ mol}^{-1}$

Discuss how the yield of ethanol will be affected by changes in temperature or pressure.

14. Ammonia is produced industrially by the Haber Process, as represented by:

$N_2(g) + 3H_2(g) \rightleftharpoons 2NH_3(g)$

a. Write the equilibrium constant expression for the reaction.

b. Describe how increasing the pressure of the system at equilibrium affects the amount of NH_3 in the system. Explain your answer.

c. What effect does liquefying the ammonia as it is formed have on the position of the equilibrium? Explain your answer.

d. The percentage of NH_3 present in equilibrium mixtures at different temperatures and at constant pressure is shown in the table below:

Temperature (°C)	Percentage NH_3 present in equilibrium mixture
200	63.6
300	27.4
400	8.7
500	2.9

Decide whether the reaction in which NH_3 is formed is endothermic or exothermic. Justify your answer.

15. A product, X, is the result of a slow, exothermic reaction. Neither product nor reactants are gases. What is the best method of speeding up the formation of X without altering the equilibrium concentrations of reactants or product?

Unit 11.4 Energy and Reaction Rates

Topic 1: Rates of reaction

Different reactions take different lengths of time: dynamite can explode in an instant; bread takes perhaps an hour to bake; a pawpaw takes weeks to ripen. In Unit 11.3 we saw that some reactions are much faster than others; in Unit 11.4 we develop our understanding about why reactions have different rates. Topic 1 looks at:

- Collision theory.
- Factors that affect the rate of a reaction.
- Measuring the rate of a reaction.

Collision theory

Chemical reactions occur by particles colliding:

- The greater the number of particles colliding, the greater the rate of a chemical reaction.
- Not all collisions result in a chemical change because colliding particles may not have enough energy for a reaction to take place – the energy of a particle can be increased by raising the temperature of the substance. Also, some collisions are glancing blows.
- Solids do not react with other solids because the particles are locked into a crystal lattice and therefore cannot move and collide.

Factors affecting the rate of reaction

Surface area

The greater the surface area of the reagents, the greater the rate of a chemical reaction.

Example

Solid food is mashed into pulp before being fed to a baby – finely ground food has a larger surface area to react with the digestive juices in the stomach and intestines than lumps of food, and so mashed food is easier for the baby to digest. Later in life, the same effect is achieved by chewing the food in the mouth with teeth, since this also increases the food's surface area.

Example

Increasing surface area

Consider a 2 cm × 2 cm × 2 cm cube:

Surface area = 6 faces × area of one face
= 6 × (2 cm × 2 cm)
= 24 cm^2

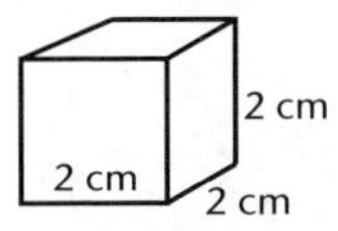

Cutting the cube in half across each section gives eight cubes:

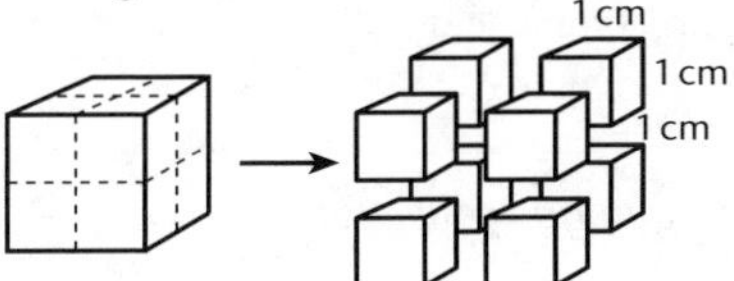

Surface area of eight small cubes is now:

8 × (6 × area of one face) = 8 × 6 × 1 × 1

= 48 cm^2

Breaking up the cube has *increased* (doubled) the surface area.

Grinding solids into a powder (eg with a pestle and mortar) will increase the rate of a reaction.

Example

Any finely divided material that is combustible (even cooking flour) can explode when mixed with air in the correct proportion. In coal mines, explosions can occur from coal dust and air mixed together. Coal seams are often covered with whitewash to improve visibility in the mine, but also to prevent coal dust getting into the atmosphere.

Concentration of chemical reagents

The higher (greater) the concentration of the reagents, the higher (greater) the rate of a chemical reaction.

Example

Oxygen gas (vital for life) is supplied to the face and mouth of a patient who has a breathing difficulty – pure oxygen has a higher concentration of oxygen gas than air does.

Example

Concentration and reaction rate

The rate of a reaction between a metal and an acid, eg

$Zn(s) + 2HCl(aq) \rightarrow ZnCl_2(s) + H_2(g)$,

can be measured by recording the volume of hydrogen produced at fixed time intervals.

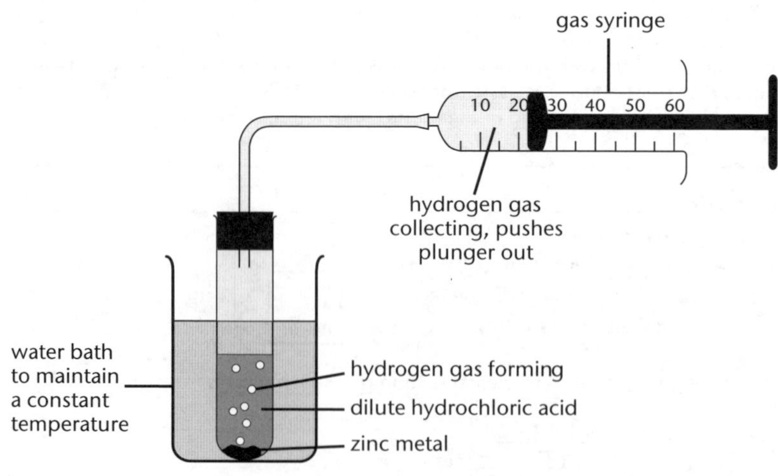

Experimental set-up for hydrogen gas collection

Results from one experiment using the apparatus on page 192 were:

Time from start (s)	0	5	10	15	20	25	30	35	40	45	50	55	60	65
Volume of gas collected (mL)	0	15	26	34	40	45	50	53	56	57	59	60	60	60

Volumes of hydrogen gas collected

Results are best presented on a graph.

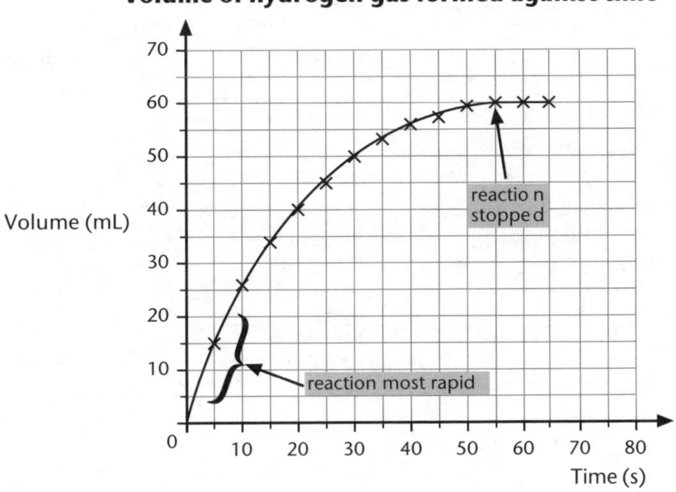

Hydrogen gas production over time

The effect of altering the concentration of the acid on the rate of the reaction can be studied by repeating the above experiment with different concentrations of acid.

Example

Increasing the concentration of the acid increases the number of acid particles in a given volume of acid – this *increases* the number of collisions between acid particles and metal particles, which *increases* the rate of the reaction.

Temperature

Temperature is a scale of hotness and coldness. The higher the temperature of the reagents, the higher the rate of a chemical reaction. At higher temperatures, particles have more energy, which means:

- Particles travel faster, and therefore create more collisions in a given time.
- When these particles collide, the impact is greater and reaction is more likely.

Example

Healthy human beings have a body temperature of 37 °C. If the body temperature rises, say by 2 °C, the reactions speed up, but not all by the same amount. The reactions are then 'out of balance', and the result can be death. A similar situation occurs if the body temperature falls by one or two degrees Celsius.

Example

Temperature and reaction rate

The rate of a reaction between a metal and an acid can be measured at different temperatures using the experimental set-up shown on page 192. Several experiments would be carried out with the same chemicals with the same particle size (for solids) and the same concentration (for aqueous solutions) – ie only temperature is changing.

Unit 11.4 Activity 1A: Rates of reaction

1. a. State three factors that can affect the rate of a chemical reaction.

b. Explain why each factor affects the rate of the reaction.

2. Below are some results for a 'metal + acid' reaction using the experimental set-up shown on page 192. The volume of the gas released in the reaction was measured and recorded every 30 seconds. The results were plotted on a piece of graph paper.

Results for metal + acid reaction

Time (min)	Volume of gas (mL)
0	0
0.5	12
1.0	23
1.5	32
2.0	39
2.5	44
3.0	46
3.5	48
4.0	50
4.5	50
5.0	50

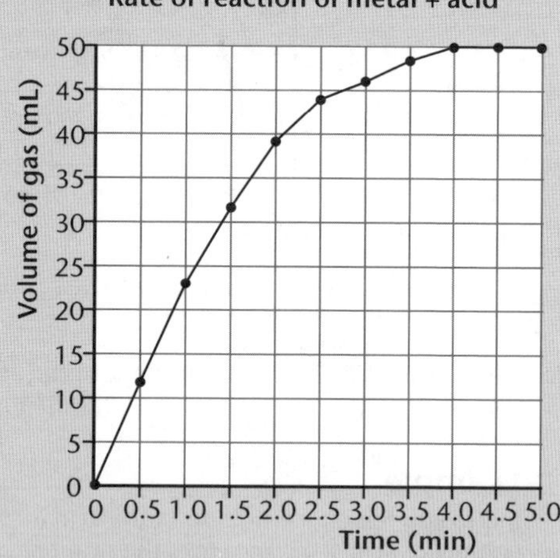

a. Explain why the total volume of gas collected reaches a maximum value (50 mL).

b. Deduce from the graph the volume of gas collected in the following time intervals:

i. 0–1 minutes.

ii. 1–2 minutes.

iii. Explain why the two volumes are not the same.

c. The curve for the results supplied is plotted on the graph. Redraw this graph and add the curves that would be obtained when the same quantities of chemicals were used *but with the following changes*:

   i. The metal was in powder form.

   ii. The temperature of the reaction was 10 °C lower than in the original experiment.

   iii. The acid concentration was double that of the original experiment.

3. The rate of the reaction between marble chips (calcium carbonate) and hydrochloric acid can be measured using the apparatus in the diagram. The contents of the flask lose mass as the reaction proceeds.

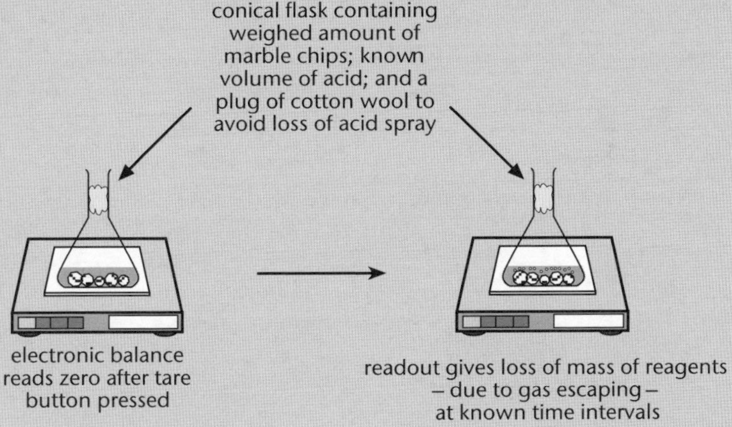

a. Write the word equation for the reaction between calcium carbonate and hydrochloric acid.

b. Name the chemical escaping from the flask during the reaction.

c. The experiment was repeated using the same quantity of chemicals but with powdered marble instead of lumps. State what difference would be noticed on the readout of the balance.

d. If powdered marble were used in the experiment, state what other change could be made to the chemicals, or to the conditions, so that the rate of loss of mass was the same as in the original experiment.

Unit 11.4 Energy and Reaction Rates

Topic 2: Energy changes

In Topic 2 we extend our understanding of chemical reactions and why reactions take place over different rates of time. We now consider heat energy and its role in chemical reactions: whether heat energy is produced during a reaction (exothermic) or absorbed during the reaction (endothermic). Topic 2 deals with:

- Classifying reactions as exothermic and endothermic.
- Determining enthalpy changes.

Energy changes in chemical reactions

Chemical reactions involve energy changes. During a reaction, energy changes from one form to another.

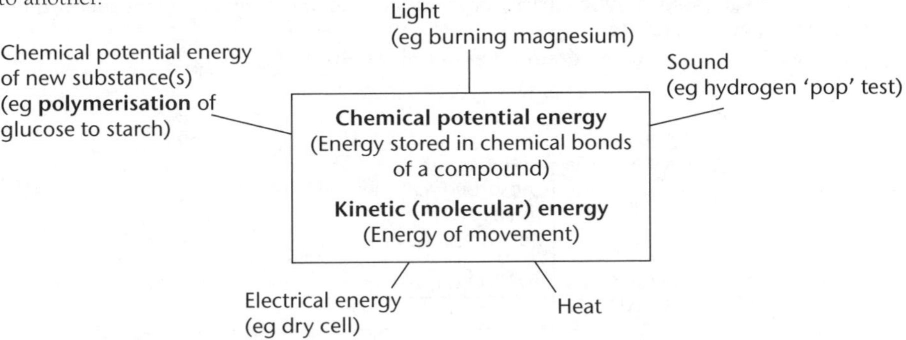

Most, if not all, chemical reactions involve changes in heat energy.

Exothermic reactions

If heat energy is *produced* during a reaction, the temperature of the surroundings *increases*, and the reaction is described as being exothermic.

Example

Dissolving sodium hydroxide

When solid sodium hydroxide pellets, NaOH(*s*), are dissolved in a beaker of water, the water becomes warm. This is an *exothermic* reaction since heat energy has been produced and transferred to the water, increasing its temperature.

Dissolved sodium hydroxide (Na^+(*aq*) and OH^-(*aq*) ions) contains *less* energy than the solid sodium hydroxide, NaOH(s). This energy difference can be shown using an **energy diagram**.

Energy diagram for sodium hydroxide dissolving

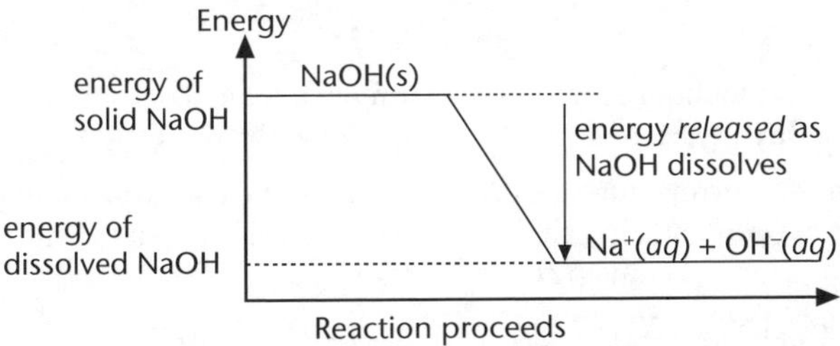

Endothermic reactions

If heat energy is *absorbed* from the surroundings during a reaction, the temperature of the surroundings *decreases* and the reaction is described as being endothermic.

When methylated spirits ('meths') is rubbed on the skin it quickly evaporates. This is an endothermic process – the meths absorbs heat energy from the skin as it evaporates (producing a 'cold feeling').

Example

Dissolving ammonium nitrate

When solid ammonium nitrate, $NH_4NO_3(s)$, is dissolved in a beaker of water, the water becomes colder as ammonium nitrate absorbs energy from the water. This is an *endothermic* process.

The dissolved ammonium nitrate ($NH_4^+(aq)$ and $NO_3^-(aq)$ ions) contains more energy than the solid ammonium nitrate.

Energy diagram for ammonium nitrate dissolving

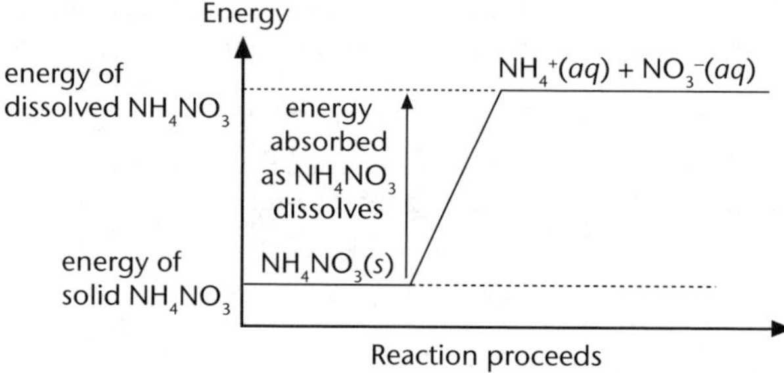

If a reaction is endothermic in one direction, it will be exothermic in the opposite direction:

$A + B \longrightarrow C + D$ exothermic

$C + D \longrightarrow A + B$ endothermic

Example

Heating hydrated copper sulfate

Heat + $CuSO_4.5H_2O(s) \longrightarrow CuSO_4(s) + 5H_2O(\ell)$ *endothermic*

blue crystals → white powder

For the 'opposite' direction:

$CuSO_4(s) + 5H_2O(\ell) \longrightarrow CuSO_4.5H_2O(s)$ + heat *exothermic*

Enthalpy

Substances have an amount of energy, in the form of either kinetic or potential energy, which is called the **enthalpy** or *heat content*. The symbol for enthalpy is H.

The energy change occurring during a reaction is called the **enthalpy change** (symbol $\Delta_r H$) or the **heat of reaction**. Enthalpy change is the difference between the enthalpy of the products ($H_{products}$) and the enthalpy of the reactants ($H_{reactants}$):

$$\text{Enthalpy change } (\Delta_r H) = H_{products} - H_{reactants}$$

Enthalpy change is:

- Negative ($\Delta_r H < 0$) for exothermic reactions.

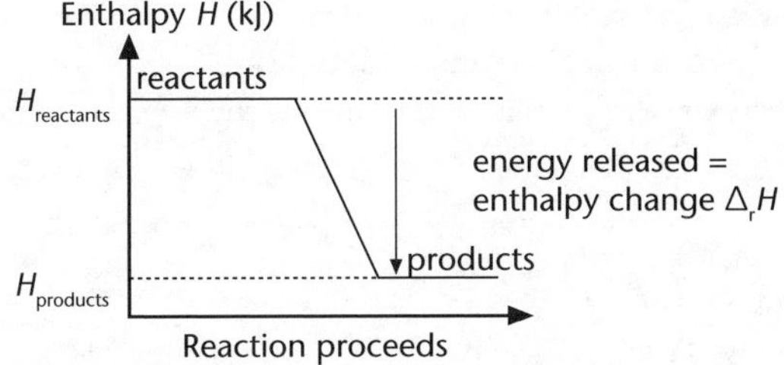

Enthalpy change is negative $\Delta_r H < 0$

- Positive ($\Delta_r H > 0$) for endothermic reactons.

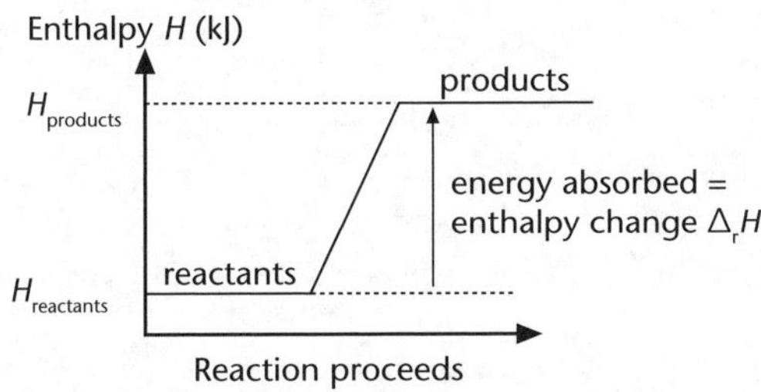

Enthalpy change is positive $\Delta_r H > 0$

Enthalpy change in a reaction depends upon the energy involved in the making and breaking of bonds:

- Bond-forming is exothermic.
- Bond-breaking is endothermic.

In a reaction that is *overall* exothermic, more energy is given out as the bonds of the products form than is taken in as the bonds of the reactants break.

In a reaction that is *overall* endothermic, more energy is taken in as the bonds of the reactants break than is given out as the bonds of the products form.

Example

Carbon dioxide subliming

Carbon dioxide subliming is overall an endothermic process.

$$CO_2(s) \longrightarrow CO_2(g)$$

Dry ice

To break the weak intermolecular bonds between CO_2 molecules in the solid state requires energy (*'bonds breaking, energy in'*).

No bonds form between CO_2 molecules in the gaseous state (*'bonds forming, energy out'* – since no bonds formed, no energy out).

Since energy in > energy out, reaction is overall endothermic.

Enthalpy is measured in joules (J) or, more commonly, in kilojoules (kJ).

Unit 11.4 Activity 2A: Enthalpy changes

1. Identify each of the following reactions as endothermic or exothermic:
   - **a.** Zinc metal is added to dilute hydrochloric acid. Hydrogen gas evolves and the test tube becomes warm. Water is vaporised to steam.
   - **b.** Ammonium nitrate dissolves in water, causing the water temperature to drop.
   - **c.** Two chlorine atoms combine to form a chlorine molecule.
   - **d.** Ice melts.
   - **e.** Some iodine sublimes (changes from solid to gas).
   - **f.** Some wood is burnt.
   - **g.** Concentrated sulfuric acid is added (carefully) to water, causing the water to heat up.
   - **h.** Photosynthesis.
   - **i.** Coal dust explodes in a mine.
   - **j.** A gas ignites.
   - **k.** Respiration.
2. The reaction of zinc metal with dilute hydrochloric acid gives out heat:

   $$Zn(s) + 2HCl(aq) \longrightarrow ZnCl_2(aq) + H_2(g)$$

   - **a.** What is the sign of $\Delta_r H$?
   - **b.** Sketch an energy diagram for the reaction. Label products and reactants on the diagram.
3. Draw and label an energy diagram for **sublimation** of dry ice, $CO_2(s)$, and decide on the sign of $\Delta_r H$.
4. Classify each of the following processes as either exothermic or endothermic:
   - **a.** The formation of snow from water vapour.
   - **b.** Dehydration of blue copper sulfate crystals, $CuSO_4.5H_2O$, to form **anhydrous** white copper sulfate.
   - **c.** Iodine subliming.
   - **d.** $H_2O_2(\ell) \longrightarrow H_2O(\ell) + \frac{1}{2}O_2(g) \quad \Delta_r H = -98.2\ \text{kJ mol}^{-1}$
5. Draw labelled energy diagrams for each of the following reactions:
   - **a.** The condensation of steam.
   - **b.** The decomposition of magnesium carbonate to form MgO and CO_2, which uses 117 kJ per mol.
   - **c.** $Fe(s) + CO_2(g) \longrightarrow FeO(s) + CO(g) \quad \Delta_r H = -11\ \text{kJ mol}^{-1}$
   - **d.** $2NH_3(g) \longrightarrow N_2(g) + 3H_2(g) \quad \Delta_r H = 92\ \text{kJ mol}^{-1}$

Determining enthalpy changes

Although enthalpy *change*, $\Delta_r H$, can be measured for a reaction, the *actual* enthalpy, H, for the reactant(s) and/or product(s) *cannot* be measured.

Data from an experiment can be manipulated to calculate values of $\Delta_r H$ for different quantities reacting.

Example

Expressing heat change information for one mole of substance

When one mole (40 g) of NaOH dissolves in water, 20 200 J (20.2 kJ) of energy is released. This can be written in any of the following ways:

- The enthalpy change for 1 mol of sodium hydroxide dissolving in water is $\Delta_r H = -20\ 200$ J. (The enthalpy change has a *negative* sign because the reaction is *exothermic*.)
- $NaOH(s) \longrightarrow Na^+(aq) + OH^-(aq) + 20.2\ kJ\ mol^{-1}$ (Emphasising that the heat energy is a product.)
- $NaOH(s) \longrightarrow Na^+(aq) + OH^-(aq)$, $\Delta_r H = -20.2\ kJ\ mol^{-1}$.
- For NaOH dissolving, $\Delta_r H = -20.2\ kJ\ mol^{-1}$.

Example

Manipulating heat change information

Since 504 J of energy is transferred to the water when 1 g of sodium hydroxide dissolves:

- 2 × 504 J = 1008 J would be transferred if 2 g dissolved.
- 10 × 504 J = 5040 J would be transferred if 10 g dissolved.

The change in temperature of a reaction mixture depends upon the amount of the substances involved.

A chemical equation for which ΔH is given is called a **thermochemical equation**.

ΔH is given in kJ mol^{-1}, where the *mol* refers to the amounts indicated in the equation by the coefficient.

Example

Thermochemical equations

The reaction:

$$2SO_2(g) + O_2(g) \longrightarrow 2SO_3(g),\ \Delta_r H = -188\ kJ\ mol^{-1}$$

means that when *2 moles* SO_2 reacts with *1 mole* O_2, 188 kJ of energy is released.
So, if 1 mole SO_2 reacts with oxygen, *half* the amount of heat energy is involved,

ie = 0.5 × 188
= 94 kJ is relased

Use either 'released' or '$\Delta_r H = -$' (negative sign); not both. In the above Example, '94 kJ is released' could have been written as '$\Delta_r H = -94$ kJ'.

Example

Changes in amounts of reactants

1. When 1 g of sodium hydroxide dissolves in 40 mL water, 504 J of energy is released so:

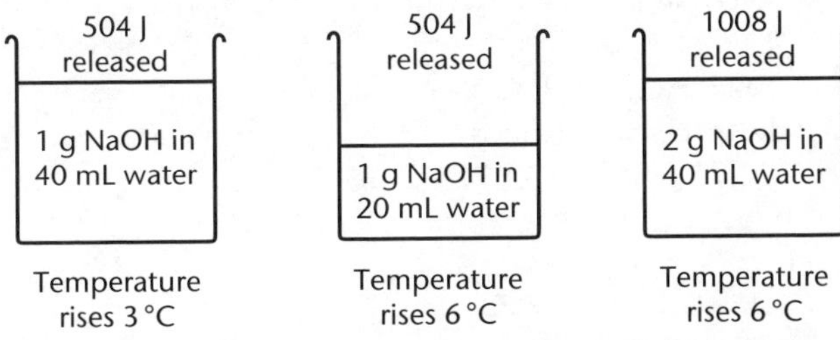

2. Calculate the heat energy released when 10 g NaOH is dissolved, given:
$NaOH(s) \longrightarrow Na^+(aq) + OH^-(aq)$, $\Delta_rH = -20.2$ kJ mol^{-1}

Amount of NaOH: $n = \frac{m}{M}$

$= \frac{10\text{ g}}{40\text{ g mol}^{-1}}$ [substituting M(NaOH) = 40 g mol^{-1}]

$= 0.25$ mol

1 mol of NaOH releases 20.2 kJ of heat energy,
∴ 0.25 mol of NaOH releases 0.25 × 20.2 = 5.05 kJ

3. Calculate the mass of octane, C_8H_{18}, needed to burn to release 10 000 kJ of energy.
$C_8H_{18}(\ell) + 12\frac{1}{2}O_2(g) \longrightarrow 8CO_2(g) + 9H_2O(g)$ $\Delta_rH = 5\,470$ kJ mol^{-1}
$M(C_8H_{18}) = 114$ g mol^{-1}
5470 kJ is released by 1 mol
10 000 kJ is released by $\frac{10\,000\text{ kJ}}{5470\text{ kJ mol}^{-1}} = 1.83$ mol

mass of C_8H_{18} is:
$m = nM$
$= 1.83\text{ mol} \times 114\text{ g mol}^{-1}$
$= 209$ g

Unit 11.4 Activity 2B: Calculating enthalpy changes

1. When 6.0 g of magnesium burns in excess oxygen, magnesium oxide forms and 150 kJ of energy is released.
Find the heat released if 2 mol of magnesium burns [M(Mg) = 24.3 g mol^{-1}].
2. The equation for the burning of carbon is:
$C(s) + O_2(g) \longrightarrow CO_2(g)$; $\Delta_rH = -400$ kJ mol^{-1}
[M(C) = 12 g mol^{-1}; M(O) = 16 g mol^{-1}].
Calculate Δ_rH when 6 g of carbon is burned.
3. The reaction between hydrochloric acid and sodium hydroxide is represented by:
$H^+(aq) + OH^-(aq) \longrightarrow$ H2O(ℓ); $\Delta_rH = -57.4$ kJ mol^{-1}.
   a. Is the reaction exothermic or endothermic? Explain your answer.
   b. 0.5 mol of hydrochloric acid reacts with 0.5 mol of sodium hydroxide. Calculate the enthalpy of reaction (Δ_rH).
4. For the reaction $N_2(g) + 3H_2(g) \longrightarrow$ 2NH3(g), ΔrH = –92 kJ mol–1:
   a. Find the heat energy released when 2 mol of nitrogen react.
   b. Find the energy released when 0.5 mol of ammonia form.
   c. How much nitrogen must react to produce 1840 kJ of energy?
5. For the reaction $2C(s) + O_2(g) \longrightarrow 2CO(g)$, $\Delta_rH = -230$ kJ mol^{-1}:
   a. Find the heat energy released when 8 moles of oxygen react.
   b. Find the heat energy released when 6 g of carbon burns [M(C) = 12 g mol^{-}].

6.

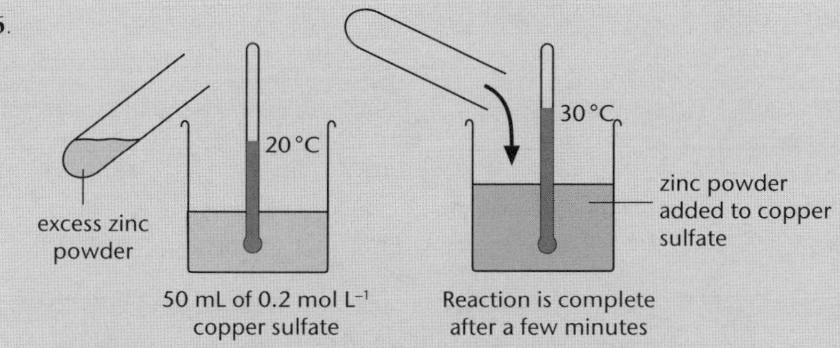

$Zn(s) + Cu^{2+}(aq) \longrightarrow Cu(s) + Zn^{2+}(aq)$; $\Delta H = -216$ kJ mol^{-1}

a. Is this reaction exothermic or endothermic? Explain your answer.

b. Calculate the heat released in this experiment.

7. Hydrogen peroxide, a common bleaching agent, decomposes as follows:

$$H_2O_2(\ell) \longrightarrow H_2O(\ell) + \frac{1}{2}O_2(g) \qquad \Delta_r H = -98.2 \text{ kJ mol}^{-1}$$

a. Calculate how much energy is released when 5 moles of hydrogen peroxide decompose.

b. Calculate how much energy is released when 1.0 g of oxygen is formed by the decomposition of hydrogen peroxide.

c. Calculate the mass of hydrogen peroxide that must decompose to produce 600 kJ of energy.

8. Propane is used in some central heating systems. It burns according to the following equation:

$$C_3H_8(g) + 5O_2(g) \longrightarrow 3CO_2(g) + 4H_2O(\ell) \qquad \Delta_r H = -2220 \text{ kJ mol}^{-1}$$

a. Draw a labelled energy diagram for the above reaction.

b. Calculate the heat energy produced when 11 g of propane is burned.

c. Calculate the mass of propane required to produce 10 000 kJ of energy.

$M(C) = 12.0$ g mol^{-1}, $M(H) = 1.00$ g mol^{-1}

9. When 1 mol of ethanol, C_2H_6O, is burned in oxygen to produce carbon dioxide and water, 1367 kJ of energy is released.

a. Write a thermochemical equation for the burning of ethanol.

b. Calculate the heat released when 4.71 g of ethanol is burnt.

$M(C_2H_6O) = 46.0$ g mol^{-1}

10. Methane, CH_4, and butane, C_4H_{10}, can both be used as fuels. When 1 mol of methane burns, 890 kJ of energy is released. When 1 mol of butane burns, 3000 kJ is released.

a. Write thermochemical equations for the reactions above.

b. Carry out calculations to determine which fuel releases the greatest amount of energy per gram.

$M(CH_4) = 16.0$ g mol^{-1} $\qquad$ $M(C_4H_{10}) = 58.0$ g mol^{-1}

11. The equation for the thermal decomposition of mercury(II) oxide is:

$$2HgO(s) \longrightarrow 2Hg(\ell) + O_2(g) \quad \Delta H = 182 \text{ kJ mol}^{-1}$$

a. Calculate the energy needed to decompose 277 g of the oxide.

b. How many grams of Hg form if 555 kJ of energy are absorbed?

12. The conversion of sulfur dioxide to sulfur trioxide is an important step in the manufacture of sulfuric acid:

$2SO_2(g) \quad + \quad O_2(g) \quad \longrightarrow \quad 2SO_3(g) \quad \Delta H = -188 \text{ kJ mol}^{-1}$

a. Calculate the energy released when 32.0 g of sulfur dioxide is burnt in excess oxygen.
$M(SO_2) = 64.0 \text{ g mol}^{-1}$.

b. Calculate the mass of SO_2 needed to produce 3000 kJ of energy.

13. Heat packs are used by skiers trapped in snowstorms. They work on the following reaction:

$4Fe(s) \quad + \quad 3O_2(g) \quad \longrightarrow \quad 2Fe_2O_3(s) \quad \Delta H = -1648 \text{ kJ mol}^{-1}$

The heat pack is sealed in an airtight plastic film. Inside this is a packet of finely divided iron. When the seal is broken, oxygen penetrates the paper packet and heat is released. A pack contains 448 g of iron.

How much energy would be released when one pack containing 448 g of iron is used?

14. Pentaborane, B_5H_9, is a highly reactive substance once considered a potential rocket fuel. It burns in oxygen according to the reaction:

$2B_5H_9 \quad + \quad 12O_2 \quad \longrightarrow \quad 5B_2O_3 \quad + \quad 9H_2O \quad \Delta H = -9036 \text{ kJ mol}^{-1}$

Calculate the heat released per gram of B_5H_9 in the reaction with oxygen.

$M(B_5H_9) = 63.0 \text{ g mol}^{-1}$

15. The first step in the production of nitric acid, HNO_3, involves the combustion of ammonia to form nitric oxide, NO:

$4NH_3(g) \quad + \quad 5O_2(g) \quad \longrightarrow \quad 6H_2O(g) \quad + \quad 4NO(g) \quad \Delta H = -912 \text{ kJ mol}^{-1}$

Calculate the energy released when 600 kg of NH_3 are used. $M(NH_3) = 17.0 \text{ g mol}^{-1}$

16. Nitrogen gas combines with oxygen gas according to the following equation:

$N_2(g) \quad + \quad 2O_2(g) \quad \longrightarrow \quad 2NO_2(g) \quad \Delta H = +68 \text{ kJ mol}^{-1}$

a. Calculate the energy needed to produce 500 g of $NO_2(g)$.

b. Calculate the mass of $N_2(g)$ reacted when 350 kJ of energy are used.

17. Hydrazine is a powerful rocket fuel, which reacts according to the equation:

$2N_2H_4(\ell) \quad + \quad N_2O_4(\ell) \quad \longrightarrow \quad 3N_2(g) \quad + \quad 4H_2O(g) \quad \Delta H = -1048 \text{ kJ mol}^{-1}$

Calculate the energy released when 100 kg of hydrazine, N_2H_4, are reacted.
$M(N_2H_4) = 32.0 \text{ g mol}^{-1}$

18. The equation for the oxidation of glucose that occurs during respiration is:

$C_6H_{12}O_6(s) + 6O_2(g) \quad \longrightarrow \quad 6H_2O(\ell) + 6CO_2(g) \quad \Delta H = -2803 \text{ kJ mol}^{-1}$

a. Calculate the energy released when 10.0 g of glucose is oxidised.

b. The energy released when 1.00 mol of sucrose, $C_{12}H_{22}O_{11}$, is oxidised, is 5640 kJ. Carry out calculations to determine which sugar – glucose or sucrose – produces more energy per gram during the oxidation process.

$M(C_6H_{12}O_6) = 180 \text{ g mol}^{-1}$ $\qquad M(C_{12}H_{22}O_{11}) = 342 \text{ g mol}^{-1}$

Unit 11.4 Energy and Reaction Rates

Topic 3: Rates of reaction (extension)

This Topic reviews and extends our understanding of chemical reactions and the factors that influence the reasons why some reactions occur rapidly (in seconds) and others occur over slowly (in days or weeks or years). Topic 3 deals with:

- The factors (changes in concentration, temperature, surface area and the presence of a catalyst) that affect rates of chemical reactions.
- Explaining changes in reaction rate.

Rates of reactions

Some chemical reactions occur *rapidly* (eg if a solution containing calcium ions is added to a solution containing carbonate ions, a white precipitate usually forms within one or two seconds).

Other chemical reactions (such as the rusting of iron), occur *slowly*, over days or even years.

Rate of reaction means the speed at which a chemical reaction occurs.

Measuring rates of reaction

The rate at which a reaction occurs is measured either by how quickly reactants are used up or how quickly products form. The rate of a reaction is easy to determine if the reaction involves a colour change or the formation of a gas.

Reactions that involve a change in colour

The rate of reaction is monitored by determining how quickly the colour changes.

Example

Measuring rate by colour change

When copper metal reacts with concentrated nitric acid, a brown gas, nitrogen dioxide, NO_2, forms:

$$Cu(s) + 4H^+(aq) + 2NO_3^-(aq) \longrightarrow Cu^{2+}(aq) + 2H_2O(\ell) + 2NO_2(g).$$

Rate of reaction is measured by observing how quickly the brown gas appears.

Reactions when one of the products is a gas

The rate of reaction can be determined by measuring how much gas is produced after a certain time interval.

Example

Rate by volume change

The reaction of zinc metal with dilute hydrochloric acid is:

$Zn(s) + 2H^+(aq) \longrightarrow Zn^{2+}(aq) + H_2(g)$

The hydrogen gas produced can be collected and its volume measured after certain intervals of time.

Experimental set-up for hydrogen gas collection

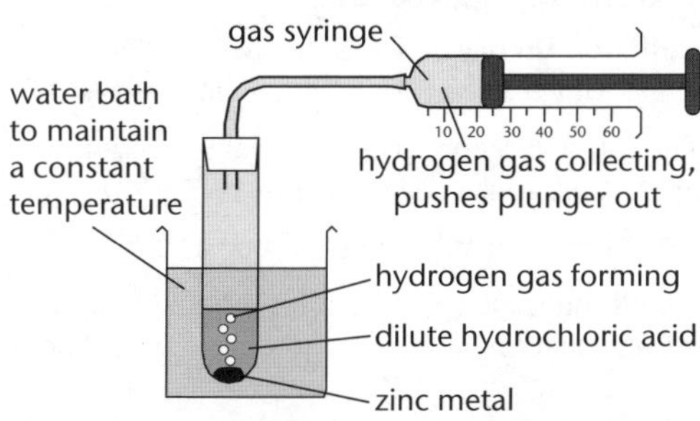

Results from one experiment using the apparatus were:

Time from start (s)	0	5	10	15	20	25	30	35	40	45	50	55	60	65
Volume of gas collected (mL)	0	15	26	34	40	45	50	53	56	57	59	60	60	60

Results are best presented on a graph:

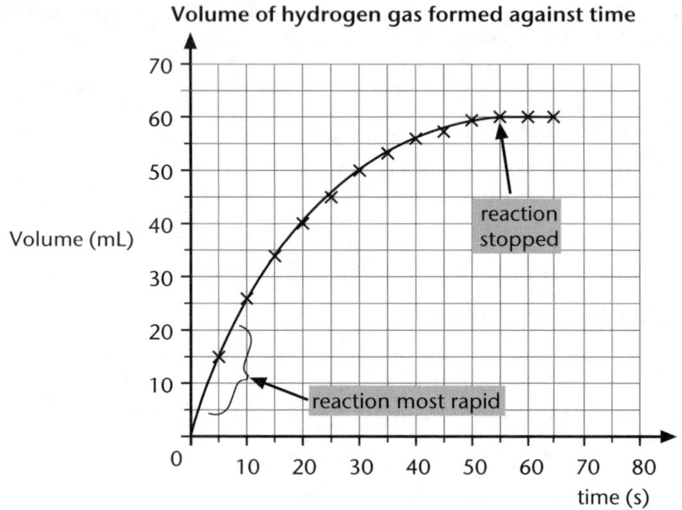

A typical rates of reaction graph (see example above) shows the reaction:

- *Begins rapidly* – initial slope of the graph is steep.
- *Slows down* as time passes – slope of the graph becomes less steep as the reaction proceeds.
- *Stops* after a certain time – graph levels out as one (or both) reactants have been used up.

Rates of reaction that do *not* involve a colour change or gas formation can also be measured, but different techniques must be used.

Sometimes, there is a delay before reactions begin because there is a protective layer or coating on one of the reactants. This delay does not necessarily mean the rate of reaction is slow, as the reaction may proceed rapidly once the coating is removed.

Collision theory

Collision theory explains the mechanism by which a chemical reaction occurs and why the rate at which a reaction occurs might change.

The collision theory states that 'for a reaction between two particles to occur, the particles must *collide* and a collision must be *effective*'.

The more frequently collisions occur, the faster a reaction occurs.

A collision is effective when reacting particles collide *with enough kinetic energy* and the *correct orientation* to break existing forces (or *bonds*) between particles so that new bonds can form.

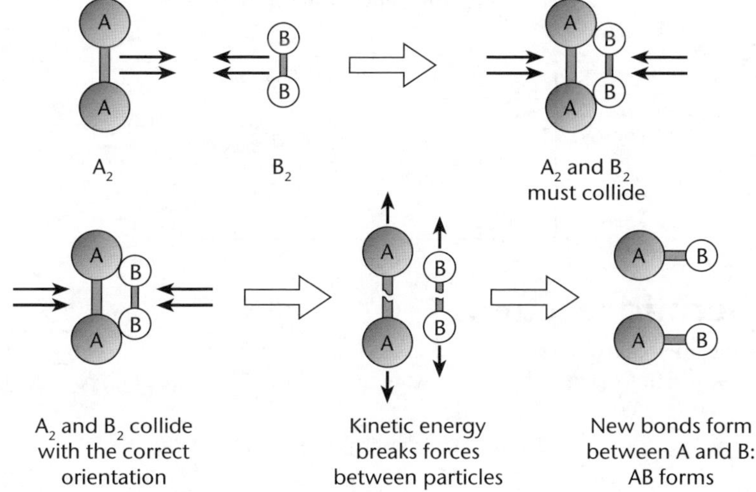

Activation energy (E_A)

Activation energy (E_A) is the 'minimum amount of energy needed for a reaction to occur between two colliding particles', ie the amount of energy needed to initiate a reaction. In collisions that do not have enough activation energy, particles bounce apart without reacting.

Activation energy can be displayed on an energy diagram, together with the enthalpy change, $\Delta_r H$, for a reaction. Activation energies are always positive, since energy must be *supplied* to start the reaction.

Example

Hydrogen peroxide, H_2O_2, decomposes to form water and oxygen:

$$2H_2O_2(aq) \longrightarrow 2H_2O(\ell) + O_2(g) \quad \Delta H = -189 \text{ kJ}, E_A = 75 \text{ kJ}$$

Energy diagram for decomposition of hydrogen peroxide

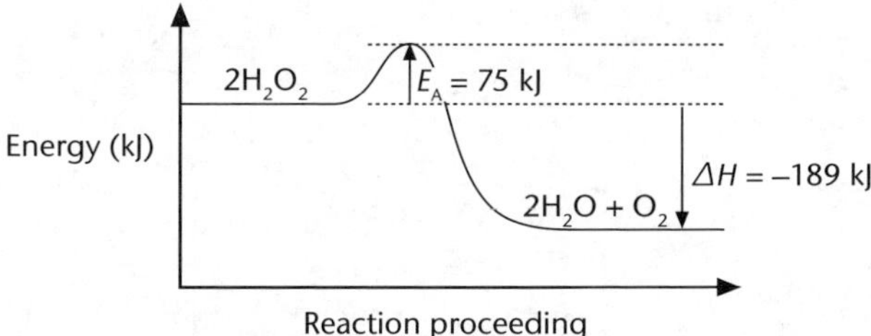

Exothermic reactions

In an **exothermic reaction**, the total energy released is greater than the activation energy for the reaction. The energy released by the exothermic reaction of just one particle reacting is used to provide the activation energy for other particles to react. Once an exothermic reaction begins, it will continue until one of the reactants is used up or some other change occurs (eg once lit, a piece of paper will burn completely unless extinguished by some other means).

Exothermic energy diagram

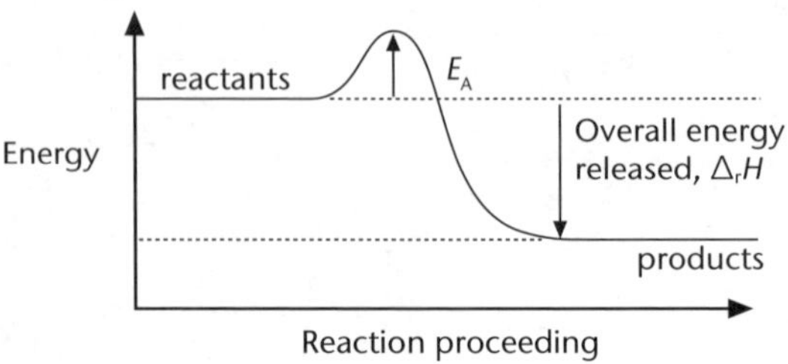

Endothermic reactions

In an **endothermic reaction**, the energy released is less than the activation energy for the reaction. These reactions have a different driving force which allows colliding particles to gain sufficient activation enery.

Endothermic energy diagram

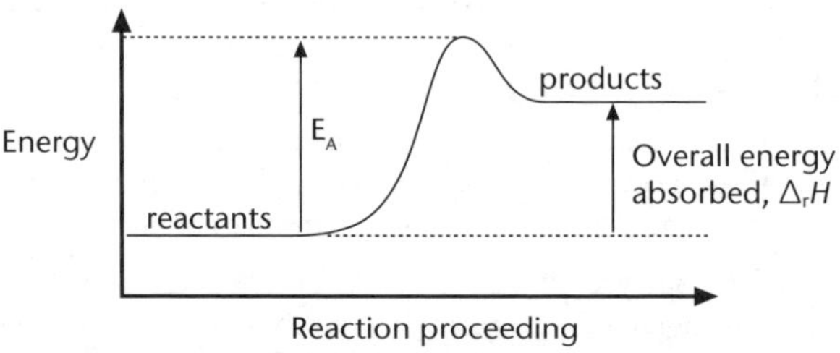

Unit 11.4 Activity 3A: Reaction rates and energy diagrams

1. Draw and label energy diagrams to illustrate each of the following situations:
   - **a.** When reactants A and B form products C and D, the overall energy released is 150 kJ and the activation energy is 55 kJ.
   - **b.** When W and X react to form products Y and Z, 42 kJ of energy are absorbed. The activation energy is 120 kJ.
2. The following diagram represents the energy curve for the reaction:

$$C(s) + O_2(g) \longrightarrow CO_2(g)$$

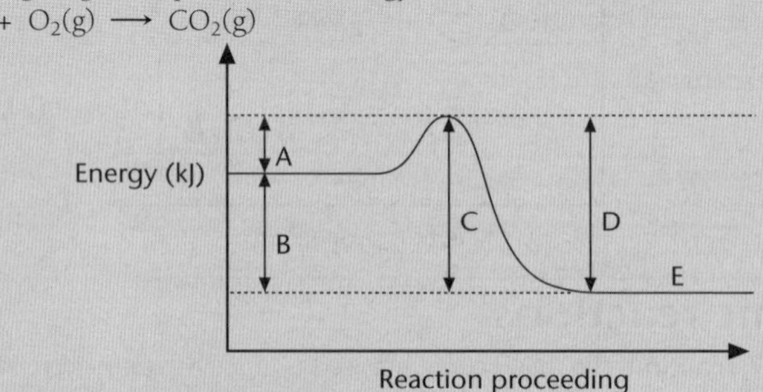

   - **a.** Is the reaction exothermic or endothermic?
   - **b.** Which letter shows the *net energy* released in the reaction?
   - **c.** Which letter shows the activation energy?

3. The diagram following represents the energy diagram for the reaction:

$N_2(g) + 2O_2(g) \longrightarrow 2NO_2(g)$

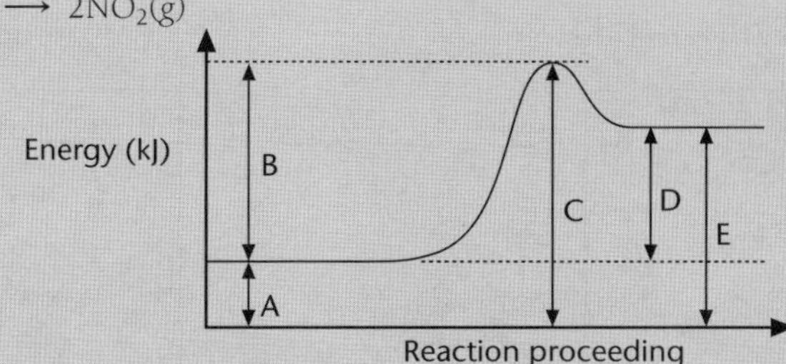

a. Is the reaction exothermic or endothermic?

b. Which letter corresponds to the activation energy for this reaction?

Factors affecting the rate of a reaction

The rate of a reaction can be increased by:

- Increasing the frequency at which collisions occur.
- Increasing the energy of the collisions so that more collisions are effective because they have energy greater than the activation energy.

The frequency of collisions can be increased by increasing the concentration, increasing the surface area and/or increasing the temperature. More effective collisions occur when temperature is increased or a catalyst is used to lower the activation energy.

Concentration and reaction rate

Increasing the concentration of a reactant increases the rate of reaction. This is because the increased concentration means an increase in the frequency of collisions, so more particles react to give products.

Example

Rates and concentration

1. Sulfur burns in oxygen according to the following equation:

$S(s) + O_2(g) \longrightarrow SO_2(g)$

There are more frequent collisions between sulfur and oxygen when the reaction occurs in *pure* oxygen, rather than in the same volume of air (which is only 20% oxygen).

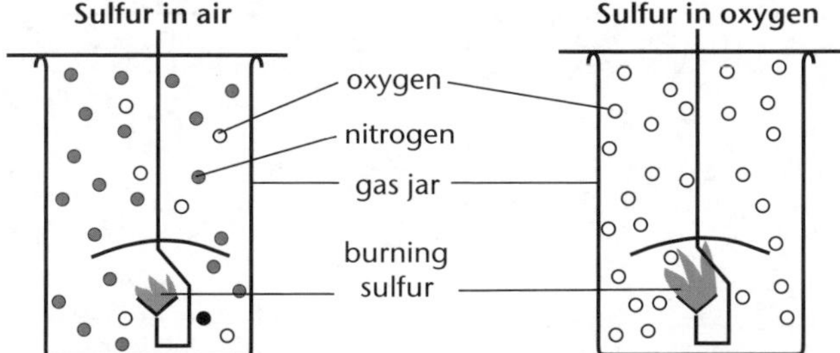

In air, only one in five particles colliding with sulfur particles is an oxygen molecule.

In pure oxygen, all particles colliding with the sulfur particles are oxygen molecules. Reaction is *faster* than in air.

2. Magnesium metal reacts faster in dilute hydrochloric acid than in the same volume of *very* dilute hydrochloric aid.

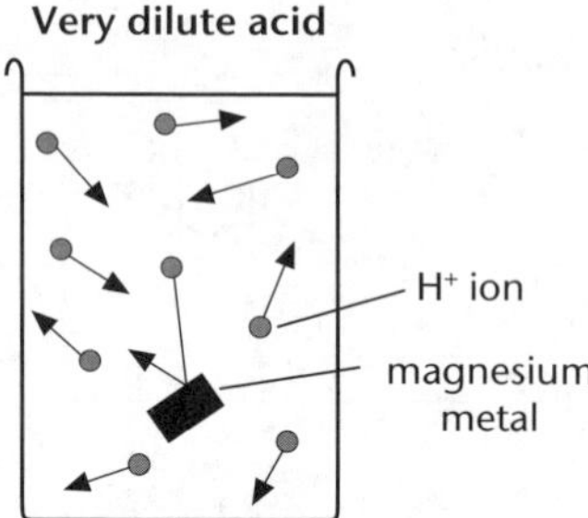

Fewer H^+ ions in the same volume of solution, so fewer collisions occur between $H^+(aq)$ and $Mg(s)$.

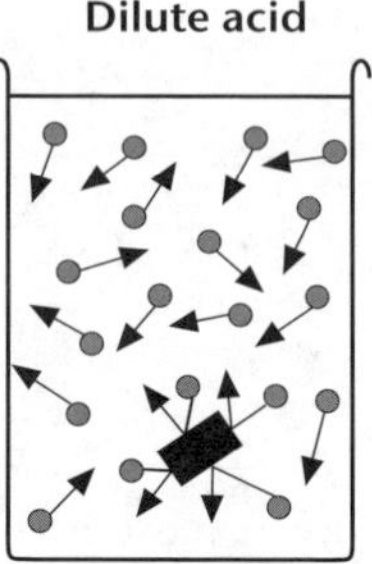

More H^+ ions in the solution, so more collisions occur between $H^+(aq)$ and $Mg(s)$. Reaction is faster than in very dilute acid.

Surface area and reaction rate

Increasing the surface area of a reactant will increase the rate of reaction. This is because an increased surface area means more reactant particles are exposed and can take part in the reaction.

Surface area is increased by breaking up pieces of reactant.

Consider a 2 cm × 2 cm × 2 cm cube:

Surface area = 6 faces × area of one face

= 6 × (2 cm × 2 cm)

= 24 cm^2

Cutting the cube in half across each section gives eight cubes:

Surface area of eight small cubes is now:

8 × (6 × area of one face) = 8 × 6 × 1 × 1

= 48 cm^2

Breaking up the cube has *increased* (doubled) the surface area.

Example

Rate and surface areas

Powdered calcium carbonate reacts faster with the same volume of dilute hydrochloric acid than a lump of calcium carbonate.

Lump of calcium carbonate

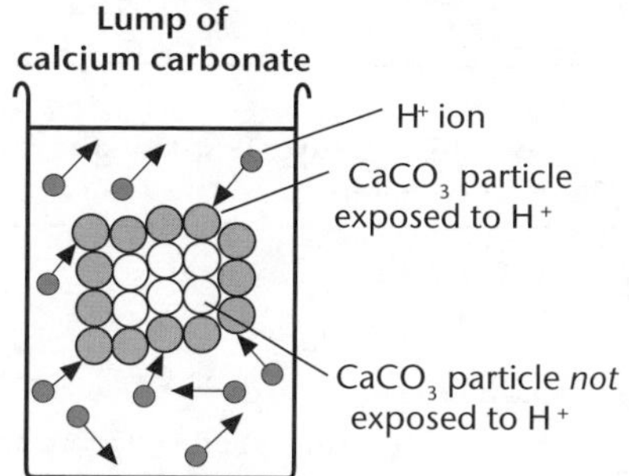

2 g lump of calcium carbonate, fewer particles exposed to H^+.

Powdered calcium carbonate

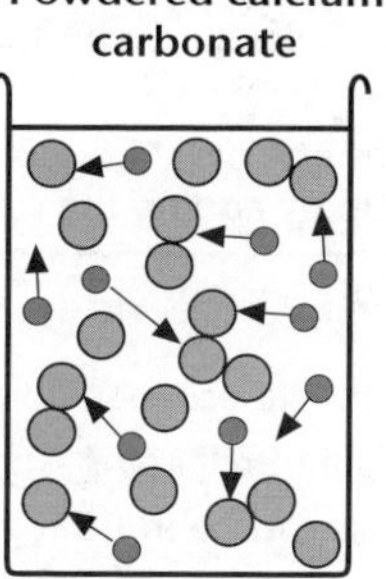

2 g of powdered calcium carbonate, more particles exposed to $H^+(aq)$. Reaction is *faster* when calcium carbonate is powdered.

Reactions in solutions are often rapid because large surface areas are involved, ie particles are separated and free to move.

Example

Forming precipitates

Solid lead nitrate does not to react at all when mixed with solid potassium iodide, yet when solutions of each are mixed, a bright yellow precipitate forms immediately:

$$Pb^{2+}(aq) + 2I^-(aq) \longrightarrow \underset{\text{yellow}}{PbI_2(s)}$$

Stirring to dissolve a solid in a liquid keeps a 'fresh' surface of the solid exposed to the liquid. This increases the frequency of collisions between the solid and liquid particles, so dissolving occurs more rapidly. Two **miscible** liquids can be *shaken* together to increase the area of contact between the two liquids and speed up the rate of mixing.

Some solids that are quite unreactive in the form of lumps can be highly reactive as powders. Coal dust explosions have occurred in mines, where the fine particles of coal spark a sudden reaction with oxygen.

Temperature and reaction rate

The rate of reaction increases with increasing temperature. This occurs because when temperature increases, energy is being added, so the kinetic energy and hence the speed of reacting particles increases. This has two effects on the colliding particles:

- Particles collide more frequently.
- Particles collide with more energy and hence a larger proportion of colliding particles have sufficient energy for a reaction to occur (ie more particles have energy > E_A).

Catalysts and reaction rate

A catalyst is a substance that alters the rate of a reaction by changing the activation energy required. Usually, catalysts are used to increase the rate of a reaction. When the activation energy is lowered, more collisions become effective.

Example

Rate and catalysts

The decomposition of hydrogen peroxide, H_2O_2, to water and oxygen is slow. The reaction can be sped up by adding the catalyst, manganese dioxide, MnO_2.

Catalysts do not appear directly in the equations for reactions, because catalysts:

- Do not change the products formed.
- Are not *used up* in reactions – they are usually recovered and reused.

The use of catalysts can be indicated by writing them above the arrow in an equation, eg:

$$2H_2O_2(\ell) \xrightarrow{MnO_2} 2H_2O(\ell) + O_2(g)$$

In biological reactions, catalysts are usually protein molecules called **enzymes**.

Example

Enzymes

The enzyme amylase breaks starch down to glucose:

$$\text{starch} \xrightarrow{\text{amylase}} \text{glucose}$$

Without amylase, the reaction rate would be too slow to be detected.

Unit 11.4 Activity 3B: Collision theory and reaction rates

1. A piece of calcium carbonate (marble) is added to some dilute hydrochloric acid. The carbon dioxide produced is collected in a graduated tube and the volume of gas collected is recorded each minute.

The results are:

Time (min)	0	1	2	3	4	5	6	7
Volume (mL)	0	3	10	29	48	51	52	52

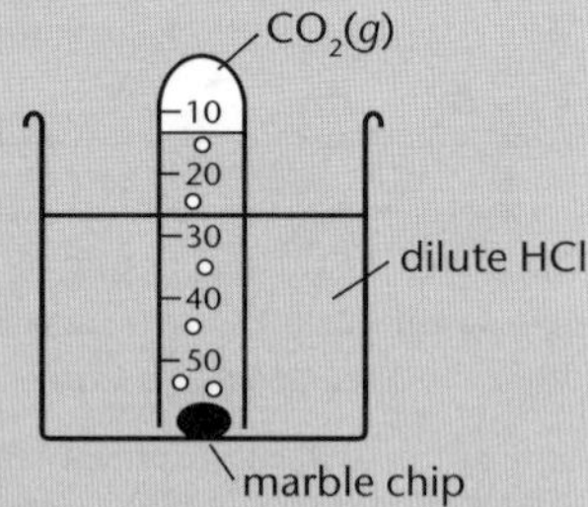

a. Plot a graph of volume versus time. Put time on the horizontal axis.

b. Label the region on your graph which represents where the reaction is most rapid.

c. After how much time does the reaction stop?

d. Sketch the curve you would expect if the marble chip was replaced by powdered marble.

e. Briefly describe two ways to alter conditions to make a similar marble chip react more slowly. Explain how these conditions work using collision theory.

2. Use the collision theory to explain the following observations:

a. Fine coal dust in coal mines has caused explosions when accidentally sparked.

b. A piece of wood will burn faster when splintered.

c. Powdered zinc decolourises a solution of copper sulfate faster than zinc granules.
d. 100 mL 0.1 mol L^{-1} hydrochloric acid reacts with magnesium metal more slowly than 50 mL 0.2 mol L^{-1} hydrochloric acid.
e. Magnesium metal burns violently in a gas jar of oxygen. In air it burns, but not explosively.
f. Despite its position on the activity series, aluminium metal does not react with water.
g. A piece of iron placed in a Bunsen flame gets hot. Iron wool (fine threads of iron) placed in a flame burns to produce iron oxides.

3. a. Describe three ways to slow down the rate of gas production when zinc granules are allowed to react with dilute hydrochloric acid.
b. Use collision theory to explain each of your responses in **a**.

4. Two grams of zinc granules were completely covered by 100 mL of 1.0 mol L^{-1} hydrochloric acid. Give the effect each of the following changes would have on the *initial* rate of hydrogen gas production and explain the effect in terms of the collision theory:
a. Using 200 mL of 1.0 mol L^{-1} hydrochloric acid.
b. Using 100 mL of 0.25 mol L^{-1} hydrochloric acid.
c. Using 2 g of powdered zinc.

5. A student used the reaction between a lump of calcium carbonate and hydrochloric acid to investigate the rate of a chemical reaction. The reaction was:

$CaCO_3(s) + 2HCl(aq) \longrightarrow CaCl_2(aq) + H_2O(\ell) + CO_2(g)$

Three experiments were carried out at constant temperature, using the quantities and concentrations set out in the following table:

Experiment number	Mass of $CaCO_3$(g)	Volume of HCl (mL)	Concentration of HCl (mol L^{-1})
1	5	20	0.1
2	5	20	1.0
3	5	20	2.0

a. In which experiment would the reaction rate be the fastest?
b. Which experiment would produce the smallest volume of carbon dioxide? Give a reason.
c. State *two* other ways, other than changing the concentration, to increase the rate of the reaction.

6. A 1 g lump of calcium carbonate was added to a 250 mL beaker containing 100 mL of 1.0 mol L^{-1} hydrochloric acid solution, at room temperature (25 °C). Bubbles of carbon dioxide were produced.
Explain how each of the following changes would affect the reaction rate. Justify your answer by referring to collisions between the particles.
a. The temperature of the reaction mixture was increased to 40 °C.
b. 100 mL of water was added to the acid. 100 mL of this diluted acid solution was added to a 1 g lump of calcium carbonate.
c. The 1 g lump of calcium carbonate was ground to form a powder, and then 100 mL of the 1.0 mol L^{-1} hydrochloric acid was added.
d. A 500 mL beaker was used instead of the 250 mL beaker, but the same amounts of reactants were used.

7. The reaction between 20.0 mL of 0.500 mol L^{-1} hydrochloric acid and 20.0 mL of 0.250 mol L^{-1} sodium thiosulfate solution at room temperature (25 °C) produces a precipitate of sulfur that makes the solution go cloudy after about 5 minutes.

a. How would the *time taken* for the solution to go cloudy be affected if:

i. The reaction were carried out in an ice bath at a temperature of 2 °C?

ii. The concentration of the hydrochloric acid was changed to 0.250 mol L^{-1}?

iii. 10 mL of 0.500 mol L^{-1} sodium thiosulfate was used?

iv. 20 mL of water was added at the same time as the solutions were mixed?

b. With reference to the collisions of particles, explain why each reaction (**i.**–**iv.**) is affected in the ways described in **a**.

8. Explain why the reaction between $SO_2(g)$ and $O_2(g)$, which is usually slow, proceeds much faster when a small amount of V_2O_5 is added.

Unit 11.5 Metals and Non-metals

Topic 1: Physical properties and uses of metals

Many substances are categorised generally as metal or non-metal – but how can it be proved that a substance is metal or non-metal? What is it that distinguishes metals from non-metals? We have learned about the basic properties of metals and non-metals and their position on the periodic table. In Unit 11.5 we study the physical and chemical properties of metals and non-metals and their uses. Topic 1 looks at:

- Physical properties of metals.
- Uses of metals based on their physical properties.
- Alloys – their composition and uses.

Physical properties of metals

Physical properties of a substance are those properties that enable it to be recognised as a separate substance but do *not* alter the chemical nature of the substance.

Common physical properties, and their characteristics for metals, include the following.

- *Physical state* – solid, liquid or gas at room temperature.
  Most metals are solids at room temperature – the exception is mercury (a liquid at room temperature). Lithium and sodium are soft enough to be squeezed between the fingers – make sure your fingers are *dry* if you ever try this.
- *Melting point* – a precise melting point indicates a pure substance.
  The melting points of metals vary from below 100 °C to above 1000 °C.
- *Boiling point* – a precise boiling point indicates a pure substance.
  Most metals have high boiling points, > 1000 °C.
- *Density* – **density** is mass per unit volume; the common unit for density is g mL^{-1}.
  Most metals have high densities, > 3 g mL^{-1}.

Example

A lump of lead is much heavier than a lump of rock of the same volume because the lump of lead has a greater mass per unit volume.

- *Colour* – related to the ability of a substance to reflect visible light.
  Most metals are silver-grey in colour.
- *Lustre* – describes how well the substance shines (reflects light).
  Metals are capable of being polished to a high lustre – and therefore shine. (Non-metals tend to be non-lustrous, ie dull.)
- *Electrical conductivity* – the ability of a substance to allow an electric current to pass through it. Substances have either high electrical resistance and are therefore poor conductors of electricity (**insulators**), or low electrical resistance and are therefore good conductors of electricity.
  Metals are very good conductors of electricity (have high electrical conductivity).
- *Thermal conductivity* – the ability of a substance to allow heat to pass through it.
  Metals are very good conductors of heat (have high **thermal conductivity**).
- *Malleability* – the ability of a substance to be beaten into a sheet.
  Most metals are readily hammered or pressed into sheets (high malleability).
- *Ductility* – the ability of a substance to be drawn into wire.
  Most metals are readily drawn into wire (high ductility).

Data for a selection of metals (alphabetically arranged) is set out in the table.

Metal	Melting point (°C)	Density (g mL^{-1})	Lustre; Colour
Aluminium	660	2.7	High; silver
Calcium	839	1.5	High; silver-grey
Copper	1083	8.9	High; pink
Gold	1064	18.9	High; yellow
Iron	1535	7.9	High; silver-grey
Lead	328	11.3	High; silver-grey
Lithium	181	0.5	High; silver-grey
Magnesium	649	1.7	High; silver
Silver	962	10.5	High; silver
Sodium	98	0.97	High; silver
Zinc	420	7.1	High; silver
Lustre – lithium, calcium and sodium have a dull appearance due to their interaction with oxygen, water and carbon dioxide in the atmosphere. Iron rusts in the presence of air and requires protection.			

Physical properties of metals

Atomic structure of metals

Atoms of metals are strongly attracted to each other – they form a close-packed arrangement which results in a hard crystalline material. The electrons in the outer energy level of a metal atom are mobile – ie they are loosely attached to the nucleus of the atom and can flow through the metal structure. The metallic bond is considered to be the nucleus of each atom bonding to a 'sea' of mobile **valence** electrons.

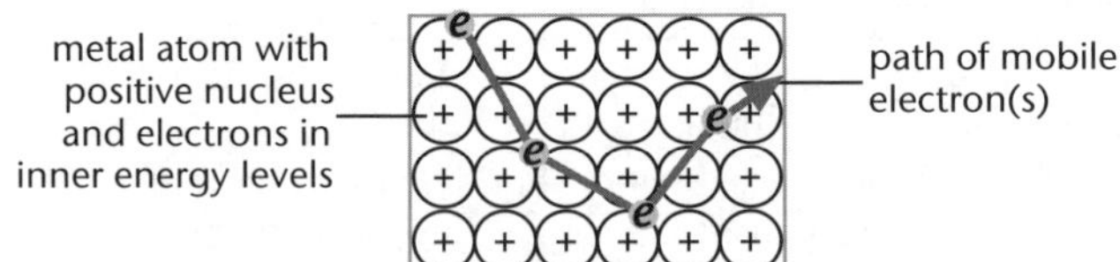

Atom arrangement in metal structure

The structure of the metal affects its physical properties, eg:

- The sea of mobile electrons acts as a 'glue', binding the nuclei together strongly – hence the high melting point of some metals.
- The mobility of the electrons explains the electrical conductivity of the metal.
- The lattice of the metal structure can be distorted by force, which accounts for the malleability and ductility of metals.

Physical properties and uses of metals

The uses of metals depend on their physical properties. Some examples are given in the table.

Property	Example and Use
Density	*Lead* – used for the keel of sailing boats because its high density gives the keel a high mass for a small volume.
Lustre	*Gold* and *silver* – used in jewellery because of their colour, lustre and their rarity.
Electrical conductivity	*Copper* – used in almost all electrical wiring (eg a typical house contains many metres of copper wire).
Thermal conductivity	*Iron* – used for cooking pots, although they are heavy to lift. *Sodium* – used in liquid form to conduct heat away from the core of a nuclear reactor.
Malleability	*Zinc* sheeting – used for weather protection of exposed wooden joinery. Zinc sheets are easily formed and can be moulded to fit awkward shapes of joinery.
Ductility	*Aluminium* – is drawn into wire to carry high-voltage electrical energy across wide valleys. Aluminium metal is preferred to copper for this use because of its low density.

Some physical properties and uses of metals

Unit 11.5 Activity 1A: Physical properties of metals

1. State which of the following behaviours of metals are dependent upon physical properties:

- **a.** Iron rusts in the presence of air.
- **b.** The metal element of a toaster gets hot when the toaster is switched on.
- **c.** A nugget (lump) of silver can be fashioned into a necklace.
- **d.** A car, in collision with another vehicle, has a door that is dented.
- **e.** A copper roof changes to a green-blue colour over a period of many years.
- **f.** Gold settles at the bottom of the pan when ground rock is mixed with water and the mixture is allowed to settle.

2. Zinc has the following physical properties: grey solid at room temperature; melting point (mp) 420 °C; silvery liquid that boils at 907 °C to become a colourless vapour. Assume room temperature is 20 °C. Describe the appearance of zinc at:

- **a.** 15 °C.
- **b.** 500 °C.
- **c.** 1000 °C.

3. Complete the following table (use words or sentences to fill the blank spaces):

Physical property	Definition	Example	Explanation of the property
Density	Mass per unit volume	Zinc has a density of 7.1 g mL^{-1}.	__________
Lustre	__________	Aluminium can be polished and used as a mirror.	The tightly packed atoms will reflect light.
Electrical conductivity	Capable of allowing an electric current to pass through readily	__________	__________
------- ------------	Capable of allowing heat to pass through readily	Saucepans often have copper bottoms.	The tightly packed copper atoms can pass heat energy from one to another.
Malleability	Ability to be beaten into a sheet	__________	__________
Ductility	__________	Steel can be used to make No. 8 wire.	Atoms of iron can slide over each other.

Alloys

Alloys are formed by mixing two or more molten metals and allowing the mixture to cool. Alloys have properties that are intermediate between the parent metals. Common alloys, and their properties, include:

- *Brass* (copper and zinc) – does not corrode as easily as copper and retains an attractive appearance; it retains the good electrical conductivity of copper.
- *Bronze* (copper and tin) – protects itself with an attractive coating that lasts for centuries.
- *Solder* (tin and lead) – melts at a low temperature and will bond with many metals.
- *Pewter* (tin and lead plus small quantities of copper and antimony) – is harder than lead and does not release lead as a poison to human beings when made into drinking vessels.
- *Duralumin* (aluminium and magnesium) – has greater strength than aluminium but retains the low density needed for aeroplane frame construction.
- *Cupronickel* (copper and nickel) – harder than copper (ie lasts longer).

Steel

The valuable alloy, **steel**, is unusual because it contains a non-metal, carbon (present in proportions of 0.5% to 2%), along with a metal, iron. Other metals are added to make specialist steels. There is a large variety of 'steels' – all of which contain iron and carbon.

Example

Stainless steel contains chromium and nickel. Acid-resistant steel contains silicon. Armour-plating steel contains manganese. Armour-piercing steel shells contain depleted uranium.

Steels have high tensile strength – ie they can be bent, and will return to their original shape without breaking. Steels resist **corrosion** better than iron.

Uses of metals and their alloys

Aluminium, copper, iron and zinc are commercially produced in very large quantities.

Aluminium is used in:

- Construction (eg window frames, ladders).
- Transport (eg aeroplane frames, small boats, caravans).
- Containers (eg cans for drinks).
- Sports equipment (eg tennis racquets).

Copper is used:

- For electrical wiring, hot-water piping and for roofing (only a thin layer of copper is needed).
- To form an alloy with tin (bronze) – can be cast into statues and retains an attractive appearance.
- To form an alloy with zinc (brass) – resistant to corrosion and is used for electrical terminals, and marine and household fittings.
- To form an alloy with nickel (cupronickel) – used for 'gold' and 'silver' coins, depending upon the proportions of copper and nickel.

Pure iron is difficult to obtain, but (apparently) is a white, soft metal. Rather than using pure iron, its alloy, steel, is used for:

- Transport (eg cars, trucks, trains, ships).
- Construction (eg reinforced concrete, bridges, scaffolding).
- Containers – large ones on ships and the small ones (eg 'tins' of food).

Example

A 'tin' can is made of mild steel, with a thin layer of metallic tin on the inside of the container to prevent corrosion of the steel by the food. Tin is non-poisonous to the human system.

- Tools and instruments (eg hammers, chisels, cutlery, scalpels, forceps, drills).
- Tubes and pipes.

Zinc, lead, magnesium, silver and gold are also commercially valuable.

Uses for zinc include:

- Protection of steel – **galvanised** iron.
- Battery casings.
- To form an alloy with copper – ie brass.

Uses for lead include:

- Roofing – can be moulded around awkward joints to provide protection from rain, etc.
- To form an alloy with tin (solder) – used for joining metals to each other, especially copper wire to other metals.

Magnesium is used:

- As the alloy, duralumin, in the construction of aeroplane frames and car wheels.
- In flares – due to its ability to burn readily and emit bright, white light.

Uses of silver and gold:

- Both metals are used for decoration and jewellery because of their high lustre and colour.

Unit 11.5 Activity 1B: Alloys, and uses of metals and alloys

1. Name:
   a. An alloy that contains carbon.
   b. Two alloys that contain tin.
   c. Three alloys that contain copper.
2. Briefly describe how alloys are made.
3. Duralumin is an alloy of aluminium and magnesium. Explain why it is used for aeroplane frames in preference to either aluminium or magnesium metal alone.
4. 24-carat gold is pure gold. 9-carat gold is used for low-cost jewellery. The 'carat' indicates the number of parts of gold present in the alloy. Calculate how much gold is used to make a 9-carat gold ring that has a mass of 50 grams.
5. Identify the following metals:

   a. 
   A cube has a side length of 1 cm. The cube has a mass of 11 g.
   The solid element can be beaten into a sheet and is used to make roofs watertight.

   b. 

   A natural substance that you have to be very lucky to find.
   The only acid that will dissolve the substance is 'aqua regia' – 3 parts conc. HCl to 1 part conc. HNO_3.

Unit 11.5 Metals and Non-metals
Topic 2: Chemical properties of metals

In the previous Topic we looked at the physical properties of metals and alloys. In Topic 2 we look at the chemical properties of metals, acids and bases and examine the:

- Activity series of metals.
- Reactivity of metals with oxygen, water and acids.
- Displacement of metals from solutions.

Metals to be studied here and in Topic 3 (in alphabetical order) are:
aluminium, calcium, copper, gold, iron, lead, lithium, magnesium, silver, sodium and zinc.

Introduction

Chemical properties are those properties that alter the chemical nature of the substance and enable the product to be recognised as a separate substance.

Metals form a group of elements in the periodic table with clearly defined similar chemical properties – much more clearly defined than the non-metals.

Activity series of metals

By observing the reaction of metals with oxygen (air), water and dilute acid, it is possible to arrange the metals in an order of activity – the **activity series**.

The reactivity of the metals with oxygen, water and acids (see table below) confirms the activity series as:

sodium > lithium > calcium > magnesium > aluminium >
zinc > iron > lead > [hydrogen] > copper > silver > gold

Hydrogen is included in the series because any metal *above* hydrogen in the series will **displace** hydrogen from dilute acid, eg

$Zn(s) + 2HCl(aq) \rightarrow ZnCl_2(aq) + H_2(g)$.

Those metals *below* hydrogen will *never* displace hydrogen from an acid, ie the following does *not* occur:

$Cu(s) + 2HCl(aq) \rightarrow CuCl_2(aq) + H_2(g)$.

Metal	Reaction with oxygen (or air)	Reaction with water (or steam)	Reaction with hydrochloric acid
Sodium	Metal must be kept under oil – burns fiercely in air	Metal must be kept under oil – reacts violently with cold water	Violent reaction with dilute acid
Lithium	Metal must be kept under oil – burns fiercely in air	Metal must be kept under oil – reacts violently with cold water	Violent reaction with dilute acid

Metal	Reaction with oxygen (or air)	Reaction with water (or steam)	Reaction with hydrochloric acid
Calcium	Metal must be kept under oil – burns fiercely in air	Metal must be kept under oil – reacts vigorously with cold water	Violent reaction with dilute acid
Magnesium	Melts when heated; then burns with an intensely bright, white light	Metal reacts slowly with cold water and burns in steam if heated	Vigorous reaction with dilute acid
Aluminium	*No reaction with air or oxygen due to protective coating of oxide	No reaction due to protective coating of oxide	Brisk reaction with dilute acid after oxide layer has been dissolved
Zinc	Metal burns in air if heated and molten metal is stirred	Metal reacts with steam when heated	Metal reacts steadily
Iron	Metal sparks in air if iron 'wool' is used. Metal burns in oxygen	Metal reacts with steam when heated	Metal reacts steadily
Lead	Oxidises on the surface when heated and stirred in air	No reaction	Very little reaction; $PbCl_2$, insoluble in water, protects metal
Copper	When heated forms a black coating, with dark pink coating underneath	No reaction	No reaction
Silver	No reaction	No reaction	No reaction
Gold	No reaction	No reaction	No reaction

** When the protective oxide coating is removed from aluminium metal, it reacts with oxygen in the air and the 'protective oxide coating' is replaced again. The melting point of aluminium is 660 °C. Once molten, aluminium will burn fiercly because the protective oxide layer escapes as smoke.*

The most active metal (sodium) is at the top of the table and the least active (gold) is at the bottom.

The reactions of metals with oxygen (air), water (or steam) and dilute hydrochloric acid

Unit 11.5 Activity 2A: The activity of metals

1. Explain the meaning of the terms:
   a. Chemical property.
   b. Activity series.
2. a. Name the metal(s) that must be kept under oil, and explain why this is required.
   b. Name the metal(s) that remain bright and shiny when in contact with air and explain why.
3. Arrange these metals in order of chemical activity, placing the most active metal first: gold, iron, lead, zinc.

Reaction of metals with oxygen

Metals will react with oxygen (oxidise) in air to varying degrees, ie

- Some metals (eg lithium, sodium) react violently in a **combustion** reaction, releasing heat and light energy.
- Some metals (eg gold) do not react with oxygen under any conditions.
- Some metals (eg aluminium, calcium) form a protective oxide layer on their surface. This oxide layer must be removed before the metal can react.

Example

Aluminium

When the protective surface oxide layer is removed from aluminium metal, it reacts with oxygen in the air, releasing large quantities of heat energy – the protective oxide 'coating' is replaced again, and the reaction ceases, unless the oxide layer is continuously removed.

The equation for the reaction is:

$4Al(s) + 3O_2(g) \rightarrow 2Al_2O_3(s)$

The experiments described below are not suitable for lithium and sodium because they melt too readily – and then they would react intensely.

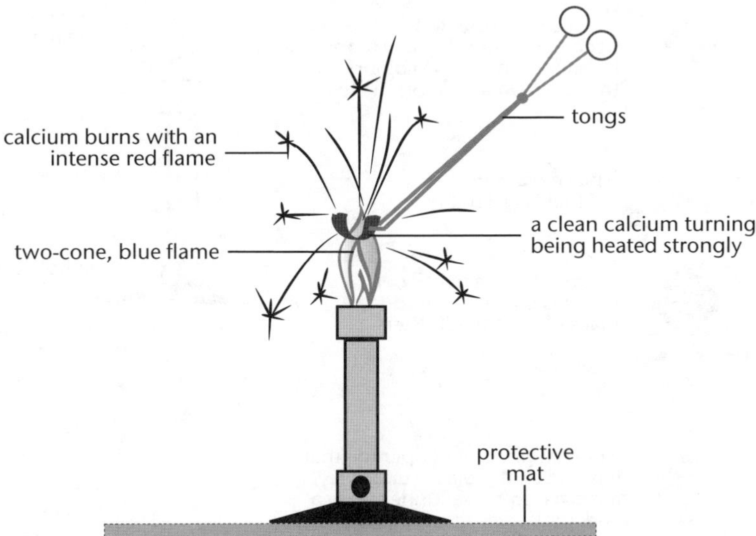

For calcium metal to oxidise, the oxide layer is first removed by heating the metal very strongly so that the metal melts - the molten metal will oxidise rapidly.

Calcium metal reacting with oxygen (air)

Copper

Copper metal **tarnishes** in air over a long period of time. Copper has a high melting point and will not melt in a Bunsen flame. When heated by a Bunsen flame, copper reacts with oxygen in air and undergoes the reactions shown by the following equations:

$2Cu(s) + O_2(g) \rightarrow 2CuO(s)$ (black copper(II) oxide)

$4Cu(s) + O_2(g) \rightarrow 2Cu_2O(s)$ (pink copper(I) oxide)

The changes that occur to a piece of tarnished copper that is cleaned and then carefully heated

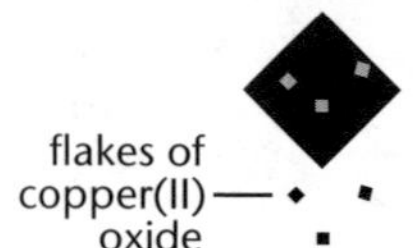

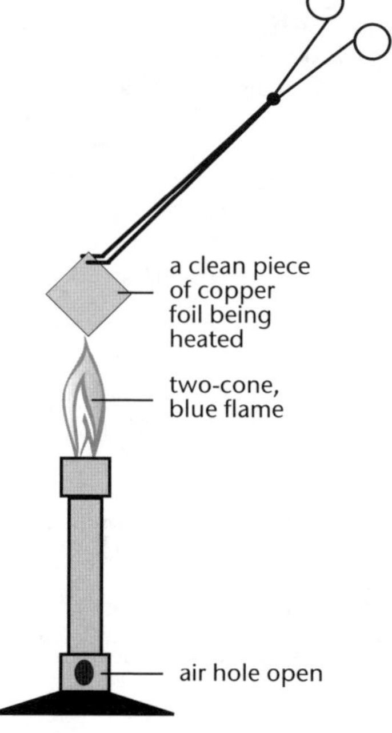

Copper metal reacting with oxygen (air)

Unit 11.5 Activity 2B: Metals reacting with oxygen

1. Name:

a. A metal that burns in air with a red flame.

b. A metal that reacts with oxygen to release a large quantity of heat, but the metal can be used for window frames, without protection from the air, and with no risk of the metal burning in the air.

c. A metal that is not silver-grey in colour and does not combine with oxygen in the air when heated.

2. Identify metal A and metal B from the following descriptions:

a. Metal A, when heated in air, burns vigorously.

b. Metal B was used for the decking of cruisers (medium-size boats used by the navy in combat) because of its low density and its resistance to atmospheric oxidation. When an Exocet missile landed on the deck of one of the cruisers, the explosion melted the metal, which then burnt extremely vigorously – the intensity of the fire was such that it was impossible to extinguish it. The cruiser was rendered ineffective.

Reaction of metals with water or steam

Lithium and sodium

Lithium and sodium react vigorously with cold water to produce lithium hydroxide solution (or sodium hydroxide solution) and hydrogen gas:

$2Li(s) + 2H_2O(\ell) \rightarrow 2LiOH(aq) + H_2(g)$

$2Na(s) + 2H_2O(\ell) \rightarrow 2NaOH(aq) + H_2(g)$.

Example

Lithium or sodium metal reacting with cold water

The experiment may be carried out by carefully placing a small piece of the metal (about the size of a rice grain) onto the surface of the water (about 75 mL of water in a 100 mL beaker). The metals are less dense than water and will float on the surface. Wear safety spectacles and stand well clear when the lithium or sodium metal is added to the water.

Calcium

Calcium metal reacts vigorously with cold water to produce calcium hydroxide and hydrogen gas:

$$Ca(s) + 2H_2O(\ell) \rightarrow Ca(OH)_2(aq) + H_2(g).$$

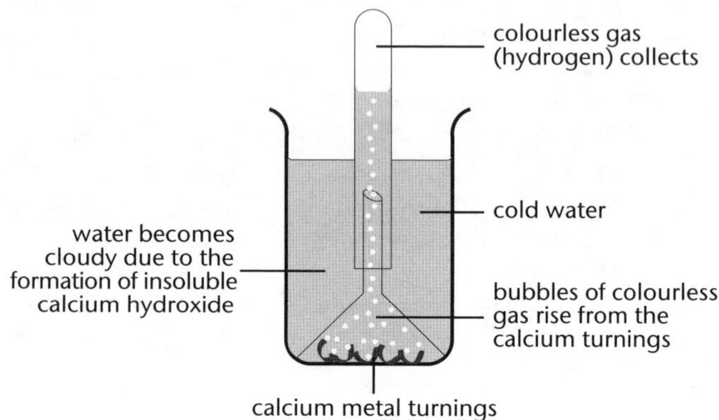

Calcium metal reacting with cold water

Magnesium

Magnesium metal reacts slowly with cold water to produce magnesium hydroxide and hydrogen gas:

$$Mg(s) + 2H_2O(\ell) \rightarrow Mg(OH)_2(aq) + H_2(g).$$

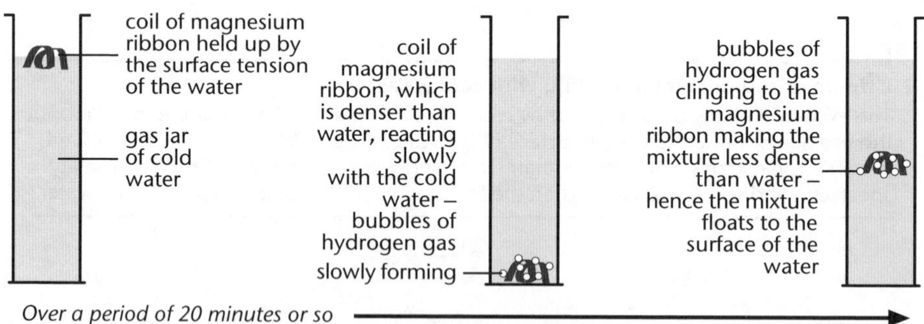

Magnesium metal reacting with cold water

Magnesium metal reacts vigorously with steam to produce magnesium oxide and hydrogen gas:

$$Mg(s) + H_2O(g) \rightarrow MgO(s) + H_2(g).$$

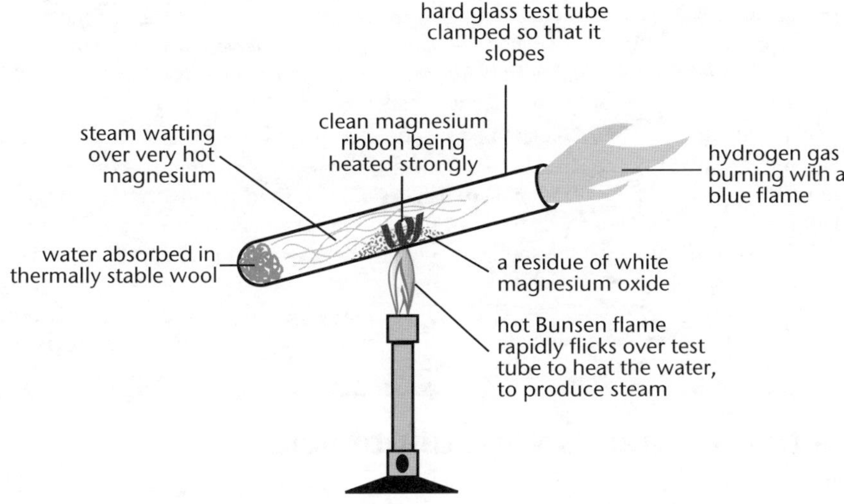

Magnesium metal reacting with steam

Other metals

- Iron, zinc (and aluminium) react with steam only when the metal is heated strongly to produce the oxide and hydrogen gas, ie:
  $3Fe(s) + 4H_2O(g) \rightarrow Fe_3O_4(s) + 4H_2(g)$,
  $Zn(s) + H_2O(g) \rightarrow ZnO(s) + H_2(g)$,
  $2Al(s) + 3H_2O(g) \rightarrow Al_2O_3(s) + 3H_2(g)$.
- Aluminium reacts with water/steam only when the oxide coating on the surface of the metal is removed.
- Lead, copper and gold do not react with water or steam at any temperature.

Unit 11.5 Activity 2C: Metals reacting with water or steam

1. Name:

   a. Three metals that will react with cold water.

   b. Three metals that will not react with cold water.

   c. One metal that will not react with cold water but will react with steam.

2. Here is a true story (mystery). *Hint*: White phosphorus – an allotrope of phosphorus – is spontaneously flammable in air and is kept out of contact with the air by storing it under water in a jar with a lid. Sodium is kept under oil to prevent it coming in contact with water.

A teacher was using metallic sodium, in the form of small lumps, in a class demonstration. When the students were dismissed at the end of the class, the teacher asked the technician to return the unused sodium (in a beaker under oil) to the store bottle in the preparation room. In the cupboard in the preparation room, the technician found a bottle with some small lumps under a colourless liquid. He poured the unused sodium and oil into the bottle. There was an immediate reaction involving flame, noise and effervescence. The technician quickly placed the bottle in the sink and turned the water tap on – there was an explosion. The technician was safe, but surprised and shocked.

Explain how the technician came to be surprised and shocked.

3. During World War II, incendiary (fire) bombs were used to set warehouses alight in London. These bombs were made of magnesium and were cylinders about 1 m in length with a diameter of about 5 cm. At the bottom end of the bomb was a detonator which, on impact with the ground, would ignite the magnesium cylinder. Explain why it was not possible to extinguish the bomb with water.
4. Draw an apparatus you could use to collect a test tube of hydrogen gas by the reaction of lithium with cold water. *Hint*: Lithium floats on water but it can be made to sink in a beaker of water if it is trapped inside a steel mesh envelope.

Reaction of metals with dilute acid

In general, metals react with acids to produce hydrogen gas and leave a **salt** in solution. The acids to be considered – along with sample equations that illustrate the reactions – are:

- Hydrochloric acid (HCl), eg
  $Li(s) + 2HCl(aq) \rightarrow LiCl_2(aq) + H_2(g)$.
- Sulfuric acid (H_2SO_4), eg
  $Zn(s) + H_2SO_4(aq) \rightarrow ZnSO_4(aq) + H_2(g)$.
- Ethanoic acid (CH_3COOH), eg
  $Mg(s) + 2CH_3COOH(aq) \rightarrow Mg(CH_3COO)_2(aq) + H_2(g)$.
- Nitric acid (HNO_3) – nitric acid is less stable than the other acids and decomposes to produce oxides of nitrogen (NO and NO_2) when it reacts with metals (rather than produce hydrogen gas). An exception is the reaction between magnesium metal and dilute (~1 mol L^{-1}) nitric acid to produce hydrogen gas:
  $Mg(s) + 2HNO_3(aq) \rightarrow Mg(NO_3)_2(aq) + H_2(g)$.

	Acids (All the acids are dilute – their concentrations are 1 mol L^{-1}.)			
Metal	**Hydrochloric, HCl(aq)**	**Nitric, HNO_3(aq)**	**Sulfuric, H_2SO_4(aq)**	**Ethanoic, CH_3COOH(aq)**
Sodium	Violent reaction; hydrogen released; sodium chloride solution remains	Violent reaction; oxides of nitrogen released; sodium nitrate solution remains	Violent reaction; hydrogen released; sodium sulfate solution remains	Vigorous reaction; hydrogen released; sodium ethanoate solution remains
Lithium	Violent reaction; hydrogen released; lithium chloride solution remains	Violent reaction; oxides of nitrogen released; lithium nitrate solution remains	Violent reaction; hydrogen released; lithium sulfate solution remains	Vigorous reaction; hydrogen released; lithium ethanoate solution remains

Acids (All the acids are dilute – their concentrations are 1 mol L^{-1}.)				
Metal	**Hydrochloric, HCl(aq)**	**Nitric, HNO_3(aq)**	**Sulfuric, H_2SO_4(aq)**	**Ethanoic, CH_3COOH(aq)**
Calcium	Vigorous reaction; hydrogen released; calcium chloride solution remains	Vigorous reaction; oxides of nitrogen released; calcium nitrate solution remains	Vigorous reaction; hydrogen released; reaction stops due to insoluble calcium sulfate being formed	Brisk reaction; hydrogen released; calcium ethanoate solution remains
Magnesium	Vigorous reaction; hydrogen released; magnesium chloride solution remains	Vigorous reaction; hydrogen released; magnesium nitrate solution remains	Vigorous reaction; hydrogen released; magnesium sulfate solution remains	Brisk reaction; hydrogen released; magnesium ethanoate solution remains
Aluminium	Delayed reaction; hydrogen gas and aluminium chloride solution produced	No reaction; aluminium oxide layer prevents acid and metal reacting	No reaction; aluminium oxide layer prevents acid and metal reacting	No reaction; aluminium oxide layer prevents acid and metal reacting
Zinc	Moderate reaction; hydrogen released; zinc chloride solution remains	Moderate reaction; oxides of nitrogen released; zinc nitrate solution remains	Moderate reaction; hydrogen released; zinc sulfate solution remains	Slow reaction; hydrogen released; zinc ethanoate solution remains
Iron	Slow reaction; hydrogen released; iron(II) chloride solution remains	No apparent reaction; iron oxide (Fe_3O_4) layer prevents acid and metal reacting	Slow reaction; hydrogen released; iron(II) sulfate solution remains	Very slow reaction; hydrogen released; iron(II) ethanoate solution remains
Lead	No visible reaction; insoluble lead chloride prevents very slow reaction	Moderate reaction; oxides of nitrogen released; lead nitrate solution remains	No visible reaction; insoluble lead sulfate prevents very slow reaction	No noticeable change; extremely slow reaction
Copper	No reaction	Moderate reaction; oxides of nitrogen released; copper nitrate solution remains	No reaction	No reaction
Silver	No reaction	Moderate reaction; oxides of nitrogen released; silver nitrate solution remains	No reaction	No reaction
Gold	No reaction	No reaction	No reaction	No reaction

Although the four acids are of the same concentration, ethanoic acid is less acidic and its reactions with metals are slower than those of the other acids.

Reactivity of metals with acids

Unit 11.5 Activity 2D: Metals reacting with acids

1. Name a metal, present as a small lump, that will:
   - **a.** React with excess dilute hydrochloric acid in a non-violent manner immediately it is added to the acid and will continue to react until it is all dissolved.
   - **b.** React with excess dilute hydrochloric acid immediately it is added to the acid but the reaction will stop quickly.
   - **c.** React with excess dilute hydrochloric acid, but with a delay, and then continue to react until it is all dissolved.
   - **d.** Not react with dilute hydrochloric acid.
2. **a.** State which of the metals – aluminium, zinc, lead, copper – would react with dilute hydrochloric acid in the following apparatus to produce hydrogen gas.

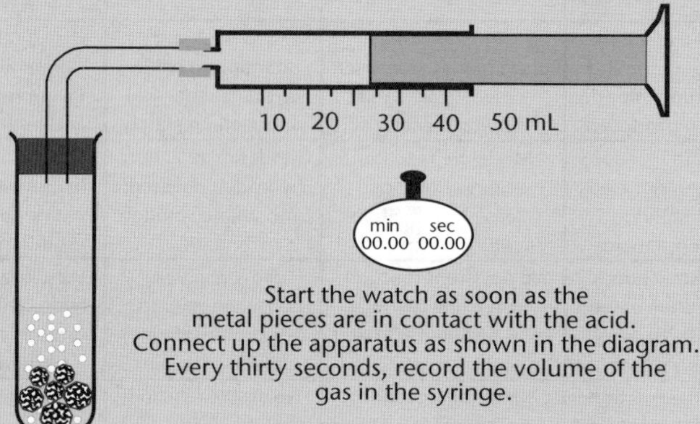

   - **b.** Explain why the other metals would not react with dilute hydrochloric acid.
3. Name a metal that will:
   - **a.** Burn in air, react slowly with cold water, and react vigorously with dilute hydrochloric acid.
   - **b.** Not react with any of oxygen, water (or steam), dilute hydrochloric acid.
   - **c.** React vigorously with oxygen, steam and dilute hydrochloric acid if the oxide layer is removed from the metal's surface.
4. Explain the following facts:
   - **a.** Lithium is stored under oil.
   - **b.** Aluminium window frames do not need to be protected from the atmosphere.
   - **c.** Copper tubing is used in hot water systems for domestic use.

Displacement of metals from solution

The activity series for metals can be confirmed by testing the proposal that 'a more reactive metal will displace a less active metal from solution'.

Example

Displacement of lead and copper from solutions

The more active zinc will displace the less active lead from solution. Similarly, the more active iron will displace the less active copper.

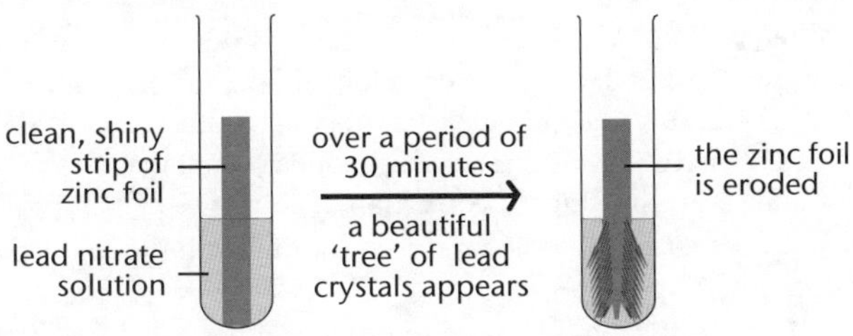

Zinc metal displaces lead from lead nitrate solution

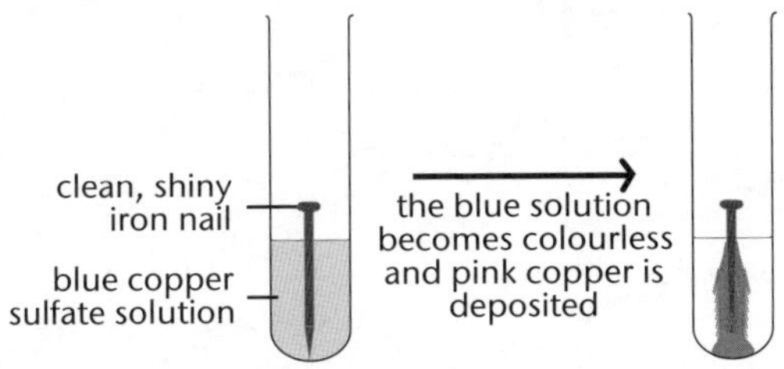

Iron metal displaces copper from copper sulfate solution

The equations can be written in either:

- Molecular form, ie
  $Zn(s) + Pb(NO_3)_2(aq) \rightarrow Pb(s) + Zn(NO_3)_2(aq)$,
  $Fe(s) + CuSO_4(aq) \rightarrow Cu(s) + FeSO_4(aq)$.

or

- Ion form – leaving out the spectator ions (nitrate or sulfate), ie
  $Zn(s) + Pb^{2+}(aq) \rightarrow Pb(s) + Zn^{2+}(aq)$,
  $Fe(s) + Cu^{2+}(aq) \rightarrow Cu(s) + Fe^{2+}(aq)$.

Example

Displacement of hydrogen gas

Sodium, lithium and calcium metals, when added to an aqueous solution of a metal salt of a lower activity metal, will displace hydrogen in preference to the metal. The reaction is:

$2Li(s) + 2H^+(aq) \rightarrow 2Li^+(aq) + H_2(g)$,

and *not*:

$2Li(s) + Zn^{2+}(aq) \rightarrow 2Li^+(aq) + Zn(s)$.

Unit 11.5 Activity 2E: Displacement of metals

1. Name a metal that will displace zinc from zinc nitrate solution.

2. Describe the observation you would make if a strip of clean, shiny zinc were placed in:

  a. Magnesium nitrate solution.

  b. Copper nitrate solution.

3. a. If you were supplied with a piece of lead foil, approximately 5 cm × 2 cm in size, copper nitrate solution and suitable laboratory apparatus, describe an experiment you could perform to produce a sample of copper metal that was pure, clean and dry. Draw a diagram showing the formation of copper to help your description.

  b. State what other metals could be produced starting with the lead foil, by a similar experiment.

4. Explain why lithium metal cannot be used to produce other metals by displacement of the metal from solution.

Unit 11.5 Metals and Non-metals

Topic 3: Extraction of metals and corrosion

In the previous two Topics we looked at the physical and chemical properties of metals. In Topic 3 we now examine how metals are extracted from the raw ore that is dug from the earth. Considering the importance of mining in PNG, this Topic is of particular relevance. It deals with:

- Reactivity of metals related to their extraction from ores.
- Methods of extraction used to obtain metals.
- Corrosion of metals.

Extraction of metals from their ores

Metals were discovered in reverse order of their chemical activity:

- The least active metals (ie gold and silver) were discovered first.
- The less active metals (ie iron, zinc, lead, copper) were discovered next.
- The highly active metals (ie sodium, lithium, calcium, magnesium, aluminium) were discovered later.

Extraction of metals from metal oxides

The less active metals either have **minerals** that are oxides, eg iron, or the oxide can be readily obtained from the mineral (often the metal sulfide), eg zinc, lead and copper. The metal is extracted by thermal reduction of the metal oxide using carbon (in the form of carbon/**coke**) as the reducing agent.

Iron

Iron oxide, present as the mineral **titanomagnetite** ($Fe_3O_4.TiO_2$), occurs in abundance in New Zealand in the **iron sand** on the west coast of the North Island.

At the local steel works, a rotary kiln is used to produce iron from iron sand. (The titanium oxide (TiO_2) in titanomagnetite is not affected by the thermal reduction process that produces iron – titanium is too reactive a metal.)

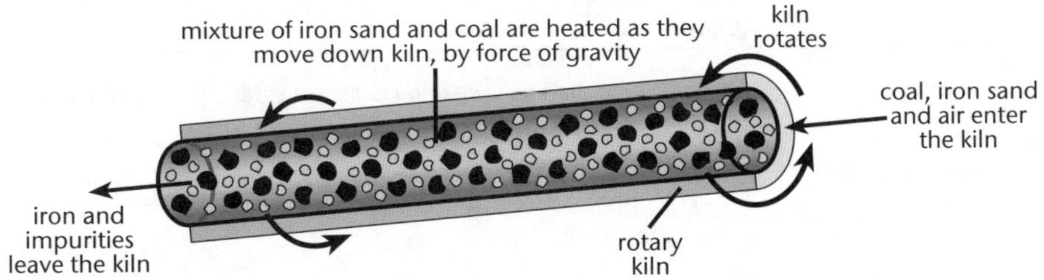

Commercial production of iron by reduction of iron oxide

The reactions occurring in the rotary kiln are:

- Oxygen, from the air, combines with the carbon in the coal to produce carbon dioxide and heat:
  $C(s) + O_2(g) \rightarrow CO_2(g)$.
- Carbon dioxide is reduced by heated carbon (coal) to produce carbon monoxide:
  $CO_2(g) + C(s) \rightarrow 2CO(g)$.
- The iron oxide is reduced by hot carbon and/or carbon monoxide to produce iron:
  $Fe_3O_4(s) + 2C(s) \rightarrow 3Fe(s) + 2CO_2(g)$ and,
  $Fe_3O_4(s) + 4CO(g) \rightarrow 3Fe(s) + 4CO_2(g)$.

The iron is then converted into steel, and the impurities (slag) are stockpiled as a source of titanium.

Zinc, lead and copper

Zinc, lead and copper metals can be obtained from their minerals by thermal reduction in a two-stage process.

Stage 1 – the mineral is roasted in air to produce the metal oxide.		
Metal	**Mineral**	**Reactions**
Zinc	Zinc blende, ZnS	$2ZnS(s) + 3O_2(g) \rightarrow 2ZnO(s) + 2SO_2(g)$
Lead	Galena, PbS	$2PbS(s) + 3O_2(g) \rightarrow 2PbO(s) + 2SO_2(g)$
Copper	Malachite, $CuCO_3.Cu(OH)_2$	$CuCO_3.Cu(OH)_2 \rightarrow 2CuO(s) + CO_2(g) + H_2O(\ell)$

Stage 2 – the metal oxide is reduced by carbon (in the form of coke), or by carbon monoxide (CO), to the metal.	
Metal oxide	**Reactions**
Zinc oxide	$ZnO(s) + C(s) \rightarrow Zn(s) + CO(g)$, or $ZnO(s) + CO(g) \rightarrow Zn(s) + CO_2(g)$.
Lead oxide	$PbO(s) + C(s) \rightarrow Pb(s) + CO(g)$, or $PbO(s) + CO(g) \rightarrow Pb(s) + CO_2(g)$.
Copper oxide	$CuO(s) + C(s) \rightarrow Cu(s) + CO(g)$, or $CuO(s) + CO(g) \rightarrow Cu(s) + CO_2(g)$.

The sulfur dioxide formed in the Stage 1 reaction for zinc and lead is converted to sulfuric acid and sold as a by-product.

The extraction of zinc, lead and copper from their ores by thermal reduction

Extraction of metals by electrolysis

The high chemical activity of sodium, lithium, calcium, magnesium and aluminium metals makes them difficult to separate from other element(s) in the mineral. As a result, these metals have been discovered only in the last two hundred years. The process of **electrolysis** – using electrical energy to produce chemicals by passing an electric current through an electrolyte – provides the high energy required to extract the metal in the mineral.

The metals sodium, lithium, calcium and magnesium are obtained commercially by electrolysis of their molten metal chloride (the electrolyte). Magnesium production is by far the largest of these metals.

Extraction of magnesium

Sea water contains a large number of dissolved minerals, including magnesium as magnesium chloride. The magnesium is separated from the other minerals by being precipitated as magnesium hydroxide. The magnesium hydroxide is converted into magnesium chloride, which is then electrolysed.

Extraction of aluminium

Aluminium occurs as a mineral, **bauxite** – a mixed oxide of aluminium and iron. The mineral is mined and then purified to produce **alumina**, Al_2O_3. The alumina is then converted to aluminium metal by electrolysis:

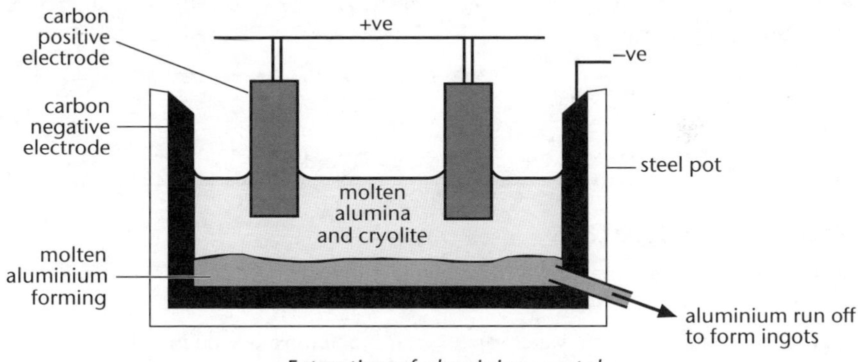

Extraction of aluminium metal

The process of electrolysis is summarised in the equation:

$2Al_2O_3(s) \rightarrow 4Al(s) + 3O_2(g)$.

The oxygen released causes the carbon positive electrodes to burn, so they then need to be replaced. The **cryolite** is present to lower the working temperature of the cell to 1000 °C (the melting point of aluminium oxide is 2072 °C).

Unit 11.5 Activity 3A: Extraction of metals

1. State the name of a metal that occurs:
   - **a.** As the element.
   - **b.** As a metal oxide.
   - **c.** As a metal sulfide.
2. State the name of a mineral that contains:
   - **a.** Aluminium.
   - **b.** Zinc.
3. Explain the meaning of the terms:
   - **a.** Thermal reduction.
   - **b.** Electrolysis.
4. Explain why copper metal can be obtained commercially by thermal reduction, whereas aluminium metal can only be obtained commercially by electrolysis.
5. Complete the following equations:
   - **a.** $C(s) + O_2(g) \rightarrow$ ____
   - **b.** ____ $+ CO_2(g) \rightarrow$ __$CO(g)$
   - **c.** $Fe_3O_4(s) +$ __$CO(g) \rightarrow$ __$Fe(s) +$ ____
   - **d.** $2PbS(s) +$ ____ $\rightarrow 2PbO(s) + 2SO_2(g)$
   - **e.** $ZnO(s) + C(s) \rightarrow$ ____ + ____
   - **f.** __$Al_2O_3(s) \rightarrow 4Al(s) +$ ____
6. **a.** Arrange the following metals in their order of discovery. Place the earliest discovery at the top of the list.
   - **i.** Aluminium.
   - **ii.** Copper.
   - **iii.** Gold.

   b. Explain why the metals were discovered in this order.

Corrosion of metals

Corrosion is a chemical process in which a metal is either 'eaten away' or its surface is changed in appearance.

The atmosphere contains oxygen, water vapour and carbon dioxide, and in industrial areas, other acidic gases (eg sulfur dioxide and nitrogen dioxide). All these chemical agents can cause metals to corrode.

Sodium, lithium and calcium

The very active metals – sodium, lithium, calcium – react so rapidly with the chemicals in the atmosphere that they are of little commercial use except as laboratory or industrial chemicals.

Example

Sodium, lithium and calcium are kept under oil to protect them from atmospheric attack.

Magnesium

The surface of magnesium is rapidly oxidised in the atmosphere to magnesium oxide, which develops into magnesium hydroxide and magnesium carbonate, and forms a protection for the metal. These compounds of magnesium protect the metal from further corrosion but make the appearance of the metal dull.

Aluminium

Aluminium metal reacts vigorously with oxygen in the air to form a strong, thin, coating of aluminium oxide on its surface. This oxide film protects the aluminium from further attack by oxygen, water and dilute acids present in air – thus, aluminium metal appears not to corrode in air. Aluminium metal does not need further protection from the chemicals in air.

Example

Aluminium window frames are often coloured for cosmetic effect, and not for protection.

Iron

Corrosion of iron – the rusting process

The commonest example of corrosion is the rusting of iron metal – the term **rust** applies only to the corrosion product(s) of the metal iron.

The corrosion of iron is dependent upon the activity of the metal and the fact that the oxide formed, iron(III) oxide, does not bond with the remaining iron underneath – the rust flakes off. Rust is hydrated iron(III) oxide. Iron will rust when in the presence of oxygen (air), water vapour *and* carbon dioxide – it will *not* rust if any *one* of these three factors (oxygen (air), water vapour or carbon dioxide) is excluded.

The rusting process can be prevented by protecting the iron from the components of air by:

- Coating the iron with zinc metal (**galvanising**), paint or tin.
- Adding to the iron a high percentage (about 25%) of chromium and nickel to produce stainless **steel**.
- Placing a thin film of oil or grease on an iron tool when the tool is not in use.

Copper

Copper is slowly corroded by the chemicals of the atmosphere to produce the oxide, the hydroxide and, finally, a green insoluble coating of basic copper carbonate.

Example

Copper roofs on European monasteries and cathedrals may be four hundred years old and still in good condition due to the insoluble copper carbonate coating protecting the copper underneath.

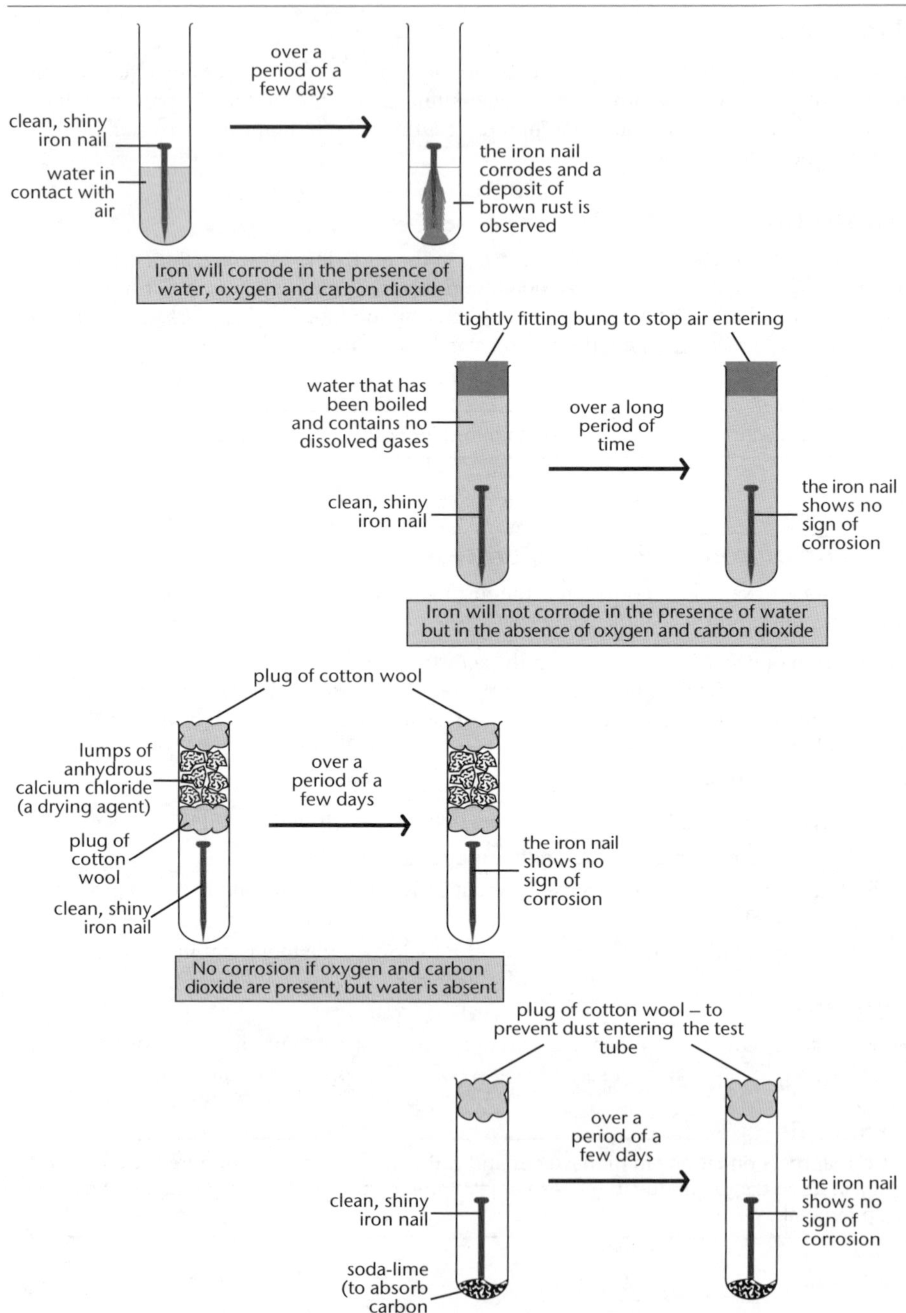

The conditions necessary for iron to corrode (rust)

Zinc and lead

Zinc and lead react with the components of air – oxygen, water and carbon dioxide – to build up a layer of metal carbonate and hydroxide which protects the metal from further corrosion. Thus, these metals do not appear to corrode in air, except that they become dull in appearance. Neither of these metals needs further protection from the chemicals in air.

Example

Zinc is alloyed with copper to form brass, which is resistant to corrosion and retains an attractive appearance.

Gold and silver

Gold has always been valued, partly because of its scarcity, but also for its yellow shiny appearance, which is not affected in any way by the chemicals in air over very long periods.

Example

When discovered, the tomb of Tutankhamen – who died in 1352 BC (3500 years ago) – contained a large quantity of shiny gold.

Example

Gold is sometimes used for teeth repair because of its appearance and its resistance to corrosion.

Silver can tarnish, especially in regions where sulfur is present – the surface of the metal becomes black silver sulfide. The high lustre of silver can be easily restored by removing the surface tarnishing, either by chemical means or by abrasion.

Unit 11.5 Activity 3B: Corrosion of metals

1. Identify the chemicals in air that can cause corrosion.

2. Identify which of the following sets of conditions would allow a shiny iron nail to rust:
- **a.** In a beaker of tap water.
- **b.** Vacuum-packed in a plastic envelope.
- **c.** Left on the moon by a visiting astronaut.
- **d.** Dropped into a shallow swimming pool.
- **e.** At the bottom of the deep blue sea.

3. Identify the change of appearance that would be noticed if each of the following metals in bright, shiny form were left in the presence of air for a year or more. For each metal, explain what new substance(s) is/are formed.
- **a.** Gold.
- **b.** Aluminium.
- **c.** Lead.
- **d.** Copper.

4. State a use for each of the following metals that involves the metal being in contact with the air. Describe the changes in appearance that would occur to the metal as it remained in contact with air for a long time (several years). Explain what chemicals are causing these changes and what new chemicals are being formed.

a. Magnesium.

b. Zinc.

c. Silver.

5. A can that contains lemonade, or some other beverage, is made from aluminium although it could be made from steel. State two reasons why aluminium is preferred to steel.

Unit 11.5 Metals and Non-metals

Topic 4: Properties, uses and production of non-metals

In the first three Topics in this Unit we looked at the properties, uses and production of metals. Now we turn our attention to non-metals and their compounds. Topic 4 deals with:

- Physical properties of selected non-metals – oxygen, chlorine, nitrogen and sulfur.
- Allotropes of oxygen and sulfur.
- Occurrence, reactions and commercial preparation of chlorine, oxygen, nitrogen and sulfur.
- Uses of chlorine and sulfur.

Physical properties

Unlike the metals, there are no clear generalisations to be made concerning the physical properties of non-metals.

Element	Physical state at room temperature	Appearance and odour	Solubility in water
Oxygen	Gas	Colourless; no smell	Very low
Chlorine	Gas	Green-yellow; strong smell	Low
Nitrogen	Gas	Colourless; no smell	Very low
Sulfur	Solid	Yellow and brittle; no smell	Insoluble

Physical properties of non-metals

Solubility

Non-metals either have low solubility in water, or are regarded as insoluble in water. The following features are of interest:

- The solubility of gases in water (the major component of blood) is much greater under high pressure.

Example

Divers who ascend too quickly to the surface from deep water (where the pressure is much greater than in the atmosphere) risk the **bends** – severe damage, or even death, caused by dissolved nitrogen gas rapidly bubbling out of the blood.

- Chlorine gas dissolves sufficiently in water to destroy bacteria that are harmful to human beings. The concentration of the gas does not make the water harmful to human beings or unpleasant to taste.
- Oxygen gas dissolves sufficiently in water to allow fish to breathe from the dissolved oxygen.

Unit 11.5 Activity 4A: Physical properties of non-metals

1. Describe the appearance of the following elements at room temperature (20 °C):

a. Chlorine.

b. Oxygen.

c. Nitrogen.

d. Sulfur.

2. a. State one physical property of each of the following elements that makes the element *different* from the other two elements:

i. Chlorine.

ii. Nitrogen.

iii. Sulfur.

b. State one physical property of each of the following elements that is the *same* or *similar* to one or more properties of the other three elements:

i. Chlorine.

ii. Oxygen.

iii. Nitrogen.

iv. Sulfur.

3. Match each of the following elements with one appropriate feature.

Element	Feature
a. Nitrogen	i. To make water fit for human consumption.
b. Oxygen	ii. Causes a problem for deep-sea divers when returning to the surface at speed.
c. Chlorine	iii. Fish can breathe underwater due to the (low) solubility of the gas in water.

4. Explain the meaning of each of the following terms:

a. Brittle.

b. Insoluble.

c. Shiny.

d. Appearance.

Allotropes

Allotropy is the existence of two or more forms of the same element in the same physical state. Each form of the element is termed an **allotrope**.

Allotropes of oxygen

Oxygen is a diatomic gas, ie the element oxygen is a gas at room temperature (20 °C) and the particles present are molecules containing two atoms. The molecule of oxygen is represented by the formula O_2.

When high energy, in the form of an electric spark, is passed through oxygen gas, some diatomic oxygen molecules are converted into triatomic molecules of **ozone**, ie:

$3O_2 \rightarrow 2O_3$.

The percentage conversion of oxygen to ozone is very low. Ozone – a colourless gas with a strong smell – is an allotrope of oxygen.

Example

Ozone's distinctive smell can be detected at beaches where it is formed by oxygen reacting with some seaweeds.

Ozonised oxygen is used to freshen air in enclosed areas where there is a high density of human beings, eg underground railway stations. The ozone destroys bacteria and decomposes to produce oxygen.

Allotropes of sulfur

There are three allotropes of sulfur:

- Alpha or **rhombic** sulfur is a pale-yellow crystalline solid.

Examples

Rhombic is a crystal form having three unequal axes – the crystal looks like a cube that has been squashed unevenly in all three dimensions.

- Beta or **monoclinic** sulfur is a pale-yellow crystalline solid.

Example

Monoclinic is a crystal form having three unequal axes, one pair of which are not at right angles to each other – the crystal is commonly seen as a needle with a cross-section that is a parallelogram.

- **Plastic sulfur** is a black, elastic, solid that has brief (one hour or so) existence.

The crystalline forms of sulfur are brittle so sulfur is commonly seen as a yellow powder or in yellow lumps (rather than as crystals).

Sulfur undergoes some unusual changes when it is heated beyond its melting point and up to its boiling point.

Observation | **Explanation** | **Structure**

Powdered sulfur melts at 119 °C and becomes a pale yellow, runny liquid.

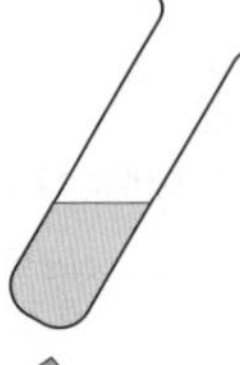

The S_8 molecules have the atoms joined in a puckered ring. At the melting point of sulfur, the molecules break away and are able to move, creating the liquid state.

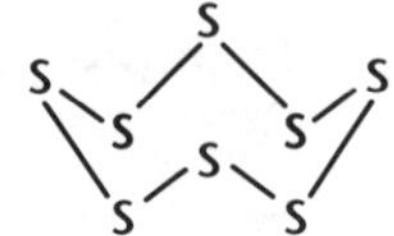

The runny yellow liquid darkens to amber and thickens slightly – 119 °C to 180 °C.

The S_8 molecules move faster, as the temperature rises, and the rings start to break into S_2, S_4, etc fragments.

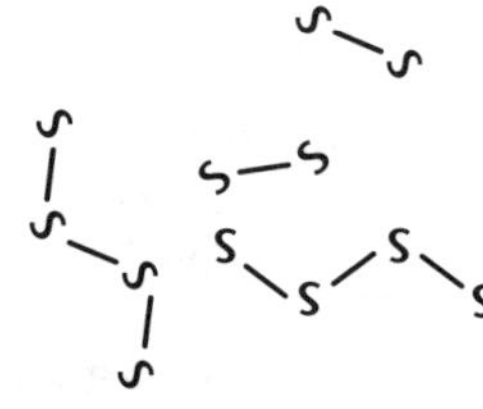

Suddenly, at 180 °C, the dark red liquid becomes so thick that the test tube can be turned upside down.

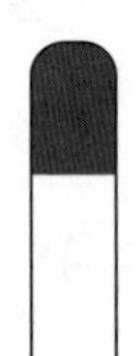

The broken S_8 molecules join up to form very long chains which interlock and move very slowly.

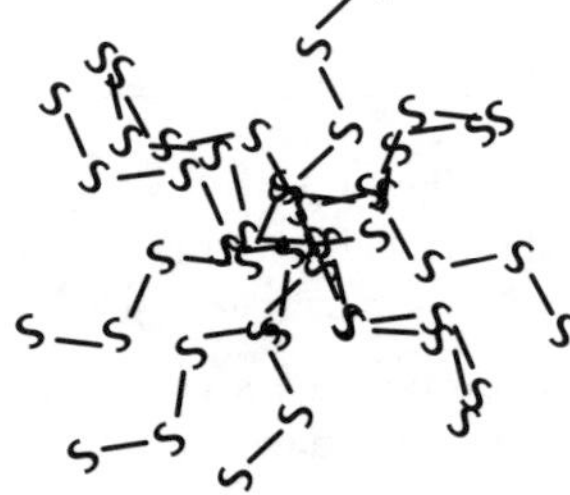

Sulfur boils at 444.4 °C and is a black, moderately runny, liquid.

The long chains have broken to form smaller chains and these particles now have enough energy to leave the liquid state – the sulfur is boiling.

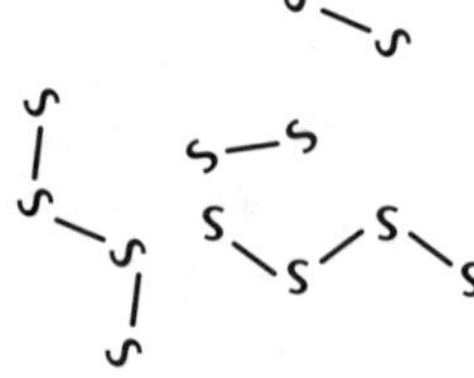

Changes occurring to sulfur as it is heated to boiling point

If sulfur at its boiling point is rapidly cooled by pouring it into a beaker of cold water, the changes illustrated in the preceding diagram are reversed but the process gets 'stuck' at the long-chain stage. The product – **plastic sulfur** – is a soft, pliable, almost-black solid that can be moulded, stretched and made into any shape you wish.

Example

The long chains of plastic sulfur can be stretched and they will return to their former position – hence the plastic nature of the solid.

Plastic sulfur is a temporary form of sulfur. Over a period of an hour or so, the long chains break and the S_8 ring molecules re-form. The sulfur is back to where it started.

Example

If you keep a piece of plastic sulfur in your pocket for an hour or so, you will find that it will lose all its plasticity and revert to a yellow brittle solid – just like ordinary sulfur lumps.

Laboratory preparation of sulfur crystals

Sulfur can be produced in the laboratory in the two different crystalline forms (rhombic and monoclinic), where the particles, S_8 molecules, are arranged in a regular pattern in three dimensions.

Rhombic sulfur

Rhombic crystals of sulfur may be prepared in the laboratory as follows:

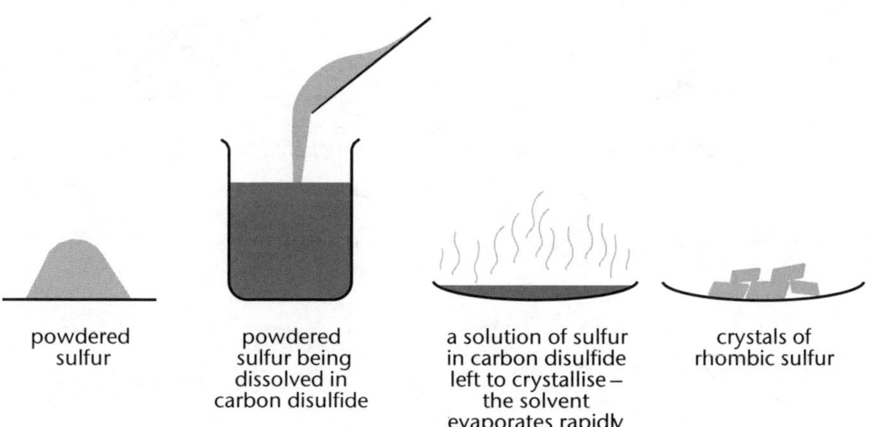

Carbon disulfide vapour is harmful if inhaled and this experiment should be carried out in a fume cupboard.

Laboratory preparation of crystals of rhombic sulfur

Monoclinic sulfur

There are two methods of making monoclinic crystals of sulfur.

- *Method 1* relies on the fact that when liquid sulfur cools and solidifies, pressure (caused by the solid enclosing some remaining liquid) increases, and under these conditions, monoclinic crystals are formed.

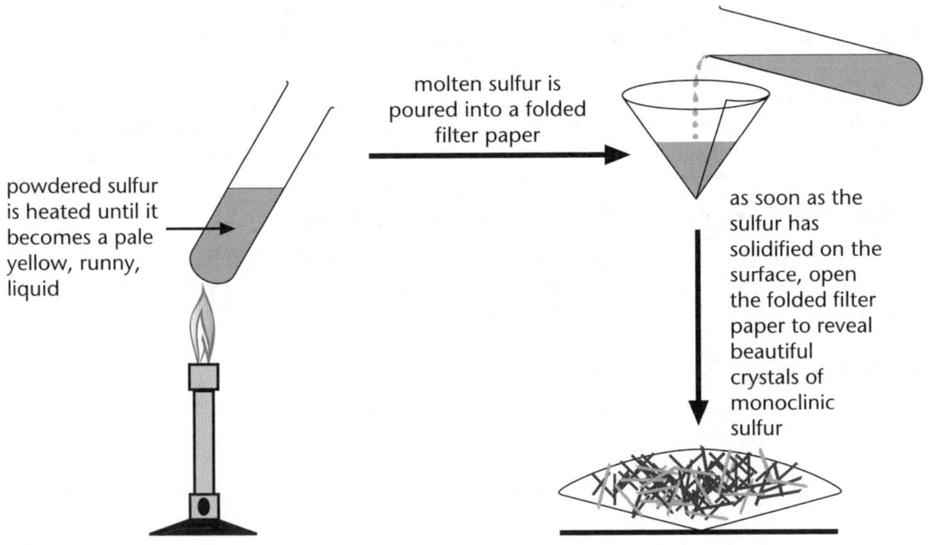

Laboratory preparation of crystals of monoclinic sulfur – method 1

- *Method 2* relies on the fact that toluene will dissolve sulfur. When the hot solution cools, the crystals first appear at a temperature at which monoclinic sulfur is the stable allotrope.

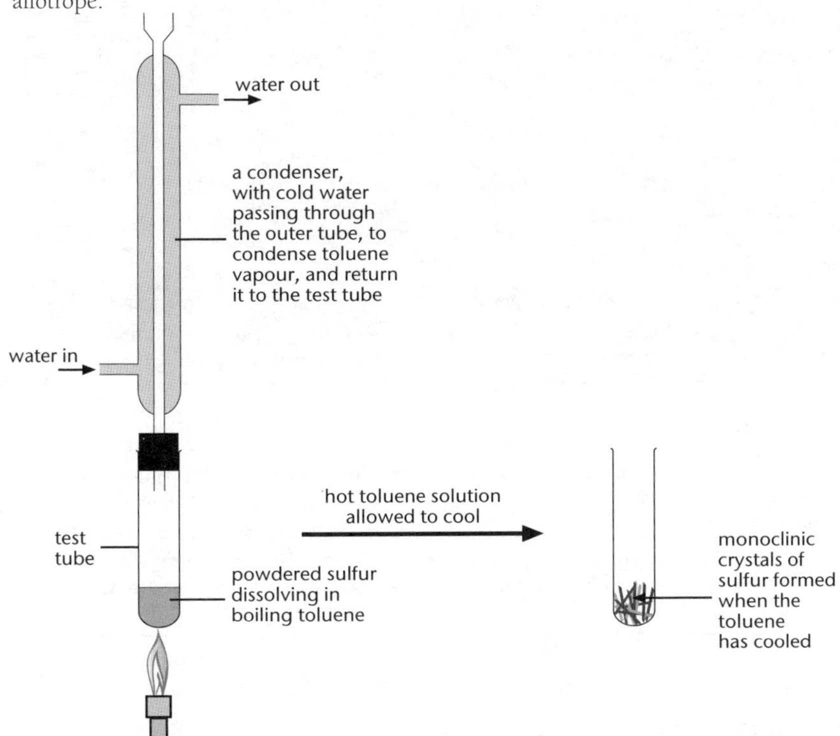

Toluene vapour is harmful if inhaled and this experiment should be carried out in a fume cupboard.

Laboratory preparation of crystals of monoclinic sulfur – method 2

Once formed, monoclinic crystals are stable over a long period – several days at least.

Unit 11.5 Activity 4B: Allotropes of oxygen and sulfur

1. Describe the appearance of:

a. Ozone.

b. Rhombic sulfur.

c. Monoclinic sulfur.

d. Plastic sulfur.

2. State and explain one use of ozone.

3. Assume room temperature is 20 °C. Sulfur has the following properties: yellow solid, mp 119 °C; yellow liquid above 119 °C, which, on heating, becomes red and, at a dark-red stage (180 °C) becomes very thick; beyond the very thick stage, the colour darkens to almost black; sulfur boils at 444 °C. Describe the appearance of sulfur at:

a. 15 °C.

b. 100 °C.

c. 181 °C.

d. 440 °C.

4. Identify the following elements:

 a. A rhomboid-shaped (parallel sides of unequal length) crystal, 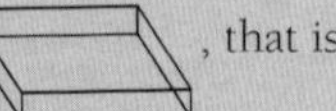, that is yellow in colour and very brittle.

 b. A gas that has a strong smell. It is produced in very low concentrations from oxygen in the air.

5. Explain why:

 a. Rhombic sulfur crystals are often seen as a yellow powder.

 b. Ozone is an allotrope of oxygen.

 c. Plastic sulfur is rubbery in texture.

6. The terms allotrope (a substance) and isotope (a particle) are often confused. Define both of the terms 'allotrope' and 'isotope' and give examples of each.

Chlorine

Chlorine is abundant in the form of sodium chloride and other metallic chlorides.

Example

Every 100 g (half cup) of sea water contains 3 g of common salt. The total amount of salt in the seas is estimated at 300 000 trillion tonnes. Some seas have dried up to produce salt flats.

Chlorine gas is a poisonous gas that attacks the lungs.

Example

Chlorine gas was used in World War I by the Germans to poison Allied troops.

Reactions of chlorine

Chemically, chlorine gas is very reactive.

- Chlorine combines directly with many elements to produce chlorides. The more active metals (eg sodium, lithium and calcium) combine very vigorously with the gas, without heating, but less reactive metals (eg zinc, iron, lead and copper) need to be heated to produce a vigorous reaction.
- When a mixture of chlorine and hydrogen gases is heated, an explosive reaction occurs to produce hydrogen chloride gas:
 $H_2(g) + Cl_2(g) \rightarrow 2HCl(g)$.

Hydrogen chloride gas is highly soluble in water and forms the solution hydrochloric acid:

$HCl(g) + H_2O(\ell) \rightarrow HCl(aq)$.

- Chlorine combines directly with other non-metals – some of the most useful products are phosphorus trichloride, phosphorus pentachloride and silicon tetrachloride.
- The reaction of chlorine with oxygen requires extreme conditions and produces a very unstable chlorine monoxide.

Uses of chlorine

Chlorine gas is used:

- To destroy bacteria.

Example

Chlorine gas is added to drinking water and to swimming pools.

Most of the chlorine ***dissolves*** in water physically.

Example

You can smell – and even taste – chlorine in swimming pool water.

Some of the chlorine ***reacts*** with water to produce a mixture of hydrochloric (HCl) and hypochlorous acids (HOCl):

$Cl_2(g) + H_2O(\ell) \rightarrow HCl(aq) + HOCl(aq)$.

The hypochlorous acid ionises to produce the hydrogen ion, H^+, and the hypochlorite ion, OCl^-. It is the hypochlorite ion that can kill bacteria.

- In the pulp and paper industry to bleach wood pulp before it is used to make paper.
- In making the **monomer** of the plastic PVC (polyvinylchloride).

Example

Environmentally, there is a problem of disposal of PVC – burning PVC produces corrosive hydrogen chloride gas.

Commercial preparation of chlorine

The high energy required to obtain chlorine from its compounds is provided by electricity in a process called electrolysis. In electrolysis, electricity is passed through an electrolyte, which undergoes chemical change. Electrolytes must contain ions and are either:

- molten ionic substances, or
- aqueous solutions containing ions.

Chlorine gas is produced in a membrane cell by the electrolysis of brine (sodium chloride solution). The membrane allows sodium ions to pass through, but not chloride ions. Chlorine gas is produced at the titanium anode and hydrogen gas at the nickel **cathode**. During the electrolysis, the electrolyte (brine) becomes sodium hydroxide solution. Sodium hydroxide and hydrogen gas are valuable by-products of the manufacture of chlorine gas.

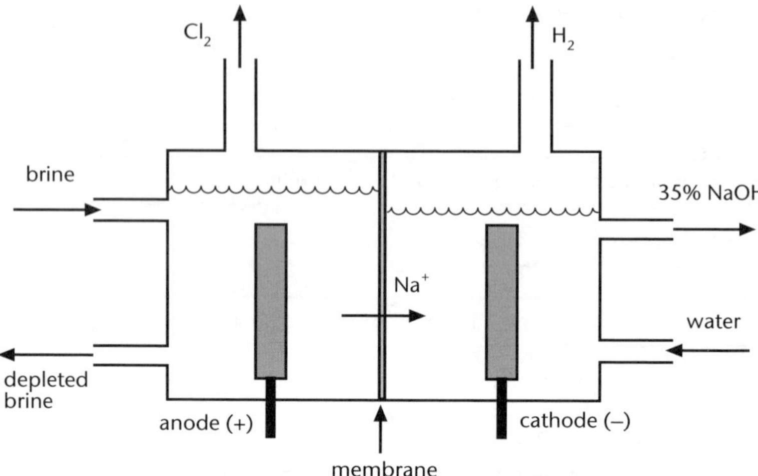

Membrane cell for electrolysis of sodium chloride solution

Unit 11.5 Activity 4C: Reactions, uses and preparation of chlorine

1. **a.** Name a source of chlorine of commercial value.
   b. Identify which of the natural substances listed in **i.** to **iv.** contains chlorine.
      i. Gypsum (calcium sulfate dihydrate).
      ii. Marble (calcium carbonate).
      iii. Carnallite (potassium magnesium chloride).
      iv. Chile saltpetre (sodium nitrate).
2. Describe the process of electrolysis using the phrases in **a.** to **f.** joined in the correct order.
   a. and hydrogen is produced at the nickel cathode.
   b. An electric current is passed through the electrolyte (sodium chloride solution)
   c. cell.
   d. whereupon chlorine is produced at the titanium anode,
   e. The electrolyte becomes a solution of sodium hydroxide.
   f. Brine, a sodium chloride solution, is placed in a
3. Explain the meaning of the following terms:
   a. Electrolysis.
   b. Electrolyte.
   c. Brine.
4. Supply the missing words in the following account of the properties of chlorine. The words may be used more than once. The dashes show the number of letters in the words or symbols.
   Chlorine will react with many _ _ _ _ _ _, some of which need to be heated. Phosphorus, silicon and _ _ _ _ _ _ _ _ _ are non-metals that will react with chlorine. The reaction with _ _ _ _ _ _ _ _ is explosive when the gases are heated.

5. Complete the following equations. Use words to complete word equations and symbols to complete formula equations.

a. Chlorine + oxygen → _ _ _ _ _ _ _ _ _ _ _ _ _ _ _ _ _ (two words).

b. $H_2(g) + Cl_2(g) \rightarrow 2$_ _ _ (g).

c. $Cl_2(g) + H_2O(\ell) \rightarrow$ _ _ _(_ _) + _ _ _ _(_ _).

6. State three uses of chlorine.

Oxygen

Oxygen is an abundant element, comprising 50% of Earth's crust and 20% of Earth's atmosphere.

Reactions of oxygen

Most elements combine directly with oxygen with the release of heat and light energy (ie they burn) in a **combustion** reaction.

Example

The burning of magnesium produces enough light for the reaction to be used in flares, flash bulbs for photography and incendiary bombs (World War II). The reaction is:

magnesium + oxygen → magnesium oxide + bright white light.

Example

Coal and wood are valuable as **fuels** because they contain carbon which burns to produce heat in the reaction:

carbon + oxygen → carbon dioxide + heat + light.

Elements that do not react with oxygen include the inert gases and the noble metals (eg gold and platinum).

Nitrogen

Nitrogen comprises 79% of Earth's atmosphere and is present in desert regions of the world, eg in Chile, as sodium nitrate.

Nitrogen is an unreactive gas chemically but, in a thunderstorm, the lightning causes a small amount of atmospheric nitrogen and oxygen to react to produce nitric oxide, ie

$N_2(g) + O_2(g) \rightarrow 2NO(g)$.

Nitric oxide combines with more oxygen to produce nitrogen dioxide,

$2NO(g) + O_2(g) \rightarrow 2NO_2(g)$.

The nitrogen cycle

The element nitrogen is continually being recycled in the living world in the **nitrogen cycle**.

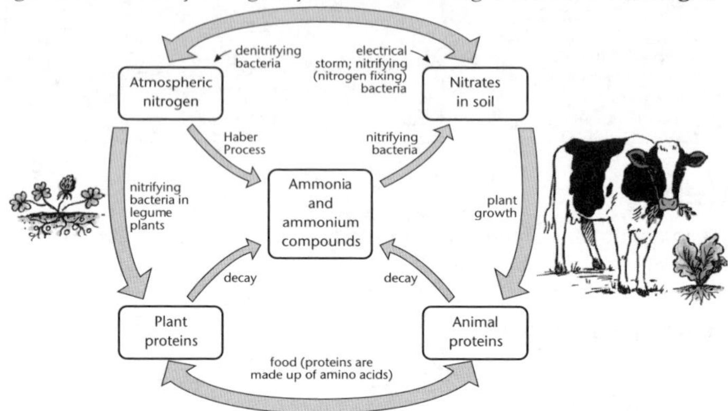

The nitrogen cycle

Various nitrogen compounds are involved in the nitrogen cycle, ie nitrites, nitrates, protein and ammonia.

Nitrites

The nitrite ion, $NO_2^-(aq)$, is obtained from atmospheric nitrogen in two ways:

- Lightning converts oxygen and nitrogen into nitric oxide gas, which is further oxidised to nitrogen dioxide which combines with water to produce the nitrite ion and the nitrate ion, $NO_3^-(aq)$.
- Nodules on some plants (eg brassica, clover, lupins) contain bacteria that will convert atmospheric nitrogen into the nitrite ion.

Nitrates

The nitrate ion, $NO_3^-(aq)$, is obtained from the nitrite ion by oxidation using atmospheric oxygen. The nitrates comprise the nitrogen compounds in the soil that plants can absorb. Denitrifying bacteria convert the nitrate ions back to nitrogen gas and thus the plant and animal worlds miss an opportunity to use them.

Protein

The nitrate ion absorbed by the roots of a plant is converted into **protein** in the plant. If this plant is eaten by animals, the protein passes to the animal.

Decomposers (organisms that occur naturally and cause decay of plants and animals) convert animal and plant protein back to the nitrate ion, which can then be absorbed by new plants to continue the cycle, or can be broken down by denitrifying bacteria to return the nitrogen to the atmosphere.

Ammonia

Ammonia, $NH_3(g)$, produced by the breakdown of animal protein (it is often present in urine) is converted to the ammonium ion, $NH_4^+(aq)$. Bacteria in the soil can convert the ammonium ion into the nitrite ion, and then into the nitrate ion.

In today's world of six billion human beings, the nitrogen in the nitrogen cycle must be supplemented with man-made ammonia.

Example

Many foodstuffs (both animals and plants) grown in one country are consumed in another country, and thus the wastes are not returned to the soil that produced the plants and animals.

The ammonia manufactured by the **Haber Process** (see Unit 11.5 Topic 8, p. 281) is converted into **fertilisers** such as ammonium sulfate, ammonium nitrate and urea. In the soil, the nitrogen component of these fertilisers is converted by bacteria into the nitrite ion and hence the cycle can continue.

Commercial preparation of nitrogen and oxygen

Nitrogen forms 79% of air, and oxygen forms 20% of air. These two gases can be obtained from air by the following process:

- Remove dust from the air using electrostatic attraction.
- Remove water vapour and carbon dioxide by cooling the air so that liquid water forms and carbon dioxide solidifies.
- Compress the air, which is free of dust, water vapour and carbon dioxide – this is an exothermic (heat energy is released) process.
- Allow the compressed air to expand – this is an endothermic (heat energy is absorbed from the gas itself) process. Sufficient cooling occurs during this expansion to liquefy the air.
- Careful warming of the liquid air allows the nitrogen to evaporate (become gas again), leaving oxygen in the liquid state. The nitrogen gas can be compressed and stored in cylinders, or returned to the liquid state if it is to be transported in bulk. Similarly, the oxygen can be transported under refrigeration as a liquid, or the gas can be formed and then compressed in cylinders for storage.

Unit 11.5 Activity 4D: Reactions and preparation of oxygen and nitrogen

1. a. Name a source of commercial value of:

i. Oxygen.

ii. Nitrogen.

b. Identify which of the natural substances listed in **i.** to **iv.** contain both oxygen and nitrogen.

i. Gypsum (calcium sulfate dihydrate).

ii. Marble (calcium carbonate).

iii. Carnallite (potassium magnesium chloride).

iv. Chile saltpetre (sodium nitrate).

2. Arrange the following statements in order to explain how nitrogen is obtained from the atmosphere.

a. The air, without dust, water vapour and carbon dioxide, is compressed.

b. The cold liquid is warmed gently to allow nitrogen to boil off, leaving liquid oxygen in the vessel.

c. Air is drawn in from the atmosphere and has dust, water vapour and carbon dioxide removed.

d. The nitrogen gas can be compressed into cylinders, or turned back into the liquid state for bulk transportation.

e. The compressed air is allowed to expand rapidly, which produces a cooling effect sufficient to liquefy the air.

3. Supply the missing words in the description of the following process. The words may be used more than once. The dashes show the number of letters in the words or symbols.

_ _ _ _ _ _ _ _ is the major component of air and lowers the activity of oxygen in the air. In a thunderstorm this gas combines with oxygen to produce _ _ _ _ _ _ _ _ _ _ _ (two words). _ _ _ _ _ _ _ _ _ _ _ (two words) then reacts with more oxygen to produce _ _ _ _ _ _ _ _ _ _ _ _ _ _ _ (two words).

4. Explain the meaning of the underlined words/terms.

Oxygen is used to support the combustion of a fuel. The reaction of a fuel with oxygen occurs exothermically. Coal is a common fuel that is mainly carbon. The principal product of combustion of coal is carbon dioxide. If gold is heated in the presence of oxygen no reaction occurs – gold is said to be inert.

5. Below is an account of the movement of element nitrogen through the living world, starting and finishing with atmospheric nitrogen. Complete the account using the words (alphabetically arranged) in the following list. The number of letters in each word is shown by the number of dashes.

Nitrogen in the _ _ _ _ _ _ _ _ _ _ can be converted to nitric oxide by combination with oxygen in the presence of _ _ _ _ _ _ _ _ _. The nitric oxide is converted to nitrogen _ _ _ _ _ _ _ by combination with more oxygen. By reacting with water, the nitrogen dioxide is converted into a mixture of nitric and nitrous _ _ _ _ _. The nitrate and nitrite ions are obtained from these acids.

Alternatively atmospheric nitrogen is converted into the _ _ _ _ _ _ _ ion in the presence of _ _ _ _ _ _ _ _ in the roots of legumes (clover, brassica, etc). The nitrite ion is _ _ _ _ _ _ _ _ to the nitrate ion and the latter is absorbed by the plant through its _ _ _ _ _. In the plant, _ _ _ _ _ _ _ is made from the nitrate ion, other minerals and water. The protein in plants can be changed into other protein in an _ _ _ _ _ _ _ _ digestive system. Eventually, the protein of the animal is _ _ _ _ _ _ _ _ _ _ by bacteria, fungi, etc and nitrogen _ _ _ _ _ _ _ to the atmosphere.

Word list: acids; animal's; atmosphere; bacteria; decomposed; dioxide; lightning; nitrite; oxidised; protein; returns; roots.

6. Fritz Haber found a way of making ammonia from readily available materials. Explain why this process is important to the nitrogen cycle.

Sulfur

Reactions of sulfur

Sulfur burns vigorously in air with a blue flame to produce sulfur dioxide, a gas with a choking smell:

$S(s) + O_2(g) \rightarrow SO_2(g)$.

Sulfur dioxide is also produced when fossil fuels (eg coal and diesel), which contain chemically bound sulfur, are burnt in air. Sulfur dioxide dissolves in water (eg rain) to produce sulfurous acid, the so-called **acid rain** that erodes stonework and corrodes metals:

$SO_2(g) + H_2O(\ell) \rightarrow H_2SO_3(aq)$.

The oxygen in the air can oxidise sulfurous acid to sulfuric acid:

$2H_2SO_3(aq) + O_2(g) \rightarrow 2H_2SO_4(aq)$

which then becomes another component of acid rain.

The oxidation of sulfurous acid to sulfuric acid is used commercially.

- Sulfur dioxide is added to fruit drinks and wine as an **antioxidant** (another name for reductant). The sulfurous acid removes any oxygen present, thus denying any harmful bacteria continued existence.
- Sulfur dioxide is used to bleach wood pulp. The sulfurous acid formed reduces the **pigment** responsible for the yellow-brown colour of natural wood to a colourless form.

Uses of sulfur

Sulfur has a number of uses.

- Sulfur is very important in the commercial production of sulfuric acid by the **Contact Process** (see Unit 11.5 Topic 8, p. 282) which produces millions of tonnes of sulfuric acid worldwide annually.
- Sulfur still plays an important role in the making of rubber, although much of the rubber used today is made synthetically using organic products from petroleum. Rubber's elastic properties allow clothing to stretch and shocks to be absorbed.

Example

Rubber tyres

Goodyear is a name associated with motor vehicle tyres. In the mid-19th century, Goodyear discovered that when latex – the white fluid obtainable from certain trees – was heated with sulfur, a solid product with elastic properties (rubber) was obtained. To make tyres, the crude rubber is filled with carbon black and plasticisers, and then molded into shapes as required.

- Many chemical compounds that can destroy harmful bacteria and fungi have a sulfur component.

Example

Lime/sulfur is a mixture used for the control of fungal disease on fruit trees. The sulfur is dispersed in a solution of calcium hydroxide and the mixture is sprayed on the leaves of citrus trees, grapevines and other fruit-bearing plants.

Commercial preparation of sulfur

When **fossil fuels** (eg natural gas or petroleum) are reacted with hydrogen in the presence of a catalyst, any sulfur in the organic compounds is converted to hydrogen sulfide gas, $H_2S(g)$. Hydrogen sulfide gas:

- Has an unpleasant smell and is poisonous to humans.
- Can be used as a fuel but one of its combustion products is sulfur dioxide gas – the main cause of acid rain.

Sulfur is recovered from hydrogen sulfide gas by dividing the gas into two parts:

- One-third of the hydrogen sulfide gas is burnt in air to produce sulfur dioxide, $H_2S(g) + 1.5O_2(g) \rightarrow H_2O(\ell) + SO_2(g)$.
- The sulfur dioxide is reacted with the remaining hydrogen sulfide (two-thirds), producing solid sulfur and water, $2H_2S(g) + SO_2(g) \rightarrow 2H_2O(\ell) + 3S(s)$.

The sulfur produced is a valuable starting material for the production of sulfuric acid (see page 234).

Unit 11.5 Activity 4E: Reactions, uses and preparation of sulfur

1. a. Name a source of sulfur of commercial value.

b. Identify which of the minerals listed in **i.** to **iv.** contains sulfur.

i. Gypsum (calcium sulfate dihydrate).

ii. Marble (calcium carbonate).

iii. Carnallite (potassium magnesium chloride).

iv. Chile saltpetre (sodium nitrate).

2. Supply the missing words or symbols in the description of the following process. Words or symbols may be used more than once. The dashes show the number of letters in the words or symbols.

Natural gas can be heated with _ _ _ _ _ _ _ _ _ gas to produce hydrogen sulfide gas. Hydrogen sulfide gas is _ _ _ _ _ _ _ _ _ _ to the human system. If hydrogen sulfide gas is burnt in air, _ _ _ (formula of gas) is formed together with water. This gas _ _ _ (formula of gas) is mixed with more hydrogen sulfide gas to produce sulfur.

3. Explain the meaning of the following terms:

a. Acid rain.

b. Reductant.

c. Antioxidant.

4. Supply the missing words in the following account of the properties of sulfur. The words may be used more than once. The dashes show the number of letters in the words or symbols.

_ _ _ _ _ _ is a component of coal and diesel oil. When these fuels are burnt in air (oxygen), _ _ _ _ _ _ _ _ _ _ _ _ _ _ (two words) is produced which, in the presence of moisture, can produce _ _ _ _ rain.

5. Complete the following equations. Use words to complete word equations and symbols to complete formula equations.

a Sulfur + oxygen → _ _ _ _ _ _ _ _ _ _ _ _ _ _ (two words).

b. $\_\,\_\,\_(g) + 1.5O_2(g) \rightarrow H_2O(\ell) + SO_2(g)$,

c. $\_H_2S(g) + SO_2(g) \rightarrow \_H_2O(\ell) + 3S(s)$.

6. State three uses of the element sulfur.

Unit 11.5 Metals and Non-metals
Topic 5: Laboratory preparation of non-metal compounds

Some of the specific learning outcomes mentioned in the Syllabus (p. 19) include an understanding of the chemistry of particular non-metals, in particular nitrogen and sulfur and some of their compounds. Topic 5 examines the following:

- Selected non-metal compounds – sulfur dioxide, nitrogen monoxide (nitric oxide), nitrogen dioxide, hydrogen chloride, ammonia.
- Properties of gases that determine method of collecting the gas, and drying agents used to remove water vapour from the gas.
- Apparatus and reagents to prepare and collect gases.

Introduction

Collection of a gas

The substances chosen for consideration (ie sulfur dioxide, nitrogen monoxide (nitric oxide), nitrogen dioxide, hydrogen chloride, ammonia) are all gases at room temperature. Laboratory preparation therefore involves the collection of a gas:

- Gases that are *insoluble in water* can be collected in a gas jar by displacement of *water*. Only one of the gases to be considered (ie nitrogen monoxide) is insoluble. The gas jar is judged to be full when there is no water in the gas jar.
- Gases that are *soluble in water* must be collected in a gas jar by displacement of *air*. A chemical test, eg **litmus** paper, is used to gauge when the jar is full of gas (place the chemical test close to the mouth of the gas jar).

The method of gas collection depends on the density of the gas relative to the density of air. If the gas is:

- *More dense than air*, the gas jar must be upright – the denser gas will fall to the bottom of the jar and displace the less dense air (the **upward displacement** of air).
- *Less dense than air*, the gas jar must be upside down – the less dense gas will rise up the jar and displace the denser air (the **downward displacement** of air).
- *Insoluble in water*, the gas jar is placed on top of a beehive shelf.

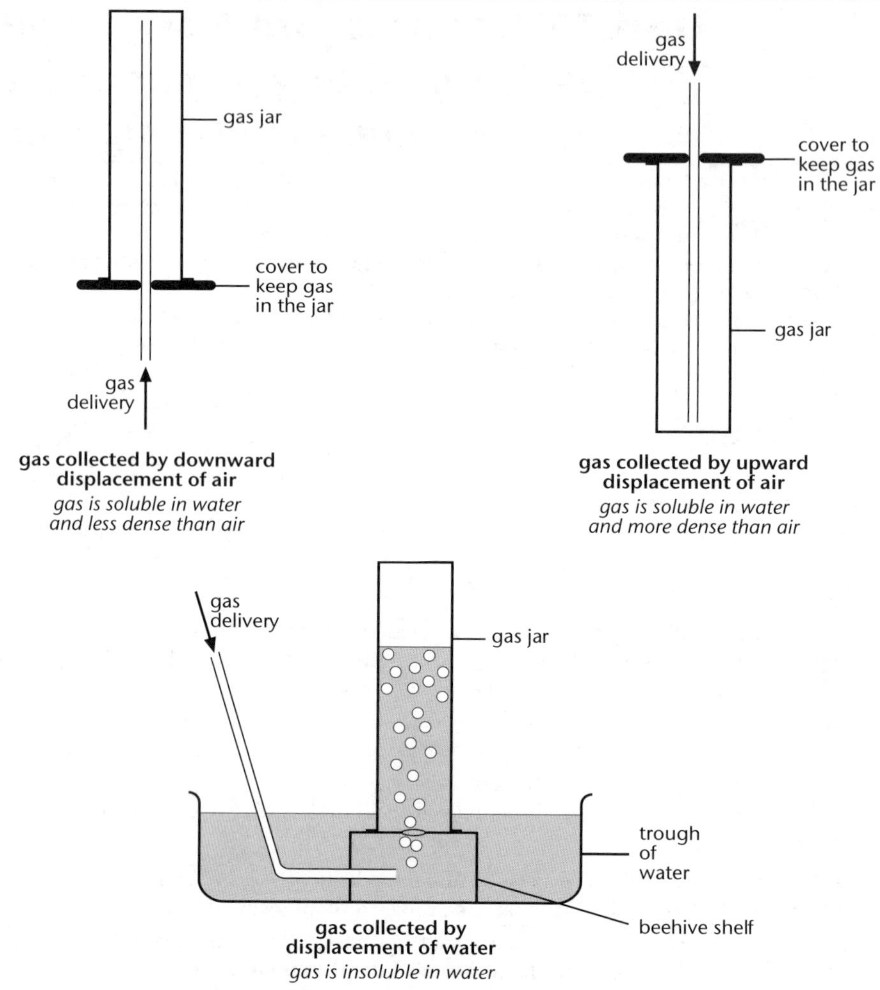

Collection of gases by displacement of air or water

A gas collected by displacement of water must contain water vapour, and can never be collected dry. For gases collected by displacement of air (which always contains water vapour), the following **drying agents** can be used:

Gas to be collected	Drying agent
Ammonia	Calcium oxide
Hydrogen chloride	Concentrated sulfuric acid (one of the reagents used to prepare the gas)
Sulfur dioxide	Concentrated sulfuric acid
Nitrogen dioxide	Anhydrous calcium chloride

Drying agents for gases

Reactions to prepare gases

The reactions to prepare and methods of collection of ammonia, hydrogen chloride, sulfur dioxide, nitrogen dioxide and nitrogen monoxide gases are summarised in the table.

Name of gas	Formula of gas	Chemicals used to prepare the gas	Equation for the preparation of the gas	Method of collection of gas
Ammonia	NH_3	Ammonium chloride + calcium hydroxide	$2NH_4Cl(s) + Ca(OH)_2(s) \rightarrow CaCl_2(s) + 2H_2O(\ell) + 2NH_3(g)$	Downward displacement of air
Hydrogen chloride	HCl	Sodium chloride + conc. sulfuric acid	$NaCl(s) + H_2SO_4(\ell) \rightarrow NaHSO_4(s) + HCl(g)$ or $2NaCl(s) + H_2SO_4(\ell) \rightarrow Na_2SO_4(s) + 2HCl(g)$	Upward displacement of air
Sulfur dioxide	SO_2	Sodium sulfite + dil. hydrochloric acid	$Na_2SO_3(s) + 2HCl(aq) \rightarrow 2NaCl(aq) + H_2O(\ell) + SO_2(g)$	Upward displacement of air
Nitrogen dioxide	NO_2	Copper + conc. nitric acid	$Cu(s) + 4HNO_3(aq) \rightarrow Cu(NO_3)_2(s) + 2H_2O(\ell) + 2NO_2(g)$	Upward displacement of air
Nitrogen monoxide (nitric oxide)	NO	Copper + 50% nitric acid	$3Cu(s) + 8HNO_3(aq) \rightarrow 3Cu(NO_3)_2(aq) + 4H_2O(\ell) + 2NO(g)$	Displacement of water

Preparation and method of collection of gases

In all the gas preparations, several jars of gas can be collected by replacing the full jar with a fresh jar containing air or water.

Unit 11.5 Activity 5A: Reactions to prepare gases, and the collection of gases

1. For the following three gases:

i. Sulfur dioxide.

ii. Ammonia.

iii. Nitrogen monoxide.

a. State the method for collecting each of the gases **i.** to **iii.**

b. State the physical property of each of the gases in **i.** to **iii.** that is used to collect the gas by this method.

2. Name the gases that can be prepared from the following reagents:

a. Dilute hydrochloric acid and potassium sulfite solid.

b. Ammonium chloride solid and sodium hydroxide solution.

3. Complete the following equations for the preparation of gases. Do not change the formulae or number of molecules for the substances provided.
   a. $2NH_4Cl(s) + Ca(OH)_2(s) \rightarrow CaCl_2(s) + 2H_2O(\ell) +$ ______.
   b. $NaCl(s) + H_2SO_4(\ell) \rightarrow NaHSO_4(s) +$ ______.
   c. $2NaCl(s) + H_2SO_4(\ell) \rightarrow Na_2SO_4(s) +$ ______.
   d. $NH_4Cl(s) +$ ______ $\rightarrow NaCl(aq) + H_2O(\ell) + NH_3(g)$.
   e. $Na_2SO_3(s) + 2HNO_3(aq) \rightarrow 2NaNO_3(aq) + H_2O(\ell) +$ ______.
4. The diagram shows a **wash bottle** that is used to absorb gases.

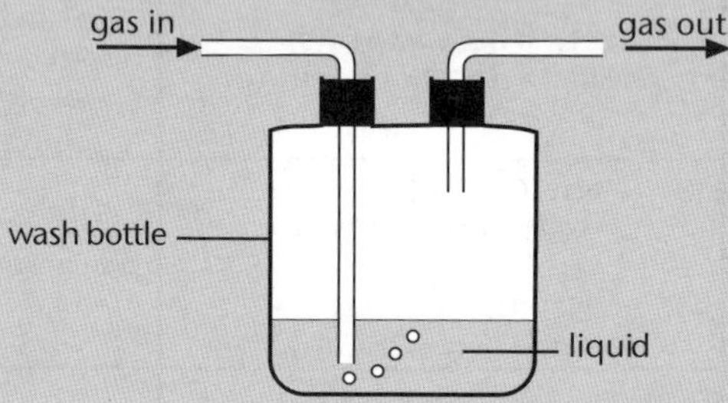

Name a suitable liquid that could be used to absorb:
   a. Water vapour that was present in a stream of hydrogen chloride gas.
   b. Nitrogen dioxide that was present in a stream of nitrogen monoxide.

Laboratory preparation of ammonia gas

The common name for the compound NH_3 is ammonia – its systematic name is nitrogen trihydride. Ammonia is a strong-smelling gas at room temperature. It is a base and is neutralised by acids to produce ammonium compounds, eg NH_4Cl (ammonium chloride), $(NH_4)_2SO_4$ (ammonium sulfate), NH_4NO_3 (ammonium nitrate). Ammonia gas is produced by reacting ammonium compounds with a strong **alkali**, eg sodium hydroxide solution (NaOH), or solid calcium hydroxide, $Ca(OH)_2(s)$:

$$2NH_4Cl(s) + Ca(OH)_2(s) \rightarrow CaCl_2(s) + 2H_2O(\ell) + 2NH_3(g).$$

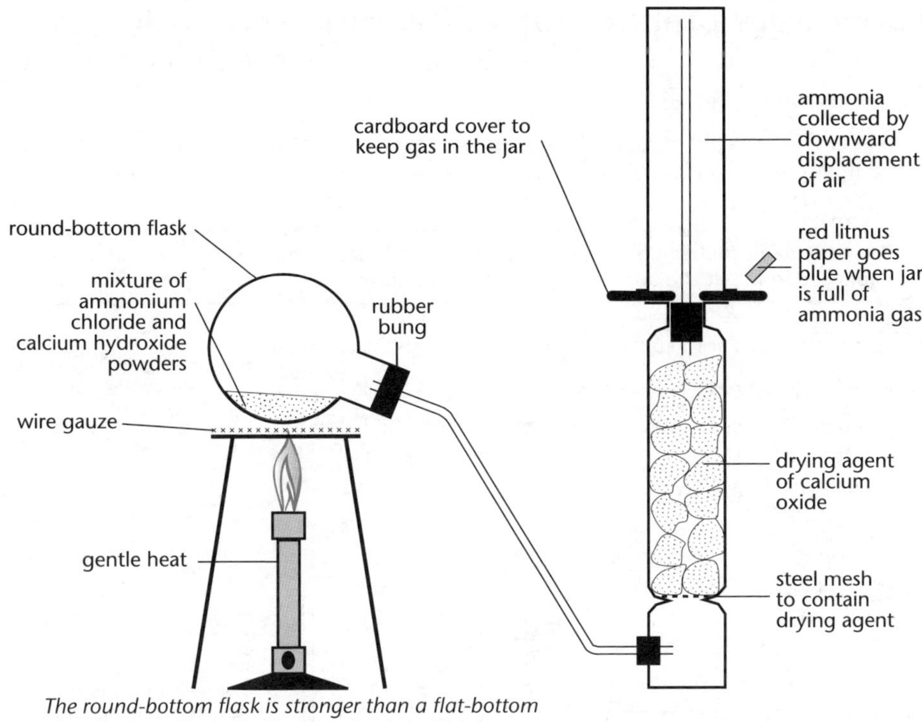

The round-bottom flask is stronger than a flat-bottom flask and is used when heating is involved.

Laboratory preparation of ammonia gas

In the experiment to prepare ammonia gas:

- The downward slope of the reaction flask is important because steam is produced in the reaction which will condense to water – if the water were to run back onto the hot glass, the glass would crack.
- The drying agent is calcium oxide, CaO – all other drying agents react with ammonia.
  Calcium oxide is converted to calcium hydroxide by water; the lumps of calcium oxide crumble to a powder, calcium hydroxide:
  $CaO(s) + H_2O(\ell) \rightarrow Ca(OH)_2(s)$.
- The steel mesh helps to contain the calcium oxide drying agent and the powdered calcium hydroxide – otherwise these would fall to the bottom of the tower and block the inlet for the gas.

Laboratory preparation of hydrogen chloride gas

Hydrogen chloride gas has the formula HCl. It reacts with water to produce hydrochloric acid:

$HCl(g) + H_2O(\ell) \rightarrow HCl(aq)$

The gas is colourless but in air appears as steamy fumes due to its reaction with water vapour in the air to produce fine droplets of hydrochloric acid.

Example

The early name for hydrogen chloride gas was 'spirits of salt' because the gas can be produced from salt and appears as a 'spirit' (wispy fumes).

The gas is produced at room temperature according to the equation:

$NaCl(s) + H_2SO_4(\ell) \rightarrow NaHSO_4(s) + HCl(g)$.

If the reaction is heated, more hydrogen chloride gas can be produced for a given quantity of sulfuric acid, ie:

$2NaCl(s) + H_2SO_4(\ell) \rightarrow Na_2SO_4(s) + 2HCl(g)$.

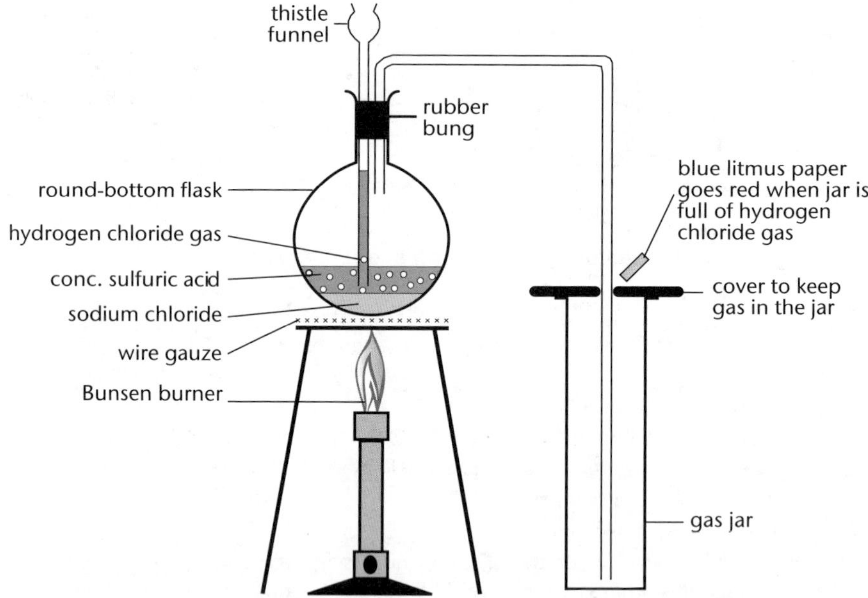

Laboratory preparation of hydrogen chloride gas

Laboratory preparation of sulfur dioxide gas

Sulfur dioxide gas, SO_2, is produced when sulfur burns in air.

A more suitable reaction for preparing sulfur dioxide gas in the laboratory is to add dilute acid to a metal sulfite compound, eg sodium sulfite (Na_2SO_3):

$Na_2SO_3(s) + 2HCl(aq) \rightarrow 2NaCl(aq) + H_2O(\ell) + SO_2(g)$.

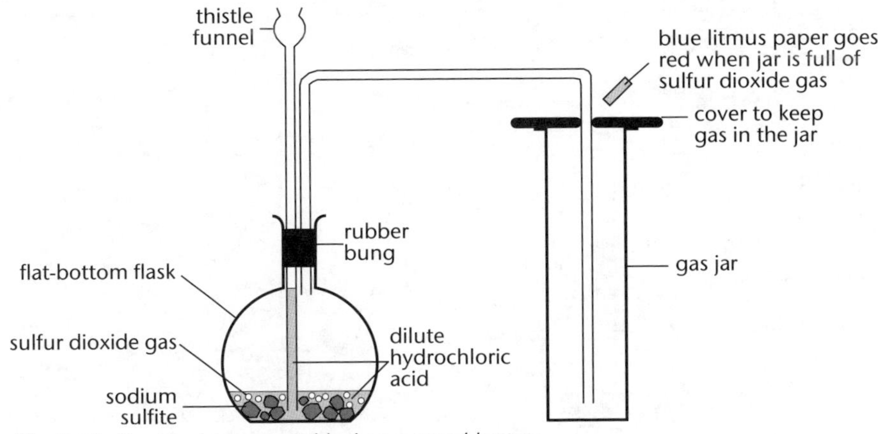

Laboratory preparation of sulfur dioxide gas

Laboratory preparation of gaseous oxides of nitrogen

Nitrogen has three oxides (N_2O, NO and NO_2). Two of these – nitrogen dioxide, NO_2, and nitric oxide, NO (also called nitrogen monoxide) – can be prepared from nitric acid, HNO_3.

Generally, an acid reacts with a metal to release hydrogen gas. There are exceptions, eg copper metal is too low in the activity series to release hydrogen from acids – the products of the reaction of copper metal with an acid are water and a nitrogen oxide.

When copper metal is added to nitric acid, the acid reacts as an **oxidising agent**.

- If the acid is in concentrated form, the product is nitrogen dioxide, NO_2:
  $Cu(s) + 4HNO_3(aq) \rightarrow Cu(NO_3)_2(aq) + 2H_2O(\ell) + 2NO_2(g)$.

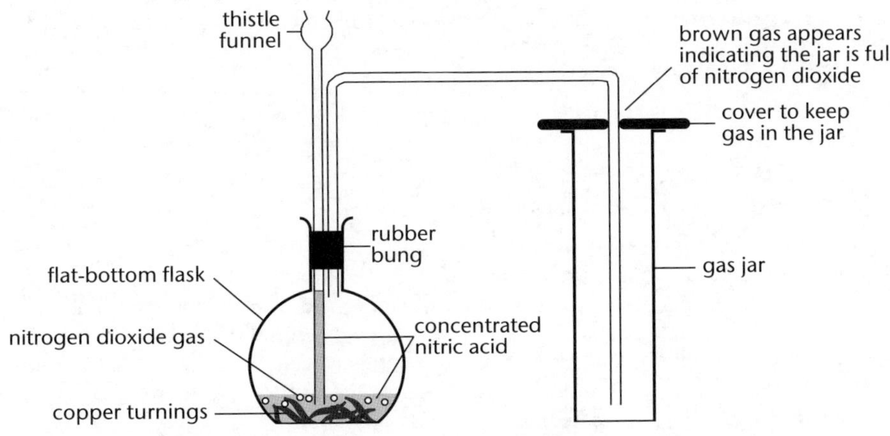

Laboratory preparation of nitrogen dioxide gas

- If the acid is of 50% concentration, the product is nitrogen monoxide (nitric oxide), NO: $3Cu(s) + 8HNO_3(aq) \rightarrow 3Cu(NO_3)_2(aq) + 4H_2O(\ell) + 2NO(g)$.

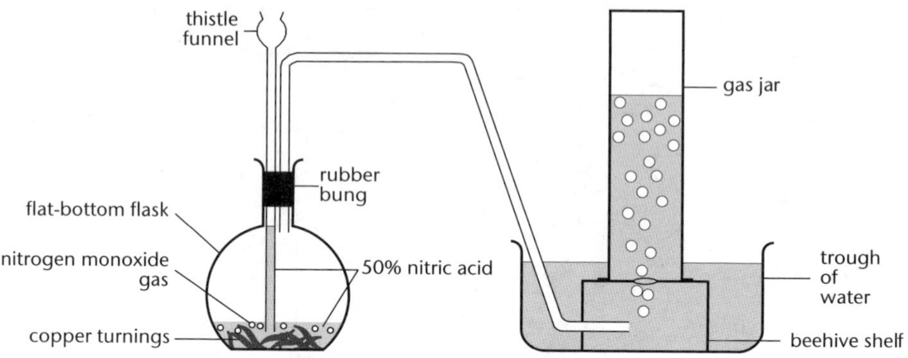

Laboratory preparation of nitrogen monoxide gas

Nitrogen monoxide reacts spontaneously with oxygen in the air, at room temperature, to produce nitrogen dioxide gas, which is brown. This reaction makes it difficult to study the properties of nitrogen monoxide.

Example

It is impossible to know the smell of nitrogen monoxide gas because to get the gas to the smell detector – in the upper regions of the nose – the gas must come in contact with air (oxygen).

Unit 11.5 Activity 5B: Apparatus for preparing and collecting gases

1. Draw diagrams of:
   - **a.** A flat-bottom flask.
   - **b.** A round-bottom flask.
   - **c.** A gas jar that would collect a gas that is less dense than air.
   - **d.** Thistle funnel.

2. Draw and label a diagram for the preparation of one gas that requires heat to be applied to the reactants. Show in the diagram how the gas is dried, if necessary, and how the gas is collected. Write an equation for the preparation of the gas.

3. State and explain one advantage of:
   - **a.** A round-bottom flask compared with a flat-bottom flask.
   - **b.** A flat-bottom flask compared with a round-bottom flask.

4. A gas called nitrous oxide is prepared in the laboratory from solid ammonium nitrate. The solid is heated and the products are nitrous oxide and steam; the steam condenses to water in the cooler regions of the flask. Nitrous oxide is insoluble in water. Draw an apparatus that would be suitable for the preparation and collection of the gas. Given that the formula of ammonium nitrate is NH_4NO_3 and the formula of nitrous oxide is N_2O, write an equation for the reaction.

Unit 11.5 Metals and Non-metals

Topic 6: Properties of non-metal compounds

The Syllabus (p. 20) makes particular mention of the chemistry of sulfur and sulfur compounds: allotropes of sulfur, sulfur dioxide, production of sulfuric acid and acid rain. In Topic 6 we extend our understanding of non-metal compounds, including sulfur and some of its compounds, by looking at:

- Selected non-metal compounds – sulfur dioxide, nitrogen dioxide, nitric acid, sulfuric acid.
- Human senses used to detect gas and liquids.
- Solubility and acidic nature of aqueous solutions.
- Impact of sulfur dioxide and nitrogen dioxide on human beings and their environment – photochemical smog and acid rain.

Physical properties of non-metal compounds

Humans possess five senses (sight, smell, taste, hearing and touch). The first three are useful in identifying and recognising chemicals by their physical properties:

- Sight – is valuable if colour is a property of a compound. Very few common gases and other non-metal compounds are coloured.

Example

Nitrogen dioxide gas (brown) is the best example of a coloured non-metal compound.

- Smell – should be used carefully.

Example

Waft the gas or the vapour towards your nose rather than place your nose into a jar full of the gas, or a beaker of liquid. Place the container of the chemical in a fume cupboard.

- Taste – should *not* be used on unknown chemicals.

Example

The sense of taste should only ever be used if your teacher specifically directs you to use it.

- Hearing – has little value for identifying chemicals unless the compound is unstable (in which case there is a danger element).
- Touch – has little value for identifying chemicals unless the compound is a skin irritant (in which case, touch should either be used very carefully or not at all).

Gases **diffuse** very rapidly (ie they can move from their source very quickly) and can enter the human system by mouth or by nose.

	Sulfur dioxide gas	Nitrogen dioxide gas
Colour	Colourless	Red-brown
Smell	Avoid inhalation – strong choking effect; throat is irritated and coughing occurs	Avoid inhalation – strong and unpleasant odour; coughing occurs

Sulfur dioxide and nitrogen dioxide gases and human senses

Liquids do not diffuse but the vapours of **volatile** liquids diffuse in the same way as gases, and may be inhaled or ingested.

	Nitric acid	Sulfuric acid
Colour / Appearance	Colourless as a pure liquid; nitrogen dioxide (NO_2) dissolved in the liquid acid causes yellow colouration.	Colourless as a pure liquid; oily liquid.
Smell	Liquid is volatile (low bp); vapour has a choking smell.	None – liquid is non-volatile (high bp).
Taste	Avoid completely.	Avoid completely.
Touch	Avoid completely – acid as liquid or vapour is extremely corrosive to skin.	Avoid completely – acid reacts violently with water, which is a large part of the human body.

Nitric acid and sulfuric acid liquids and human senses

Unit 11.5 Activity 6A: Non-metal compounds and the human senses

1. Name a substance that is a:

- **a.** Colourless gas at room temperature.
- **b.** Coloured gas at room temperature.
- **c.** Gas which irritates the throat.
- **d.** Liquid that is volatile (bp below 100 °C).
- **e.** Liquid that is non-volatile (bp above 200 °C).
- **f.** Liquid whose vapour will make you choke.
- **g.** Liquid that will dehydrate your skin.
- **h.** Skin irritant.

2. Explain how to:

- **a.** Obtain the smell (just a sniff) of a gas that attacks the throat, having been provided with the gas in a jar with a secure lid that can slide over the top of the jar.
- **b.** Transfer 10 mL of a liquid, which is volatile and corrosive to skin, from a stoppered bottle to an open beaker.
- **c.** Show that nitrogen dioxide, a brown gas which is heavier than air and dangerous to inhale, will diffuse in air. *Hint*: Assume a gas jar (with a sliding lid) of nitrogen dioxide is available, plus another gas jar that contains air.

Solubility of gases in water

The solubility of a gas in water can be tested by inverting a test tube of gas into a beaker of water.

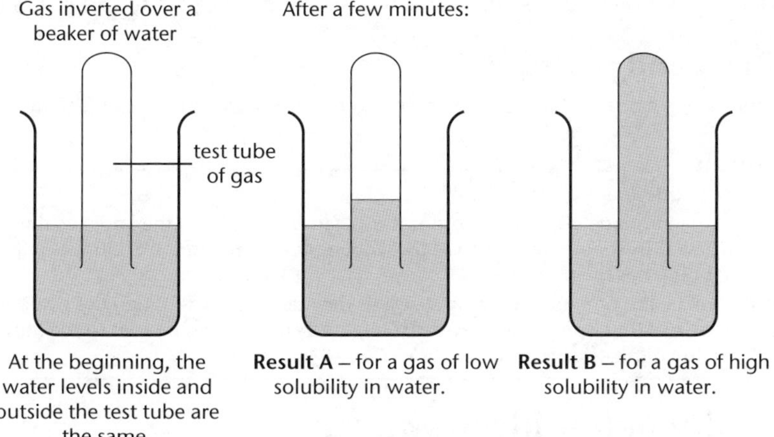

Solubility of gases in water

When the test tube of gas is placed under the water in the beaker, the pressure inside the test tube is the same as atmospheric pressure.

- If the gas is **insoluble** in water, the water level in the test tube will not change.
- If the gas dissolves in the water, the pressure inside the test tube is reduced because there is less gas. The atmospheric pressure is now greater than the pressure inside the test tube and pushes the water up the test tube.

Example

Carbon dioxide gas has low solubility in water at room temperature (Result **A** in diagram above).

- For a gas that is highly soluble in water, the test tube becomes empty of gas and thus the pressure inside the tube is zero, and the water rises to the top of the test tube.

Example

Sulfur dioxide and nitrogen dioxide gases are very soluble in water at room temperature (Result **B** in diagram above).

Solubility and temperature

All gases have lower solubility in water at higher temperature. When water is heated, the first bubbles of gas observed are bubbles of air. Once the air has been removed, and the temperature of the water reaches 100 °C, the bubbles are much bigger and are of steam.

Solubility and pressure

The solubility of a gas in water is greatly increased by raising the pressure on the gas.

Example

Carbon dioxide gas and fizzy drinks

The 'fizz' in lemonade is made by compressing carbon dioxide in the liquid as the bottle is sealed. The fizz in champagne is made by allowing **fermentation** to continue in the bottle once the bottle has been sealed.

In both drinks, the pressure of the gas (carbon dioxide) increases. When the stopper is removed from either bottle, the pressure is released and the gas escapes to the atmosphere rapidly, ie the drink **effervesces.**

Solubility of liquids in water

Both nitric acid and sulfuric acid are very soluble in water, ie the acids mix with water in all proportions. Acid solutions are described as either:

- Concentrated – aqueous solutions that contain a large number of acid particles, or
- Dilute – aqueous solutions that contain a small number of acid particles.

Dissolving nitric acid in water

Pure nitric acid is a colourless liquid which when added to water produces a solution of nitric acid – concentrated or dilute depending upon the number of acid particles present. The addition of acid to the water generates a lot of heat, and the mixing should be done with care and constant stirring of the mixture.

Dissolving sulfuric acid in water

Sulfuric acid is a colourless liquid with a boiling point of 338 °C and a density of 1.84 g mL^{-1}.

When sulfuric acid is added to water, the mixing should be done slowly and with great care. Because the acid is almost twice as dense as water, the acid will sink below the water, but as the mixing occurs the heat generated accelerates the mixing and the evolution of heat.

Example

If 50 mL of sulfuric acid is mixed with 50 mL of water and the mixture is stirred, the mixture gets hot enough to boil – the mixture can escape from a beaker and harm the person doing the mixing!

Always add the acid to the water.

Example

Sulfuric acid in the eye

Getting sulfuric acid in the eye is to be avoided at all costs. The heat generated by the acid mixing with water in the eye is extremely dangerous to the eye components. Even when the heat damage is over, the solution of sulfuric acid that remains is also very dangerous to all parts of the eye.

If sulfuric acid comes in contact with the eye, the region should be immediately flooded with water. Treatment with mild alkali (sodium hydrogen carbonate solution) can be carried out later. Get medical attention immediately.

Acidic nature of sulfur dioxide and nitrogen dioxide gases

Both sulfur dioxide and nitrogen dioxide gases are acidic:

- Damp blue litmus paper added to a gas jar containing each gas will go red.
- The gases react rapidly with alkali according to the equations:
  $SO_2(g) + 2NaOH(aq) \rightarrow Na_2SO_3(aq) + H_2O(\ell)$, and
  $2NO_2(g) + 2NaOH(aq) \rightarrow NaNO_3(aq) + NaNO_2(aq) + H_2O(\ell)$.

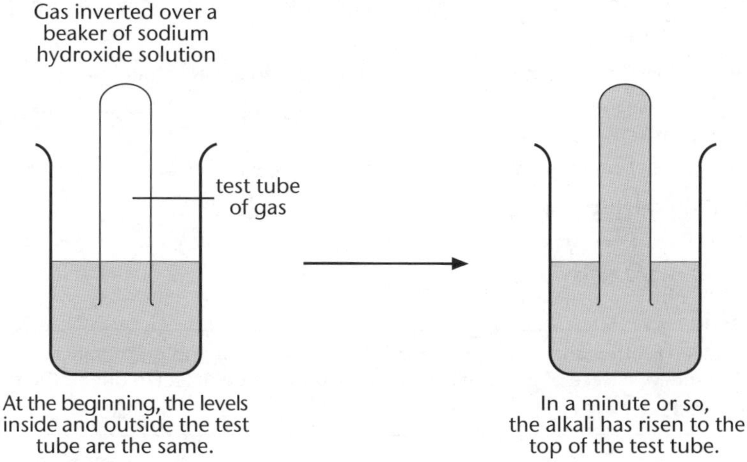

Gas inverted over a beaker of sodium hydroxide solution

test tube of gas

At the beginning, the levels inside and outside the test tube are the same.

In a minute or so, the alkali has risen to the top of the test tube.

The pressure inside the test tube is reduced as the reaction proceeds, and the external atmospheric pressure forces the alkali to fill the test tube.

Reactivity of acidic gases in alkali

Aqueous solutions of sulfur dioxide and nitrogen dioxide gases

Aqueous solutions

An aqueous solution is a solute dissolved in the solvent water.

Example

Aqueous solution of a gas

Sulfur dioxide gas dissolves in water to produce an aqueous solution of sulfur dioxide. The word equation for this process is:

sulfur dioxide gas + water → aqueous solution of sulfur dioxide

Using symbols, where $SO_2(g)$ represents sulfur dioxide gas and $SO_2(aq)$ represents sulfur dioxide gas dissolved in water, the formula equation is:

$SO_2(g) + H_2O(\ell) \rightarrow SO_2(aq)$.

Acidic nature of aqueous solutions of sulfur dioxide and nitrogen dioxide gases

Aqueous solutions of sulfur dioxide gas and nitrogen dioxide gas are acidic. They display the typical acidic properties of reacting with:

- Magnesium metal to release hydrogen gas – the gas burns with a 'pop' in air.
- Marble chips (calcium carbonate) to release carbon dioxide gas – the gas turns **limewater** milky.

There are two distinct processes when the gases are in contact with water.

- **Physical dissolving** of the gas in water, ie
  $SO_2(g) + H_2O(\ell) \rightarrow SO_2(aq)$.
  $NO_2(g) + H_2O(\ell) \rightarrow NO_2(aq)$.

Example

Sulfur dioxide is often added to a fruit drink as an antioxidant and preservative. When the bottle is opened, a trace of the smell of sulfur dioxide can be detected due to the reaction,

$SO_2(aq) \rightarrow SO_2(g) + H_2O(\ell)$.

- In all cases, the gas that dissolves in water also *reacts* with the water.

Example

Sulfur dioxide reacts with water to produce sulfurous acid, $H_2SO_3(aq)$. This can be represented by the equation,

$SO_2(g) + H_2O(\ell) \rightarrow H_2SO_3(aq)$.

Nitrogen dioxide reacts with water to produce nitric acid, $HNO_3(aq)$, and nitrous acid, $HNO_2(aq)$ according to the equation:

$NO_2(g) + H_2O(\ell) \rightarrow HNO_3(aq) + HNO_2(aq)$.

Any compound that produces an acid by reaction with water is termed an **anhydride**.

Example

Sulfur dioxide is the anhydride of sulfurous acid.

Nitrogen dioxide is termed a **mixed anhydride** because it produces a mixture of acids when it reacts with water.

Aqueous solutions of sulfuric acid and nitric acid

When samples of sulfuric acid and nitric acid are added to water, there is an exothermic reaction.

The heat released is due to the reaction between the acid and the water to produce the hydronium ion, $H_3O^+(aq)$, and the anion(s) of the acid. The reaction can be represented by the equations:

- For pure sulfuric acid,
  $H_2SO_4(\ell) + H_2O(\ell) \rightarrow H_3O^+(aq) + HSO_4^-(aq)$, then,
  $HSO_4^-(aq) + H_2O(\ell) \rightarrow H_3O^+(aq) + SO_4^{2-}(aq)$.
- For pure nitric acid,
  $HNO_3(\ell) + H_2O(\ell) \rightarrow H_3O^+(aq) + NO_3^-(aq)$.

In a similar manner, the acids produced from gases react with water to produce the hydronium ion:

- For an aqueous solution of sulfurous acid,
  $H_2SO_3(aq) + H_2O(\ell) \rightarrow H_3O^+(aq) + HSO_3^-(aq)$.
- For an aqueous solution of nitric acid and nitrous acid,
  $HNO_3(aq) + H_2O(\ell) \rightarrow H_3O^+(aq) + NO_3^-(aq)$ and
  $HNO_2(aq) + H_2O(\ell) \rightarrow H_3O^+(aq) + NO_2^-(aq)$.

The reactions with water of acids produced from gases are much less exothermic than when pure sulfuric acid or pure nitric acid is added to water.

Aqueous solutions of sulfuric and nitric acids have all the properties of an acid, ie they will:

- Turn blue litmus red.
- React with magnesium metal to release hydrogen gas – the gas burns with a 'pop' in air.
- React with marble chips (calcium carbonate) to release carbon dioxide gas – the gas turns limewater milky.

Unit 11.5 Activity 6B: Solubility and acidity of non-metal compounds

1. Name a gas that is very soluble in water at room temperature and atmospheric pressure.
2. Write the formula of the substance formed when sulfur dioxide gas dissolves in water.
3. If a test tube that is full of nitrogen dioxide gas is placed over a beaker of water, the water rises rapidly to fill the test tube. Explain fully why this observation is made. Include in your explanation what property of the gas is causing the observation and why the water rises to the top of the test tube.
4. Describe the process represented by the following equation:
   $SO_2(g) + H_2O(\ell) \rightarrow SO_2(aq)$.
5. Three different gases (X, Y and Z) were tested for their solubility in water in the apparatus below. Indicate what **Results X, Y,** and **Z** tell you about the solubility in water of the three different gases X, Y, and Z.

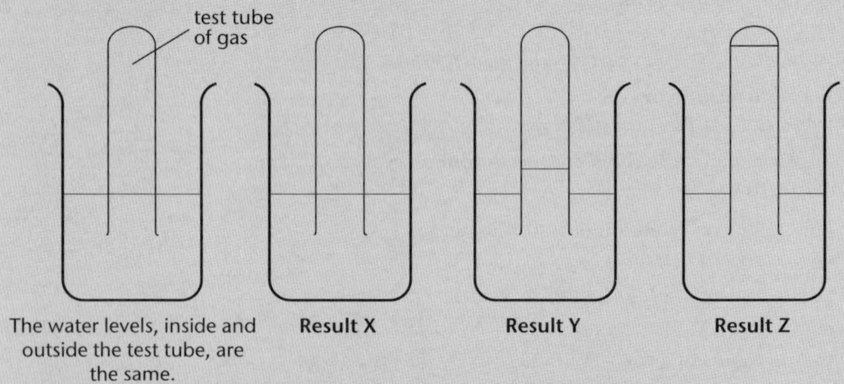

6. Explain, with an example, the meaning of each of the following terms:
   a. Solvent.
   b. Aqueous solution.
   c. Anhydride.
7. Describe four observations that you would make when 50 mL of sulfuric acid is added to 50 mL of water in a 250 mL beaker.
8. Complete the following equations:
   a. $H_2SO_4(\ell)$ + _____ $\rightarrow H_3O^+(aq) + HSO_4^-(aq)$.
   b. $HSO_4^-(aq) + H_2O(\ell) \rightarrow$ _____ $+ SO_4^{2-}(aq)$.
   c. _____ $+ H_2O(\ell) \rightarrow H_3O^+(aq) + NO_3^-(aq)$.
   d. $HNO_2(aq) + H_2O(\ell) \rightarrow H_3O^+(aq) +$ _______.
9. Write the equation in symbol form for:
   a. Sulfurous acid reacting with water to produce the hydronium ion and the hydrogen sulfite (HSO_3^-) ion.
   b. Nitric acid solution reacting with water to produce the hydronium ion and the nitrate (NO_3^-) ion.

Sulfur dioxide and nitrogen dioxide in the atmosphere

Photochemical smog

Photochemical smog is a form of local pollution caused by the internal combustion engine. Some of the chemicals formed in the internal combustion engine, and emitted in the exhaust of the vehicle (eg unburnt fuel, oxides of nitrogen, sulfur dioxide), react in the presence of sunlight with other chemicals in the atmosphere. The necessary presence of sunlight explains the use of the term photochemical (the Greek word *photo* means light).

Ozone (an allotrope of oxygen) is formed by interaction of oxides of nitrogen with oxygen in the air. Ozone, nitrogen dioxide and sulfur dioxide react with unburnt hydrocarbons from the car exhausts to produce a mixture of organic molecules. These organic molecules form an **aerosol** in the atmosphere – an aerosol consists of minute droplets of liquid suspended in the gases of the air. The result is a blue-brown haze, or photochemical smog. This process is occurring in the lower atmosphere (up to 10 km above Earth's surface).

Acid rain

The water cycle on the surface of planet Earth involves water evaporating from the seas, lakes and rivers and returning to these reservoirs (after cloud formation and then rain falling on the planet). The rain can dissolve gases (eg carbon dioxide, sulfur dioxide) in the atmosphere.

The main component of **acid rain** is sulfurous acid, formed by the reaction of sulfur dioxide with water (rain). Sulfur dioxide in the atmosphere comes from:

- The burning of fossil fuels containing small quantities of sulfur – this is particularly significant in industrial areas.
- In any thermal area, sulfur present in the rock combines with oxygen in the air to produce sulfur dioxide.

Oxides of nitrogen released in the atmosphere through car exhausts contribute to acid rain. Very small quantities of oxides of nitrogen are also formed during thunderstorms.

Carbon dioxide gas present in the atmosphere by natural processes of **photosynthesis** and **respiration** also contributes to the acidity of rain – but to a much lesser extent than sulfur dioxide.

The effects of acid rain include:

- Corrosion of stone buildings, which contain calcium carbonate, over periods of years.
- Corrosion (sometimes severe) of metals (especially steel) damages bridges, roofs and motor vehicles.
- Plant life can be severely affected to the extent of plants, even large trees, dying.

Unit 11.5 Activity 6C: Photochemical smog and acid rain

1. State one way in which each of the following gases enters the atmosphere:
   a. Sulfur dioxide.
   b. Nitrogen dioxide.

2. Give the name of an acid formed when sulfur dioxide is dissolved in water.

3. Write five sentences on each topic **a.** – **c.** to explain the features of:
   a. Acid rain.
   b. Photochemical smog.
   c. Ozone in the atmosphere.

4. Name a chemical substance that is:
   a. Produced in the internal combustion engine, and during thunderstorms.
   b. Produced when fossil fuels are burned, and has a choking effect on the throat.

Unit 11.5 Metals and Non-metals

Topic 7: Properties, uses and production of metalloids

In Topic 7 we examine metalloids, in particular:

- Elements classified as metalloids, their production and their physical properties.
- Chemical properties of metalloids and location on the periodic table.
- Uses of metalloids based on their physical and chemical properties.

What is a metalloid?

Elements can be divided into metals and non-metals, with eight elements that are neither metals nor non-metals being called metalloids. Astatine is classified as a halogen by some sources. The term 'metalloid' derives from the Greek words *metallon*, which means 'metal', and *edios*, which means 'sort of'.

The metalloids are placed in groups 13–17 in the periodic table, where they form a zigzag shape. They are separated from the non-metals by a line called the **amphoteric line**.

The metalloids are shown in the following table.

Metalloid	Atomic number	Electronic configuration	Mass number (atomic mass unit)
Boron (B)	5	[He] $2s^2 2p^1$	10.81
Silicon (Si)	14	[Ne] $3s^2 3p^2$	28.09
Germanium (Ge)	32	[Ar] $4s^2 3d^{10} 4p^2$	72.59
Arsenic (As)	33	[Ar] $4s^2 3d^{10} 4p^3$	74.92
Antimony (Sb)	51	[Kr] $5s^2 4d^{10} 5p^3$	121.75
Tellurium (Te)	52	[Kr] $5s^2 4d^{10} 5p^4$	127.60
Polonium (Po)	84	[Xe] $6s^2 5d^{10} 4f^{14} 6p^4$	209
Astatine (At) (man-made)	85	[Xe] $6s^2 5d^{10} 4f^{14} 6p^5$	210

Atomic and mass numbers of metalloids

Metalloid physical properties

Metalloids have properties between those of metals and non-metals. Metalloids can be shiny like metals or dull like non-metals. They are brittle and ductile and can be drawn into shapes. They can conduct heat and electricity, but not as well as metals can and behave as electrical insulators at room temperature.

Metalloid	Form	Structure	Density (g/cm^3)	Solubility in water	Melting point (°C)	Boiling point (°C)
Boron (B)	Crystalline solid, several allotropes	Cluster B^{12}, unit icosahedron (20 sides)	2.3	Insoluble	2200	3900
Silicon (Si)	Crystalline solid, several allotropes	Diamond cubic	2.3	Insoluble	1410	2680
Germanium (Ge)	Crystalline solid, several allotropes	Diamond shape	5.3	Insoluble	937	2830
Arsenic (As)	Crystalline grey solid	Body-centred cubic	5.7	Insoluble	81	613
Antimony (Sb)	Crystalline solid		6.7	Insoluble	631	1380
Tellurium (Te)	Crystalline solid, several allotropes	Chain	6.24	Insoluble	450	1390
Polonium (Po)	Crystalline solid	Metallic cubic, 6-coordinate	9.4	Insoluble	254	962
Astatine (At)	Halogen, more metallic	Unknown, 20 isotopes	Unknown	Unknown	302	335

Physical properties of metalloids

Metalloid synthesis and general reactions

The chemical reactivity of metalloids depends on the substance they react with. Metalloids are usually found combined with non-metals either in a molecular structure such as SiO_2 or in an oxoanion such as those found in the silicates. In such compounds, the metalloid has a lower **electronegativity** than the non-metal, so the metalloid exists in a positive **oxidation state**. Chemical reduction usually with carbon or hydrogen, at temperatures as high as 3000 °C, produces the metalloid in the elemental state.

Metalloid	Oxidation state	Reduction reaction
Boron (B)	Positive	$2BCl_3(g) + 3H_2(g) \rightarrow 2B(s) + 6HCl(g)$
Silicon (Si)	Positive	$SiO_2(s) + C(s) \rightarrow Si(s) + CO_2(g)$
Germanium (Ge)	Positive	$GeO_2(s) + C(s) \rightarrow Ge(s) + CO_2(g)$ $GeO_2(s) + 2H_2(g) \rightarrow Ge(s) + 2H_2O(g)$
Arsenic (As)	Positive	$2As_2O_3(s) + 3C(s) \rightarrow 4As(s) + 3CO_2(g)$ $2As_2O_3(s) + 3H_2(g) \rightarrow 2As(s) + 3H_2O(g)$
Antimony (Sb)	Positive	$2Sb_2O_3(s) + 3C(s) \rightarrow 4Sb(s) + 3CO_2(g)$ $2Sb_2O_3(s) + 3H_2(g) \rightarrow 2As(s) + 3H_2O(g)$
Tellurium (Te)	Positive	$2Te_2O_3(s) + 3C(s) \rightarrow 4Te(s) + 3CO_2(g)$ $2Te_2O_3(s) + 3H_2(g) \rightarrow 2Te(s) + 3H_2O(g)$
Polonium (Po) (radioactive)	Positive (+4, +2)	$2Po_2O_3(s) + 3C(s) \rightarrow 4Po(s) + 3CO_2(g)$ $2Po_2O_3(s) + 3H_2(g) \rightarrow 2Po(s) + 3H_2O(g)$ Better produced by radioactive decay of bismuth or uranium
Astatine (At)	Positive and negative (+7 to –1)	Man-made

Chemical synthesis reactions of metalloids

Many metalloids have multiple oxidation states or valences. For example, the radioactive polonium has oxidation states of +4 and +2 while astatine has **oxidation numbers** ranging from +7 to –1.

Metalloids react as non-metals with metals and as metals with non-metals. For example, when boron reacts with fluorine, it acts as a metal:

$$B + 3F \rightarrow BF_3$$

but when it reacts with sodium, it acts as a non-metal to produce sodium borohydride ($NaBH_4$).

Some metalloids, such as boron, silicon and germanium, behave as semiconductors.

Many metalloids have different allotropes – are different physical forms of the same element. One allotrope of a metalloid may behave as a metal and the other allotrope may behave as a non-metal.

Oxides of metalloids are usually amphoteric as metals oxides are generally basic and non-metals oxides are generally acidic. Boron does not form simple oxides or borates. The simplest one involves hydrogen oxoacid (H_3BO_3).

Germanium occurs occurs mostly in the oxidation state +4, although there are many germanium compounds with the oxidation state +2. Elemental germanium oxidises slowly to GeO_2 at 250 °C. Germanium is insoluble in dilute acids and alkalis but dissolves slowly in concentrated sulfuric acid and reacts violently with molten alkalis to produce germanates ($[GeO_3]^{2-}$). Germanium can form covalent bonds with itself. This is called *catenation*. The tendency for catenation is greatest for lighter metalloids.

Silicon is a relatively inert element. It reacts with halogens and dilute alkalis, but does not react with most acids, except for some very reactive combinations of nitric acid and hydrofluoric acid. Silicon has four **bonding electrons**, like carbon does, allowing for many different forms of chemical bonding. It is a semiconductor, readily either donating or sharing its four outer electrons.

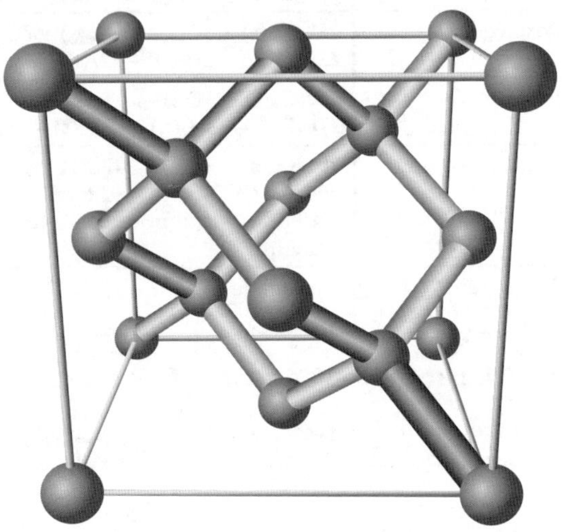

Silicon crystallises in a diamond cubic crystal structure

Silicon forms only one oxide (SiO_2). Although its formula is similar to CO_2, structurally it is very different because silicon does not form Si–O multiple bonds like carbon does. Quartz is an oxide of silicon that has a –Si–O–Si–O–spiral chain hexagonal three-dimensional network. Like carbon, silicon can combine with many other elements and compounds.

Silicon undergoes catenation, but not naturally because it has a very stable structure in oxidation state –2. It forms an organosilicon polymer. Silicon can form sigma bonds to other silicon atoms. However, it is difficult to prepare and isolate Si_nH_{2n+2} (analogous to the saturated alkane hydrocarbons) with $n > 8$, as thermal stability decreases as the number of silicon atoms increases.

In recent years a variety of double and triple bonds in silicon, germanium, arsenic and bismuth have been reported. Catenation of certain main group elements is currently the subject of research into inorganic **polymers**.

Polonium is very scarce and is usually produced in milligram amounts in a nuclear reactor by bombarding bismuth-209 with neutrons. This forms bismuth-210, which has a half-life of 5 days. Bismuth-210 decays into polonium-210 through beta decay.

Metalloid hydrides

Because of its relatively high electronegativity, hydrogen forms covalent bonds with metalloids to form *hydrides*. These have the **general formula** H_4X, with the exception of boron, which has the simplest formula H_6B_2. Many boron hydrides have the three-centre bond arrangement B–H–B.

Germanium forms the hydride germane (GeH_4), which is a compound similar in structure to methane (CH_4).

Germane (GeH_4) structure

Metalloid	Name	Formula	Structure bonding
Boron (B)	Diborane	B_2H_6	Many structures, bridging H, B–H–B two three-centre bonds
Silicon (Si)	Silicon hydride	SiH_4	Catenation between Si atoms (–S–Si–) to form silanes (Si_2H_6, Si_6H_{14})
Germanium (Ge)	Germanium hydride	GeH_4	Catenation between Ge atoms (Ge–Ge) to form Ge_2H_6 and Ge_3H_8
Arsenic (As)	Arsenic hydride	AsH_3	Catenation not observed
Antimony (Sb)	Antimonium hydride	SbH_3	Catenation not observed
Tellurium (Te)	Tellurium hydride	TeH_2	Catenation not observed

Some common metalloid hydrides

Metalloid uses

Metalloids are very useful. Silicon and germanium are central to the remarkable progress in the field of solid-state electronics. The operation of every hi-fi stereo system, TV receiver, TV camera, VCR, CD player and AM-FM radio relies on transistors made from semiconductors.

Silicon–germanium alloys are used in semiconductors in high-speed integrated circuits, where they can be much faster than silicon alone. Si–Ge alloys are starting to replace gallium arsenide (GaAs) in wireless communications devices.

Some germanium compounds are available in the US as low-dose dietary supplements in oral capsules or tablets. Some germanium compounds are quite reactive and hazardous to humans. The liquid germanium chloride and the gas germane (GeH_4) can be very irritating to the eyes, skin, lungs and throat.

Metalloid	Electrical property	Uses
Boron (B)	Semiconductor	Eye wash, tennis racquets, heat-resistant glass, regulator in nuclear plants
Silicon (Si)	Semiconductor	Microchips, solar cells, tools, quartz, cement, heat-resistant glass, greases and oils, semiconductors, stereos, TV receivers, TV cameras, VCR, CD players, transistors
Germanium (Ge)	Semiconductor	Fibre optics, infrared prisms, polymerisation catalysts, solar reflectors, wide-angle lenses, dentistry, semiconductors, stereos, TV receivers, TV cameras, VCR, CD players, transistors, phosphors, metallurgy, chemotherapy
Arsenic (As)	Semiconductor	Hardens shot, metal for mirrors, glass, lasers, light-emitting diodes (LED)
Antimony (Sb)	Semiconductor	Solder, type for printing, lead batteries, bearings, infrared detectors, cosmetics
Tellurium (Te)	Semiconductor	Percussion caps, vulcanisation of rubber, battery plate protectors, electrical resistors
Polonium (Po)	Semiconductor	Nuclear batteries, neutron source, film cleaner
Astatine (At)	Unknown	Seldom found in nature; no known uses

Uses of metalloids

Unit 11.5 Activity 7A: Metalloids

1. Why are metalloids usually recovered from their compounds by chemical reduction rather than by oxidation?
2. Write the chemical equation for the:
   a. chemical reduction of BCl_3 with hydrogen.
   b. production of Si from SiO_2 using carbon as a reducing agent.
   c. reduction of As_2O_3 with hydrogen.
3. What is the simplest hydrogen compound formed by boron? What is it called? Describe the shape of the molecule.
4. Differentiate between catenation in carbon and catenation in silicon and germanium.

Unit 11.5 Metals and Non-metals

Topic 8: Commercial preparation and uses of non-metal compounds

The Chemistry Syllabus (p. 19) mentions that students should understand the usefulness of non-metal compounds such as those used in artificial fertilisers. Topic 8 looks at:

- Commercial preparation of ammonia, sulfuric acid, superphosphate and sodium hypochlorite.
- Properties of sulfuric acid, sodium hypochlorite solution and sulfur dioxide.
- Uses of ammonia, sulfuric acid and superphosphate.
- Uses of sulfur dioxide and sodium hypochlorite related to some of their properties.

Ammonia

During World War I, Germany needed to make nitric acid in large quantities – nitric acid is used in the manufacture of all common explosives.

Example

Traditionally, nitric acid had been made from Chile saltpetre (sodium nitrate) and sulfuric acid. Chile saltpetre was unavailable to the Germans because the shipping lanes across the Atlantic Ocean were controlled by British warships.

Example

Nitroglycerine and TNT (trinitrotoluene) are two types of explosives.

In 1915, Fritz Haber invented a process for making ammonia from coal, water and air – all these substances were readily available to the Germans. Once ammonia is produced, it is commercially simple and cheap to make nitric acid.

Commercial preparation of ammonia – the Haber Process

Ammonia can be made from a modified **Haber Process** using natural gas as the source of hydrogen, and air as the nitrogen source, under carefully selected reaction conditions. The process involves several stages which are outlined in the reaction scheme:

Stage	Reaction and reaction conditions
Primary Reformer – hydrogen is produced from steam and natural gas.	$CH_4(g) + H_2O(g) \rightarrow CO(g) + 3H_2(g)$ 750 °C, using a nickel catalyst
Secondary Reformer – air is introduced as the nitrogen source and oxygen is removed.	$4N_2 + O_2 + 2H_2 \rightarrow 2H_2O + 4N_2$ 1100 °C, using a nickel catalyst
Shift Reactor – carbon monoxide is converted to carbon dioxide.	$CO(g) + H_2O(g) \rightarrow H_2(g) + CO_2(g)$ This is a two-stage reaction. • The first occurs at 400 °C, using a catalyst of Fe_2O_3. • The second occurs at 200 °C, using a catalyst of CuO.

Scrubber – carbon dioxide is removed using potassium carbonate solution, $K_2CO_3(aq)$. (Carbon dioxide is more soluble in potassium carbonate solution than in water.)	$CO_2(g) + H_2O(\ell) + K_2CO_3(aq) \rightarrow 2KHCO_3(aq)$ Room temperature, no catalyst required
Methanator – final traces of carbon monoxide are removed.	$CO + 3H_2 \rightarrow H_2O + CH_4$ 300 °C, using a nickel catalyst
Synthesiser – ammonia is produced from nitrogen and hydrogen.	$N_2 + 3H_2 \rightarrow 2NH_3$ 400 °C and 200–1000 atm, using an iron catalyst

The Haber Process

The very high gas pressures (200–1000 atmospheres, ie 20–100 MPa) in the synthesiser mean that the tubes and towers of the synthesiser are made of steel several centimetres thick. Gases under these pressures must be contained in leakproof equipment.

The conversion to ammonia, as nitrogen and hydrogen pass through the synthesiser, is approximately 10%. The ammonia is separated from the gas mixture by liquefying the ammonia or by dissolving ammonia in water, and then the unreacted nitrogen and hydrogen are returned to the synthesiser for further conversion to ammonia.

Uses of ammonia

The Haber Process has been a valuable industrial process since 1915, producing the ammonia used:

- For making nitric acid (used to manufacture high and low explosives).
- As an agricultural chemical to supplement nitrogen in the soil which has been removed by intensive plant growth (*see* the nitrogen cycle in Unit 11.5 Topic 4).
- In the manufacture of **synthetic** fibres, eg nylon.
- In refrigeration on a large scale.

Sulfuric acid

Sulfuric acid was discovered by heating green vitriol crystals (iron(II) sulfate heptahydrate, $FeSO_4.7H_2O$).

Example

The historical name of sulfuric acid is 'oil of vitriol'.

Commercial preparation of sulfuric acid – the Contact Process

Annually, millions of tonnes of sulfuric acid are produced worldwide from sulfur using the **Contact Process** – the name of the process derives from the fact that the reacting gases of sulfur dioxide and oxygen come in *contact* with the catalyst.

Stage		Process	Equation(s)	Conditions
1	Producing sulfur dioxide *either from*: Sulfur *or from*: A sulfur mineral, eg zinc blende, ZnS	 Sulfur is burnt in air *or* A sulfur mineral is roasted in air	 $S(s) + O_2(g) \rightarrow SO_2(g)$ *or* $2ZnS(s) + 3O_2(g) \rightarrow 2ZnO(s) + 2SO_2(g)$	 Burn sulfur in air *or* Roast the zinc sulfide in air
2	Converting sulfur dioxide into sulfur trioxide, SO_3	Sulfur dioxide is mixed with air (oxygen) and passed over a catalyst of vanadium pentoxide.	$2SO_2(g) + O_2(g) \rightarrow 2SO_3(g)$	A temperature of 400 °C is used to melt the catalyst to increase its efficiency.
3	Dissolving sulfur trioxide in sulfuric acid to make oleum $H_2S_2O_7$	Sulfur trioxide is absorbed in *pure* sulfuric acid*.	$SO_3(g) + H_2SO_4(\ell) \rightarrow H_2S_2O_7(\ell)$	The process is carried out at room temperature.
4	Diluting the oleum with water to produce pure sulfuric acid $H_2SO_4(\ell)$	The oleum is carefully mixed with water in the correct proportions to produce sulfuric acid.	$H_2S_2O_7(\ell) + H_2O(\ell) \rightarrow 2H_2SO_4(\ell)$	The process is exothermic, and cooling must occur.

* Pure sulfuric acid is used because it contains no water – the reaction of sulfur trioxide with water is violent, and would fill the plant with droplets of sulfuric acid.

The Contact Process

Sources of the sulfur used to produce sulfur dioxide for the Contact Process include:

- Large sulfur deposits in the United States (Texas and Louisiana).
- Petroleum and natural gas.
- Naturally occurring mineral sulfide ores – eg zinc blende (ZnS), galena (PbS), iron pyrite (FeS_2).

Sulfuric acid is manufactured in industrial plants comprising steel tubes, pipes and towers. Since *pure* sulfuric acid will not attack metals, there is no corrosion of the plant by the sulfuric acid.

Properties of sulfuric acid

Sulfuric acid is referred to as the 'king of the acids' due to its varied properties which make it an extremely useful chemical in many industries.

Example

A clear indication of the economic prosperity of a country is the production and consumption of sulfuric acid by that country.

The properties of sulfuric acid include the following:

- In aqueous solution, it has acidic properties, ie it will turn blue litmus red; release hydrogen gas when magnesium is added to it; neutralise bases to form salts.

- It will **dehydrate** (remove water from) other chemical substances.
- It will oxidise metals and non-metals.
- It will sulfonate organic molecules (eg to make detergents).
- It acts as a catalyst in many chemical reactions.

Uses of sulfuric acid

Sulfuric acid is used:

- For making fertilisers, eg superphosphate.
- For making explosives, eg nitroglycerine and TNT.
- As an acid bath, eg to remove rust from steel before a car body is painted.
- For making synthetic fibres, eg terylene.

Unit 11.5 Activity 8A: Commercial preparation, properties and uses of ammonia and sulfuric acid

1. From the word list below, choose those chemicals that could be used as starting materials to produce:

a. Sulfuric acid.

b. Ammonia gas.

Word list: air, chlorine gas, natural gas, rock phosphate, sodium hydroxide solution, sulfuric acid, water, zinc blende.

(Penalties will apply for using extra chemicals, ie chemicals that are not required for the production of the commercial product.)

2. a. Complete and balance the following equations:

i. $S(s) + O_2(g) \rightarrow$ ______.

ii. $2ZnS(s) +$ ______ $\rightarrow 2ZnO(s) + 2SO_2(g)$.

iii. $CH_4(g) + H_2O(\ell) \rightarrow CO(g) +$ ______.

iv. $K_2CO_3(aq) +$ ______ $+$ ______ $\rightarrow 2KHCO_3(aq)$.

b. Each equation in **a.** illustrates one reaction in either of two series of reactions that result in the formation of a commercial product. Name the commercial product formed as the final product of the reaction series to which each reaction belongs.

3. Write a sentence to explain why:

a. Sulfuric acid is regarded as the 'king of the acids'.

b. Nitric acid is needed to wage a war.

c. The term 'contact' is used in the manufacturing process of sulfuric acid.

4. Explain how:

a. Carbon dioxide gas can be efficiently removed from other gases such as hydrogen, carbon monoxide and nitrogen.

b. Ammonia gas can be separated from nitrogen and hydrogen gases.

5. Write the chemical formula for, and give the physical state at room temperature, of:

a. Sulfuric acid.

b. Ammonia.

Superphosphate

Phosphorus is an important element for plant growth.

Example

The root systems of many plants require phosphorus, which can be taken up only in the form of the phosphate ion, PO_4^{3-}.

The conventional source of phosphate for plants – the term commonly used for these substances is **fertiliser** – is superphosphate, which is derived from rock phosphate.

Example

Rock phosphate on Nauru

Over millions of years, the solid waste products from migrating birds that have rested on the island of Nauru on their way north or south over the Pacific Ocean has filled the hollows of the coral atoll. Over a long period of time, these deposits hardened into rock phosphate, $Ca_3(PO_4)_2$. The low solubility of rock phosphate in water means it has remained on the island for millions of years.

Rock phosphate is not used as a fertiliser as its low solubility in water makes it very slow acting, eg it can take six months to see improved plant growth. In contrast, the much greater solubility of superphosphate means improved plant growth can be seen in two weeks. The prefix 'super' came from superphosphate's superior solubility in water.

Commercial preparation of superphosphate

Rock phosphate is crushed to a fine powder. It is then mixed with sulfuric acid. The reaction that occurs is shown by the equation:

$$Ca_3(PO_4)_2(s) + 2H_2SO_4(\ell) \rightarrow Ca(H_2PO_4)_2(s) + 2CaSO_4(s).$$

The mixture of calcium dihydrogen phosphate, $Ca(H_2PO_4)_2$, and calcium sulfate is superphosphate.

Example

Initially, the calcium sulfate component of superphosphate, which was too difficult to separate from the calcium dihydrogen phosphate, was thought to be of no value to the plants. Research has shown that the sulfate ion, SO_4^{2-}, is involved in plant growth and development. Two fertilisers in one!

Example

New Zealand dairy farming

The success of dairy farming in the Waikato region of New Zealand is attributed to the regular application of superphosphate fertiliser and regular rainfall. The soils are naturally deficient in phosphate minerals and the rainfall ensures the superphosphate is quickly available to the plants. Clover in the pasture can absorb the phosphate ion through its roots, and thereby increase its growth. Bacteria in nodules on the roots of the clover then convert nitrogen in the air to the nitrate ion which makes grass growth improve. The cows love it!

Sodium hypochlorite

Commercial preparation of sodium hypochlorite

The membrane cell used to produce chlorine gas (*see* Unit 11.5 Topic 4) also produces sodium hydroxide solution and hydrogen gas.

If the sodium hydroxide solution produced in the membrane cell is reacted with chlorine gas, the reaction produces a mixed solution of sodium chloride and sodium hypochlorite. The reaction can be represented by the equation:

$2NaOH(aq) + Cl_2(g) \rightarrow NaCl(aq) + NaOCl(aq) + H_2O(\ell)$.

It is not possible to separate the sodium chloride and the sodium hypochlorite – the solution is used 'as is'.

Properties of sodium hypochlorite solution

Sodium hypochlorite solution contains the following particles:

- A large number of water molecules, H_2O.
- Sodium ions, Na^+, and chloride ions, Cl^-, from sodium chloride.
- Sodium ions, Na^+, and hypochlorite ions, OCl^-, from sodium hypochlorite.

The hypochlorite ion is unstable, especially in sunlight, and decomposes as shown in the equation:

$OCl^-(aq) \rightarrow Cl^-(aq) + O$.

The oxygen produced (O) is atomic oxygen and extremely reactive. This atomic oxygen reacts so rapidly that it exists only for an extremely short time.

Uses of sodium hypochlorite solution

Sodium hypochlorite is not stable as a solid and is always used in aqueous solution. When the solution has a concentration of 3% (3 g per 100 g of water) of sodium hypochlorite, it is commercially sold as bleach.

The major uses for sodium hypochlorite solution (bleach) are:

- As a **bleaching agent** – the atomic oxygen, produced from the hypochlorite ion in the bleaching agent, reacts with the colour agent in the stain or soiling and the oxidised product is colourless.

Example

Coffee stains, tea stains, blood stains, nappy stains can be removed by using sodium hypochlorite solution without damaging the fabric that has been soiled or stained.

- As a sterilising agent – the atomic oxygen destroys harmful bacteria that might be present.

Example

Kitchen and bathroom surfaces can be washed with a solution of sodium hypochlorite to remove harmful bacteria or fungi. Babies' drinking vessels can be sterilised by rinsing in sodium hypochlorite solution.

Example

Petri dishes containing agar jelly can be deliberately contaminated with bacteria. If the dish is covered and then warmed to 35 °C for 24 hours, the bacteria multiply (by feeding off the jelly) and colonies of bacteria become visible as white spores (growths). Patches of filter paper, soaked in sodium hypochlorite solution, placed on top of some parts of the jelly destroy the bacteria and no spores are visible in these regions.

Sulfur dioxide

Properties of sulfur dioxide

The following properties of sulfur dioxide make it useful as a commercial chemical:

- It is very soluble in water at room temperature. The process of dissolving is represented by the equation:
  $SO_2(g) + H_2O(\ell) \rightarrow SO_2(aq)$.
  This process is reversible – when sulfur dioxide gas escapes from water, the choking smell of the gas can be noticed.
- The SO_2 gas can combine with oxygen slowly at room temperature – ie it can act as an **antioxidant**. An antioxidant can prevent the oxidation of other substances because it has removed the oxygen that would be needed for oxidation.

Uses of sulfur dioxide

The major uses for sulfur dioxide are:

- As a preservative in drinks.

Example

Sulfur dioxide is dissolved in the water of fruit juices and wines to remove microbes that can alter the beverages' taste and produce toxins. Like all other living organisms, microbes require oxygen for respiration. Sulfur dioxide in the beverages removes the oxygen present and thus the microbe is destroyed.

- As a bleaching agent.

Example

Wood pulp is pale yellow or straw-coloured in its natural state. When the pulp is made into paper, eg newsprint, there is a demand for the paper to be white. Treating the wood pulp, which is in the form of a slurry (fine particles of solid suspended in water), with sulfur dioxide gas removes oxygen from the chemical in the pulp that is causing the yellow colour – ie the pulp is bleached (rendered colourless). (The yellow colour of newsprint reappears if the newsprint is kept for some time, especially in sunlight, due to oxygen in the atmosphere combining with the bleached chemical in the newsprint – the paper is said to fade.)

Unit 11.5 Activity 8B: Commercial preparation and uses of superphosphate, sodium hypochlorite and sulfur dioxide

1. From the word list below, choose those chemicals that could be used as starting materials to produce:

- **a.** Superphosphate
- **b.** Sodium hypochlorite solution.

Word list: air, chlorine gas, natural gas, rock phosphate, sodium hydroxide solution, sulfuric acid, water, zinc blende.

(Penalties will apply for using extra chemicals, ie chemicals that are not required for the production of the commercial product.)

2. Write a sentence to explain why:

- **a.** In the membrane cell to produce chlorine (see diagram on p. 250) the sodium hydroxide solution produced is not regarded as a waste product.
- **b.** Superphosphate has the prefix 'super'.

3. Explain how superphosphate fertiliser can be regarded as a double fertiliser.

4. Write the chemical formula for, and give the physical state at room temperature, of:

- **a.** Sodium hypochlorite.
- **b.** Calcium phosphate.
- **c.** Calcium sulfate.
- **d.** Calcium dihydrogen phosphate.

5. Name suitable chemicals, either gases or aqueous solutions, that will:

- **a.** Restore the white colour to a piece of old newsprint that has gone yellow.
- **b.** Remove a coffee stain from a white T-shirt.

6. Write the formula for:

- **a.** The hypochlorite ion.
- **b.** Superphosphate.

7. Explain the following:

- **a.** Why sulfur dioxide gas, added to a fruit drink, will make the drink suitable for human consumption over a period of months.
- **b.** Why sodium hypochlorite solution is used to wash down the walls of a shower box at a gymnasium.

8. Describe what difference you would notice, and explain why you notice the difference, between the appearance of today's newspaper and the appearance of one that has been kept in a library for two years.

9. A petri dish was contaminated with bacteria from a bathroom wall, covered with a plate, and warmed to 35 °C for 24 hours. Another dish was treated in the same manner but drops of sodium hypochlorite solution were placed in two regions of the dish before the dish was warmed. Describe and explain any differences you would observe in the appearance of the two dishes after they had been warmed for 24 hours.

Supplementary Unit
Practical chemical investigations

The Chemistry Teacher Guide (p. 23) states that an 'investigation' involves students in a study of an issue or a problem. The emphasis in the assessment is on the student's investigation of the issue in its context by researching, identifying the issues or problems, collecting, analysing and commenting on secondary data and information. Students should consider and explore a variety of perspectives as they develop and state their position on the issue. Students may present the final investigation in a variety of forms, including one or a combination of the following: a written scientific report, an oral presentation, a website, linked documents, multimedia, a video or audio recording.
This Supplementary Unit offers advice on the following:

- Planning the investigation.
- Providing a statement of the purpose of the investigation.
- Identifying a range for the independent variable.
- Collecting, recording and processing data.
- Interpreting the data by writing a conclusion.

Having an idea

You may need some help to get started. Some ideas (numbered **1** to **6**) are shown below – there are many more available.

Some ideas for a practical chemistry investigation

Stating the purpose of the investigation

The purpose of the investigation may be stated as any of:

- *Aim* – the intention to solve, or attempt to solve, a problem.
- *Testable question* – to pose a problem that can be, or may be, answered by practical measurement, or by practical observation.
- *Prediction* – to make a forecast; to suggest something that is likely to be true from known information.
- *Hypothesis* – a proposition made as a basis for reasoning, without the assumption of its truth.

From the ideas in the drawing on p. 289, you should be able to form a statement of the purpose of the investigation, eg,

Some examples of a statement of the purpose of an investigation

Independent and dependent variables

Variables in a chemical experiment can include:

- Amount of one chemical.
- Concentration of one chemical.
- Temperature of the experiment.

A variable affects the extent or the rate of the reaction.

An independent variable is a factor that is varied (eg in a reaction), producing a series of related experiments. It is the variable that you will change from one experiment to the next.

A dependent variable is a factor (eg in a reaction) that is measured as the reaction proceeds. It is the variable that you will measure or observe.

Example

Independent and dependent variables

The following examples of independent and dependent variables are based on the ideas **1** to **6** in the drawing on p. 289.

Idea	Independent variable	Dependent variable
1	The concentration of ammonia in a series of ammonia solutions.	The volume of acid used to neutralise ammonia solution in a series of experiments.
2	The temperature at which the experiments are carried out.	The volume of gas collected at known time intervals.
3	The amount of alkali present in one antacid tablet.	The volume of acid solution used to neutralise the alkali in an antacid tablet.
4	The concentration of ethanoic acid in a series of ethanoic acid solutions	The volume of alkali solution used to neutralise ethanoic acid solutions.
5	The mass (or volume) of alcohol burnt.	The temperature measured at given time intervals as the alcohol burns.
6	The concentration of a reagent used to form a precipitate.	The particle size and appearance of the precipitate formed in a precipitation reaction.

Independent and dependent variables

Investigation plan

When you have selected the idea, it is useful to complete an Investigation Plan.

Each of the items in the Investigation Plan is discussed below, using Idea **1** ('How much ammonia is in a commercial cleaner?') from the drawing on p. 289.

1. **Purpose of investigation**

   To measure the concentration of ammonia in a solution of a commercial cleaner.

   Your teacher will advise you that you need to prepare a series of ammonia solutions from a supply of one solution of ammonia whose concentration is given to you.

2. **Which variable will be changed (ie which is the independent variable)?**

   The concentration of ammonia solution.

   You will be provided with a volume of ammonia solution and told its concentration is 2 g of ammonia per litre of solution. Your teacher will advise you to prepare solutions of ammonia whose range of concentrations is, eg 2.0 g L^{-1}, 1.5 g L^{-1}, 1.0 g L^{-1}, 0.5 g L^{-1}.

1. Purpose of investigation? (This may be an aim; a testable question; a prediction; or an hypothesis.)	
2. Which variable will be changed (the *independent variable*)?	
How will the independent variable be changed?	
Give a suitable range of values for this variable.	
3. Which variable will be measured or observed (the *dependent variable*) in order to get some data or information from this investigation?	
How will the dependent variable be measured or observed?	
4. Other variables that need to be controlled to make the results more accurate.	
Other variables.	**Describe how this variable will be controlled or measured.**
5. How do you show the results are reliable?	
6. Notes from the trials.	

Investigation plan

3. Which variable will be measured or observed (ie which is the dependent variable)?

The volume of hydrochloric acid used in each experiment.

You will be supplied with a solution of hydrochloric acid and a coloured acid/alkali indicator. You will use the hydrochloric acid solution to find the volume of acid that will neutralise a known volume (say 10 mL) of each of the ammonia solutions you have available – the one you were given, plus the solutions you have prepared. You should use a **burette** to add the acid to the ammonia solution. By this method, you can accurately (to the nearest 0.1 mL) find the volume of acid needed to neutralise the alkali.

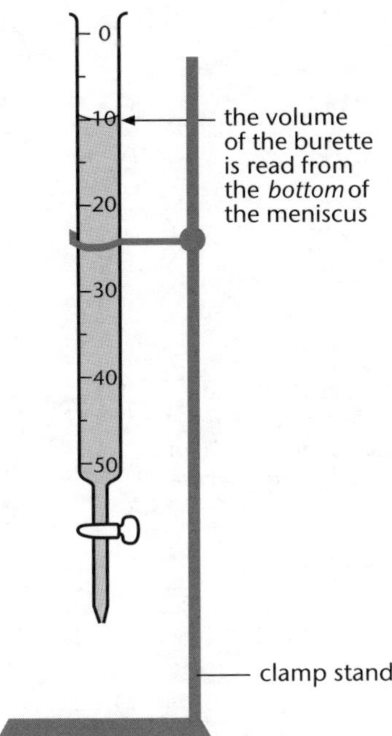

The volume in the burette should be recorded to one decimal place, eg 10.0 mL, 21.3 mL.

The burette

4. Other variables that need to be controlled

- *The volume of each sample of ammonia solution*. The volume of ammonia solution must be the same in each experiment. This is best achieved by using a **pipette** – a glass instrument designed to contain and deliver a known volume (say 10 mL) of solution.
- *The volume of indicator used in each experiment*. The ammonia will be neutralised by the acid and the indicator will change colour when neutralisation is just complete. For comparison of results, the same volume (usually 3 drops) of indicator will give the same intensity of colour change.
- *The temperature of the experiment*. The solubility of ammonia gas in water changes with change in temperature. If the series of experiments is carried out in the space of a few minutes, then the temperature should be constant.

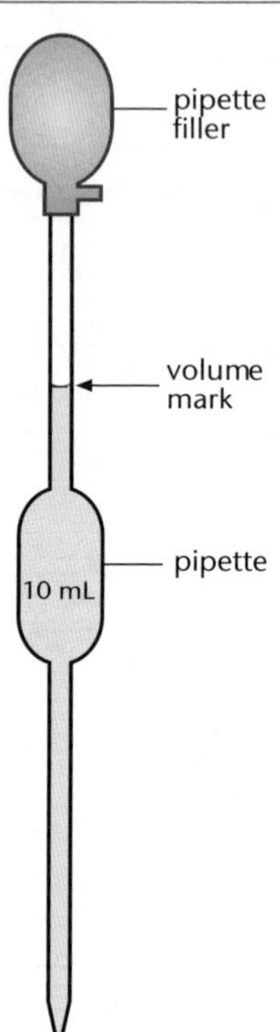

*A pipette is a glass tube **calibrated** to dispense an exact volume of solution, typically 10.00 mL, 20.00 mL or 25.00 mL. A pipette filler is used to fill and dispense liquid from the pipette.*

The pipette

5. **Reliability of results**

   Repeat the experiments to show that the results obtained are consistent. For each solution, at least two results should agree to within 0.2 mL of acid solution used.

6. **Notes from the trials**

   You will need to carry out trials to:

   - Be familiar with the equipment you are using.
   - Be familiar with the colour changes of the indicator.
   - Discover whether the commercial product needs to be diluted in order to obtain a satisfactory range of results. Your teacher may advise you of the dilution of the commercial product that you should consider. The acid you will be supplied with should neutralise 10 mL of the ammonia solutions when using acid solution volumes in the range of 2 mL to 15 mL.

Example

Dilution of commercial ammonia

If your trial shows that 25 mL of acid are needed to neutralise 1 mL of the commercial ammonia, then 250 mL of acid will be needed to neutralise a 10 mL sample of the commercial product.

This is far too much acid, because:

- The colour change of the indicator would be difficult to observe in a large volume and there would be a loss of accuracy.
- The exercise would take too long.
- The cost of the acid would be high.

In this illustration, dilute the commercial product by a factor of one hundred before you neutralise a 10 mL sample of the diluted product with the acid you have been provided with.

Method

Use the information you have gathered from your Investigation Plan to write a method for carrying out your investigation. The method should be sufficiently detailed so that someone else could carry out your experiment. Diagrams should be used to help explain the method you will use, especially of apparatus that is unfamiliar.

Dilution of ammonia solution

It will be useful to have 100 mL of each ammonia solution. The 100 mL will be enough solution for several experiments using 10 mL samples of solution, including some repeats of experiments that go wrong.

To make up 100 mL solutions, use either:

- A measuring cylinder – calibrations should be indicated in 1 mL divisions.
- A volumetric flask – has a guaranteed volume of 100 mL if filled to the volume mark.

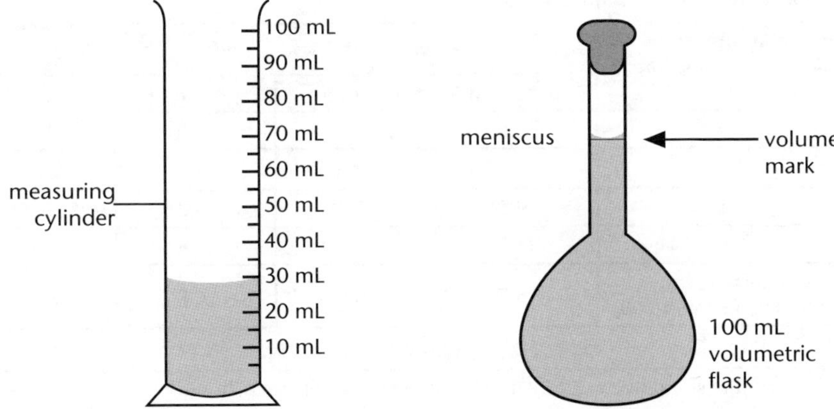

Measuring cylinder and volumetric flask

The table below indicates how much water is added to the original ammonia solution (eg of concentration 2 g L^{-1}) to prepare a solution of the required concentration.

Volume (mL) of ammonia solution provided (conc 2 g L^{-1})	Volume of water (mL) to add to make 100 mL of solution	Total volume of solution (mL)	Concentration of prepared solution (g L^{-1})
100	0	100	2
75	25	100	1.5
50	50	100	1.0
25	75	100	0.5

Preparing solutions

Results

In this investigation, the following details should be recorded:

- Dilution of commercial ammonia, say 100:1.
- Concentration of ammonia solution provided, say 2 g L^{-1}.
- Concentration of ammonia solutions prepared, say 1.5 g L^{-1}, 1.0 g L^{-1}, 0.5 g L^{-1}.
- **Indicator** used and colour of indicator in acid and alkali solution.
- A table of results showing the volume of acid used in each experiment.

Example

Recording experimental results

Volume of ammonia solution used in each experiment = 10 mL.
Indicator is methyl orange: red in acid solution/yellow-orange in alkali solution.
The table of results for this investigation is:

Concentration of ammonia solution (g L^{-1})	Volume of acid used (mL)			
	Experiment 1	Experiment 2	Experiment 3	Average
0.5	2.8	3.0	3.0	2.9
1.0	(5.3)	5.6	5.7	5.65
1.5	9.0	8.8	8.8	8.87
2.0	11.7	11.7	11.8	11.73
Commercial sample (diluted by factor of 100)	6.7	6.9	6.8	6.8

Record of experimental results

Your table of results should be examined to identify any results that should be rejected.

Example

Rejected results

In this experiment, the volume of acid used can be read from the burette, to the nearest 0.1 mL. All the results for the volume of acid used should be within 0.2 mL of each other. The value of 5.3 mL (circled) in the table of results in the above example has been rejected when calculating the average volume of acid used in this experiment.

Interpretation of results

A graph of results, with the independent variable along the horizontal axis and the dependent variable along the vertical axis, is useful if the results are numerical. *Note*: If the results are not numerical, a graph is not appropriate.

Decide what shape the graph is. If you think it is 'straight line', then draw the best straight line – the line may not go through any of the points plotted.

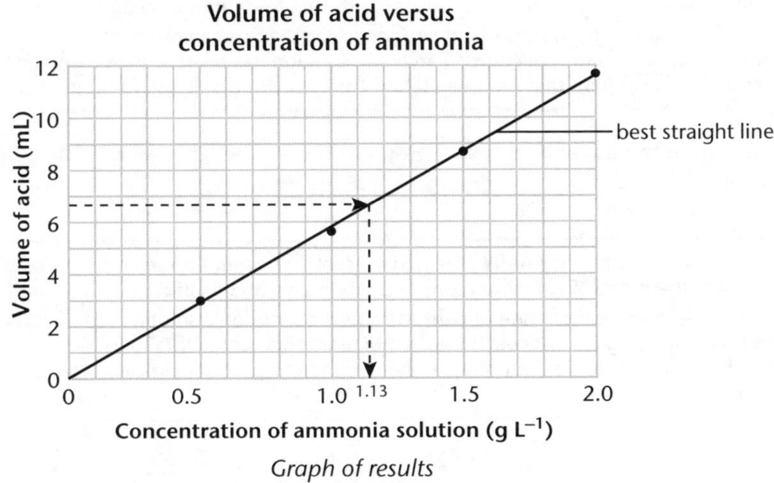

Graph of results

Read from the graph the concentration of the diluted ammonia solution.

Example

The average volume of acid used was 6.8 mL, so the concentration of the diluted ammonia solution = 1.13 g L^{-1} for the sample used.

Concentration of the original commercial product = 1.13 × 100 = 113.0 g L^{-1}.

Conclusion

Make a statement related to the aim / testable question / prediction / hypothesis.

Example

'The concentration of ammonia in the commercial product has been measured to be 113.0 g L^{-1}.'

Evaluation

In your evaluation, you should ask, and answer, the following queries.

- What problems did you encounter during the experiment? How did you overcome these difficulties?

Example

Problem – Using unfamiliar equipment and practical techniques.

Solution – Practise using the equipment and carrying out the practical processes.

- Where there any systematic errors? How did you measure them?

Example

Errors – Ammonia solution, on exposure to the air, loses ammonia gas.

Solution – With a given solution of ammonia, carrying out an estimation of the concentration of ammonia immediately the solution has been prepared and comparing the result with a solution that has been left in contact with the air for 10 minutes or so.

- Were your results comprehensive? How could you check that your results were comprehensive?

Example

Query – Were sufficient results obtained for each concentration of ammonia solution used and was the range of concentrations of ammonia solution sufficient?

Solution – Increase the range of concentrations of ammonia solutions used and carry out more investigations of each solution of ammonia used. Obtaining the same value for the concentration of ammonia in the commercial product indicates your results were comprehensive.

- What science ideas explain the trend in results?

  Response – A straight-line graph indicates that the two variables studied (independent and dependent) are directly proportional, ie if one variable doubles its value, then the other variable doubles its value.

Supplementary Unit Activity A: Practical investigation

1. Explain the terms:
   - **a.** Testable question.
   - **b.** Hypothesis.

2. State which term (aim, testable question, prediction, hypothesis) is being used in each of the following statements.
   - **a.** To measure the concentration of common salt in sea water.
   - **b.** The Dead Sea has a higher concentration of common salt than the Pacific Ocean.

3. For each of the investigations **1** to **6** in the drawing on p. 289, identify:
   - **a.** An independent variable.
   - **b.** A dependent variable.

4. Complete the following table, which shows how to prepare a series of acid solutions of varying concentration (independent variable), from one solution that is supplied:

Volume of acid solution taken (concentration is 2 g L^{-1}) (mL)	Volume of water to add to make 100 mL of solution (mL)	Total volume of solution (mL)	Concentration of prepared solution (g L^{-1})
a. 100	0	100	
b. 80		100	
c.		100	0.8
d. 5		100	

5. The table below contains results for the neutralisation reaction between different concentrations of ethanoic acid (independent variable) and a solution of sodium hydroxide (dependent variable). The aim of the experiment is to measure the concentration of ethanoic acid in a sample of vinegar.

Concentration of ethanoic acid solution (g L^{-1})	Volume of ethanoic acid solution taken (mL)	Volumes of sodium hydroxide solution taken (mL)	Average volume of sodium hydroxide solution used (mL)
5	10	8.3; 8.2; 8.4	
2.5	10	4.0; 4.4; 4.1	
1	10	2.1; 2.1; 2.2	
Diluted vinegar solution	10	4.6; 4.5; 4.6	

a. State which result(s) should be not used. Explain your decision.

b. Calculate the average volume of sodium hydroxide used for each concentration of ethanoic acid.

c. Plot a graph of the dependent variable (vertical axis) against the independent variable.

d. Estimate the concentration of the diluted vinegar solution. Show on the graph how you obtained this estimate.

e. If the diluted vinegar solution was obtained from a sample of vinegar by diluting the sample of vinegar by a factor of ten, calculate the concentration of the sample of vinegar.

6. State the similarities and the differences between a pipette and a burette.

7. A graph is drawn to show the relationship between an independent variable and a dependent variable. The graph is a straight line. State what you learn from this result.

8. Explain why:

a. The temperature should be kept constant when estimating the amount of ammonia in a series of ammonia solutions by neutralising the ammonia solution samples with acid.

b. The same volume (3 drops) of indicator should be used when adding acid to alkali, (say 10 mL), to find the volume of acid required to neutralise the alkali.

c. A pipette filler should be used when filling a pipette with ammonia solution.

APPENDIX: PERIODIC TABLE OF THE ELEMENTS

Atomic Number 1 **H** 1.0 Atomic Mass

1	*2*	*3*	*4*	*5*	*6*	*7*	*8*	*9*	*10*	*11*	*12*	*13*	*14*	*15*	*16*	*17*	*18*
																	2 **He** 4.0
3 **Li** 6.9	4 **Be** 9.0											5 **B** 10.8	6 **C** 12.0	7 **N** 14.0	8 **O** 16.0	9 **F** 19.0	10 **Ne** 20.2
11 **Na** 23.0	12 **Mg** 24.3											13 **Al** 27.0	14 **Si** 28.1	15 **P** 31.0	16 **S** 32.0	17 **Cl** 35.5	18 **Ar** 40.0
19 **K** 39.1	20 **Ca** 40.1	21 **Sc** 45.0	22 **Ti** 47.9	23 **V** 50.9	24 **Cr** 52.0	25 **Mn** 54.9	26 **Fe** 55.9	27 **Co** 58.9	28 **Ni** 58.7	29 **Cu** 63.6	30 **Zn** 65.4	31 **Ga** 69.7	32 **Ge** 72.6	33 **As** 74.9	34 **Se** 78.9	35 **Br** 79.9	36 **Kr** 83.8
37 **Rb** 85.5	38 **Sr** 87.6	39 **Y** 88.9	40 **Zr** 91.2	41 **Nb** 92.9	42 **Mo** 95.9	43 **Tc** (98)	44 **Ru** 101.1	45 **Rh** 102.9	46 **Pd** 106.4	47 **Ag** 107.9	48 **Cd** 112.4	49 **In** 114.8	50 **Sn** 118.7	51 **Sb** 121.8	52 **Te** 127.6	53 **I** 126.9	54 **Xe** 131.3
55 **Cs** 132.9	56 **Ba** 137.3	71 **Lu** 175.0	72 **Hf** 178.5	73 **Ta** 180.9	74 **W** 183.9	75 **Re** 186.2	76 **Os** 190.2	77 **Ir** 192.2	78 **Pt** 195.1	79 **Au** 197.0	80 **Hg** 200.6	81 **Tl** 204.4	82 **Pb** 207.2	83 **Bi** 209.0	84 **Po** (209)	85 **At** (210)	86 **Rn** (222)
87 **Fr** (223)	88 **Ra** 226.0	103 **Lr** 262.1	104 **Rf**	105 **Db**	106 **Sg**	107 **Bh**	108 **Hs**	109 **Mt**									

Lanthanide Series	57 **La*** 138.9	58 **Ce** 149.1	59 **Pr** 140.9	60 **Nd** 144.2	61 **Pm** 146.9	62 **Sm** 150.4	63 **Eu** 152.0	64 **Gd** 157.3	65 **Tb** 159.0	66 **Dy** 162.5	67 **Ho** 164.9	68 **Er** 167.3	69 **Tm** 168.9	70 **Yb** 173.0
Actinide Series	89 **Ac** 227.0	90 **Th** 232.0	91 **Pa** 231.0	92 **U** 238.0	93 **Np** 237.1	94 **Pu** 239.1	95 **Am** 241.1	96 **Cm** 247.1	97 **Bk** 249.1	98 **Cf** 251.1	99 **Es** 254.1	100 **Fm** 257.1	101 **Md** 258.1	102 **No** 255

Answers

Answers for many questions include a 'Marking Guide':

- **A** ('Achievement', meaning 'satisfactory achievement').
- **B** ('Merit', meaning 'high achievement').
- **E** ('Excellence', meaning 'very high achievement').

The Marking Guide has been made by the authors and the publishers and is not an official guide but we hope it will be a help to students who are striving for the best possible results.

Preliminary Unit Activity A: Processing information involving chemistry and technology (page 5)

1. **a.** The branch of science concerned with the composition, properties and reactions of substances. (***A***)

b. The application of practical or mechanical sciences to industry or commerce.

This explanation is the one most commonly given. (***A***)

2. **a.** Reliable data or facts. (***A***)

b. Drawing up data, etc in a table. (***A***)

c. Making a brief account; dispensing with needless detail. (***A***)

3. **a.** Processing. (***A***)

b. Collecting. (***A***)

c. Collating. (***A***)

4. The following 'hits' were obtained using the *Google.com* search engine – they were current at the time of search, and they will be different now.

a. 159 000 000 (***A***)

b. 79 800 000 (***A***)

c. 6 540 000 (***A***)

d. 6 280 000 (***A***)

e. 117 000 000 (***A***)

f. Generally the more hits, the more significant the entry.

It is most likely that the early entries in the response from the search engine will be the most useful. You will not have to examine every entry! (***A***)

5. The following 'hits' were obtained using the *Google.com* search engine – they were current at the time of search, and they will be different now.

a. 421 000 (***A***)

b. 23 200 (***A***)

c. Generally the more hits, the more significant the entry. (***A***)

6. **a.** Smog is a mixture of fog and smoke. (***A***)

b. Acid rain is rain that contains gases that are acidic (***A***) and these gases are generated from burning fossil fuels. (***M***)

c. Plastic is a material that can be moulded (***A***) and that occurs either naturally, or is produced by humans. (***M***)

7. **a.** Water, petroleum, salt. (***A*** – all three correct)

b. Testing electrical conductivity. (***A***)

c. Production of drinking water from sea water. (***A***)

d. *Either:* The cost is too high (***A***) of providing heat energy for the process of distillation (***M***),

or: Rain water is an alternative (***A***) that is collected in tanks and is suitable for drinking. (***M***)

Unit 11.1 Activity 1A: Nature and states of matter (page 10)

1. 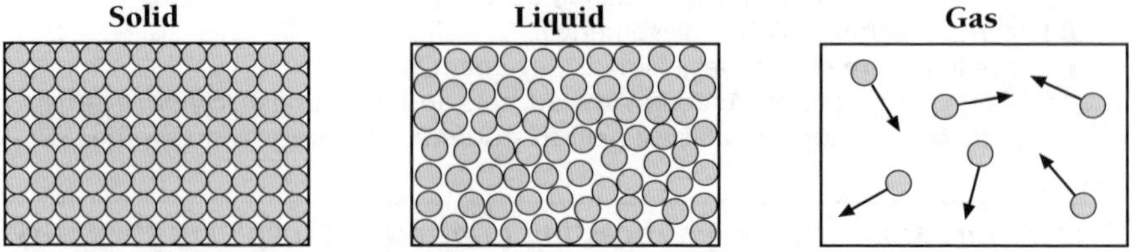 **Solid** **Liquid** **Gas**

(***A*** – three different boxes, each showing particles as circles; ***M*** – one box shows regular arrangement, or sparse regular arrangement; ***E*** – all three boxes show correct arrangement)

2. **a.** Particles in the gaseous state have high energy. Large distances between the particles mean the particles do not affect each other (the particles have already escaped from each other). Due to their high energy, the particles impact on the walls of the container and create pressure.
(***A*** – large distances apart and high energy of particles; ***M*** – impact on walls causes pressure)

b. Particles in the liquid state have medium energy. Particles are in close contact and so the particles do affect (attract) each other, but they can move slowly. Particles on the liquid's surface can escape from the other particles. Due to the particles attracting each other, they do not impact on the walls of the container to create pressure.
(***A*** – medium energy of particles; ***M*** – close together but capable of movement; ***E*** – strong attraction between particles prevents escape and pressure on walls of container)

c. Particles in the solid state have low energy. Particles are arranged regularly and in close contact, so the particles do affect (attract) each other. Particles cannot escape from the other particles. Due to the particles attracting each other, they do not impact on the walls of the container to create pressure.
(***A*** – low energy of particles; ***M*** – close together and regular arrangement so no movement; ***E*** – strong attraction between particles prevents escape and pressure on walls of container)

3. Hydrogen gas creates a pressure inside the balloon that is slightly less than the air pressure outside. As the balloon rises, the external pressure reduces because the air mass reduces, but the internal pressure remains constant – this results in the balloon expanding. A greater volume of air is displaced and the balloon continues to rise. If the balloon contained a lot of hydrogen gas (was at high internal pressure) when it was released, then the balloon would continue to expand as it rose in the air and would burst.

(***A*** – air pressure reduces as balloon rises; ***M*** – pressure in the balloon is constant; ***E*** – lower pressure outside the balloon allows the balloon to expand and thus displace more air, causing balloon to rise further)

4. **a.** As the particles get closer to each other, they reach a point where they attract each other – this bond-forming process releases heat energy. (***M***)

b. The heat energy released is due to the particles getting close enough to be attracted to each other and form bonds. (***M***)

c. In the cool part of the refrigerator, a liquid under pressure changes to its gaseous state by reducing the pressure on the liquid. (***A***) The energy needed to break the bonds between the liquid particles (***M***) is taken from the refrigerator and its contents, so the temperature of the refrigerator and its contents drops. (***E***)

d. As soon as the lighter is flicked, the liquid is exposed to the (lower) outside pressure and it boils. (***M***) Gas is released from the lighter and this gas, in contact with oxygen in air, is ignited by the spark. (***E***)

Unit 11.1 Activity 1B: Types of matter (page 13)

1. a.

Element	Compound	Mixture
Silver, chlorine, mercury	Ice, common salt, sugar	Sea water, butter

(***A*** – 6 of 8 correctly placed)

b. All elements and compounds are pure, ie have a constant composition. (***A***) Mixtures can never be pure – their composition varies.

2. a. True. The composition of air is continually changing due to, for example, water vapour increasing or decreasing; dust settling or being added to the air; industrial and animal waste adding gases to the atmosphere.

b. False. A sharp and precise melting point is an indication of a pure substance – the substance could be an element or a compound.

c. False. River water is collected from hills and mountains where the rain has dissolved small parts of the solid matter in the hills and mountains. Silt forms in the river from small particles of insoluble solid matter from the hills and mountains. The soluble and insoluble matter in the river adds to the cocktail of chemicals already in the sea.

(***A*** – three correct true/false answers; ***M*** – ***A*** plus one explanation correct; ***E*** – ***A*** plus three explanations correct)

Unit 11.1 Activity 1C: Separating mixtures (page 19)

1. a. A solute will dissolve in a solvent to produce a solution, eg sugar is a solute that will dissolve in the solvent water to produce sugar solution.

b. Volatile means that a liquid has a low boiling point and is easily vaporised, eg petrol is a volatile liquid because it evaporates rapidly at room temperature.

c. Evaporation is the process of a liquid changing to a vapour below the boiling point of the liquid, eg a puddle of water that has dried up has undergone evaporation.

d. Insoluble means that a substance will not dissolve in a solvent, eg sand is insoluble in water.

(***A*** – one explanation with example; ***M*** – two explanations with examples; ***E*** – three explanations with examples)

2. a. Water.

b. White crystals.

c. Yellow solid.

d. Evaporation.

Water dissolves the white crystals to produce a clear solution. The white crystals dissolve in water and then are recovered when the water evaporates. The yellow solid does not dissolve in water. Evaporation of the liquid occurs when it is left in contact with warm air. (***A*** – two identifications correct; ***M*** – all identifications correct)

3. **a.** matches with **v.**; **b.** matches with **iii.**; **c.** matches with **i.**; **d.** matches with **ii.** or **iv.**; **e.** matches with **iv.** or **ii.** (***A*** – three matches correct; ***M*** – five matches correct)

Pure water is obtained by distillation – the impurities remain in the distillation flask. Sea water, if left in the open air, will become salt because the water will evaporate. Chromatography can separate similar colours – if the blue dye is only one pigment (colour), then only one colour patch will show up on the chromatogram (the result of the chromatography process). The denser sand will settle to the bottom of the container – the petrol can then be decanted from the sand. Excess copper sulfate will not pass through filter paper – the copper sulfate solution will.

4. **Separating salt and sand**

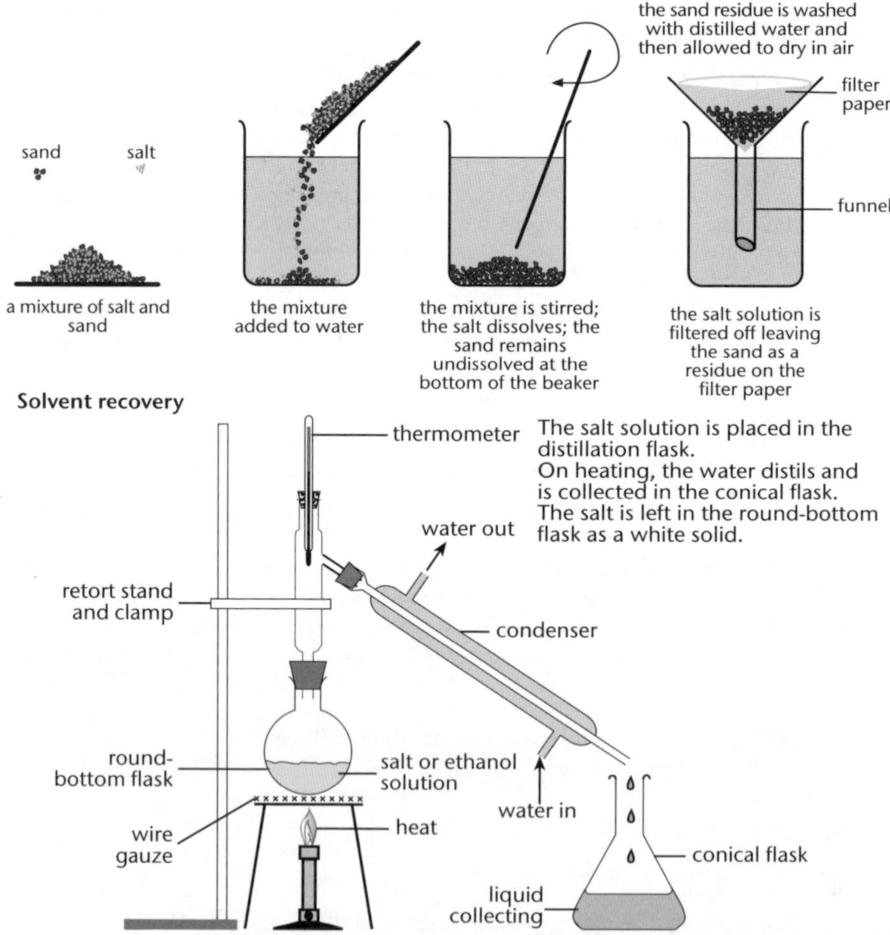

(***A*** – recognising that salt is soluble in water and sand is not; ***M*** – description of processes, and diagrams that show dissolving, filtration and distillation processes; ***E*** – complete description of the three processes, and complete, correct and well-labelled diagrams

5. **a.** Fractional distillation. (***A***)

   Petrol, diesel, fuel oil, bitumen are fractions of petroleum and all have different boiling points. When petroleum is heated to a high temperature, it will all vaporise, with the exception of bitumen – which has a very high boiling point. On careful cooling, fuel oil will condense at a higher temperature than other fractions and thus can be collected as a liquid. On further controlled cooling, other fractions will condense and can be collected as separate samples of petroleum.

   b. The process is evaporation. Sea water is trapped in shallow lakes as the tide comes in and the water is lost by the action of the Sun. Salt is left as a solid residue in the dry lake. (***A*** – correct process identified; ***M*** – Sun's action explained)

Unit 11.1 Activity 1D: Matter and separations (page 25)

Answers may vary – discuss with your teacher.

Unit 11.1 Activity 2A: Diffusion of solids, liquids and gases (page 29)

1. b.
2. b.
3. True.
4. False.
5. True.
6. True.
7. True.

Unit 11.1 Activity 3A: Behaviour of gases (page 37)

1. **a.** $V \propto T$

   b. $P \propto \frac{1}{V}$

   c. $P \propto T$
2. A gas system in which the molecules are in constant motion, all collisions are elastic and the volume of the molecules is insignificant.
3. PV = nRT
4. 0.0821 L atm mol^{-1} K^{-1}
5. 62.3637 L torr mol mol^{-1} K^{-1}
6. 8.314 m^3 Pa mol^{-1} K^{-1}
7. **a.** Oxygen is made up of individual molecules of the formula O_2.

   b. 1 mole of oxygen molecules consists of 6.022×10^{23} molecules. This is a fixed amount and is the same for any substance. 1 moles of any gas occupies a volume of 22.4 L at STP.

   c. The molar mass of oxygen is the mass of 1 mole of oxygen molecules.

   d. The molar volume of oxygen is the volume occupied by 1 mole of oxygen. At STP (0 °C and 1 atm) this is 22.4 litres.

8. 1 mole of hydrogen molecules = 6.022×10^{23} molecules

9. The molecules of a gas are in constant motion and collide with each other and the side of the container, which causes its pressure. Increasing the temperature of the gas gives the molecules more energy and results in more collisions with the side of the container. Hence, the pressure increases.

10. According to Dalton's law of partial pressure, each gas in a mixture of non-reacting gases exerts its own pressure because each gas behaves independently of the other gases.

11. Collisions are not perfectly elastic, so some kinetic energy is lost. The molecules of real gases do have some forces between them. In a real gas, the volume of the particles is not negligible compared to the total volume.

12. According to Graham's law, the smaller the molecular mass, the greater the rate of diffusion.

13. According to Boyle's law, $P_1V_1 = P_2V_2$.

$755 \times 225 = 365 \times V_2$

$$V_2 = \frac{755 \times 225}{365} = 465 \text{ mL}$$

Unit 11.1 Activity 4A: Solubility (page 42)

Answers will vary – discuss with your teacher.

Unit 11.1 Activity 4B: Conditions affecting solubility (page 43)

Observations will vary – discuss with your teacher.

Unit 11.1 Activity 4C: Solubility of substances (page 47)

1. Oxygen, carbon dioxide, water vapour, neon and other gases.

2. Ethanol, carbon dioxide, sugar and flavourings.

3. Increasing the temperature causes the bonds holding the solid molecules together to break, and more solid molecules go into solution. Gases tend to become less soluble because heating the solution enables the gas molecules to escape the solution and move to the gas phase.

4. **a.** Soluble

b. Insoluble

c. Soluble

5. $P = kc$

6. B

7. False – a solution will have a higher boiling point than the pure solvent.

Unit 11.2 Activity 1A: Atomic theory (page 51)

1. **a.** J. Dalton. (***A***)
   b. Electron. (***A***)
   c. Neutron. The particle has a neutral charge. (***M***)
   d. Radioactivity. (***A***)
   e. Rutherford showed that an atom was mainly space with a very dense and small nucleus. (***M***)
2. **a.** matches with **ii.**, Electron. **b.** matches with **iii.**, Neutron. **c.** matches with **i.**, Proton. (***A*** – all answers correct)
3. The *nucleus* of an atom contains nearly *all* the mass of the atom and is surrounded by *electrons* in energy *levels*. Although it is not possible to see atoms – clusters of about one hundred atoms can be seen under an electron *microscope* – every student of chemistry believes that they *exist*. *Atomic* bombs have been exploded. So this is the atomic *age* or the *nuclear* age; or maybe by now we have moved on to the *IT* age! (***A*** – eight correct; ***M*** – all correct)
4. **X** – the symbol for the element and represents one atom of the element. **Z** – the atomic number for the element and indicates the number of protons in the nucleus of an atom and the number of electrons in the energy levels of the atom. **A** – the mass number for an atom of the element and indicates the number of protons plus neutrons in the nucleus of an atom. (***A*** – two correct explanations; ***M*** – all three correct)

Unit 11.2 Activity 1B: Isotopes (page 53)

1. **a.** Hydrogen-1, $^{1}_{1}H$. (***A***)
   b. The mass of the proton and the mass of the neutron are the same. (***A***) The sum of protons plus neutrons in both atoms is the same, ie 14. (***M***)
   c.

Isotope	Symbol	Number of protons	Number of electrons	Number of neutrons
Carbon-12	$^{12}_{6}C$	6	6	6
Carbon-14	$^{14}_{6}C$	6	6	8
Iodine-127	$^{127}_{53}I$	53	53	74
Iodine-131	$^{131}_{53}I$	53	53	78
Strontium-88	$^{88}_{38}Sr$	38	38	50
Strontium-90	$^{90}_{38}Sr$	38	38	52

   (***A*** – symbols correct for carbon-14, iodine-131 and strontium-90; ***M*** – ***A*** plus correct number of protons, electrons and neutrons for all six isotopes)
2. **a.** 100 kg of natural uranium contains 0.7 kg of uranium-235. 100 ÷ 0.7 = 142.86 kg of natural uranium contains 1 kg of uranium-235. (***M*** – working required)
   b. $^{235}_{92}U$ and $^{238}_{92}U$. (***A*** – one symbol correct; ***M*** – both symbols correct)
3. Each atom has one proton in its nucleus. Each atom has one electron in the first energy level of the atom. The deuterium atom has twice the mass of the hydrogen atom. The tritium atom has three times the mass of the hydrogen atom. (***A*** – two points correct; ***M*** – all points correct)

Unit 11.2 Activity 1C: Electron configuration (page 55)

1.

Electron configuration	Element	Atomic number
a. 2	Helium	2
b. 2, 2	Beryllium	4
c. 2, 8, 2	Magnesium	12
d. 2, 8, 8, 2	Calcium	20
e. 2, 5	Nitrogen	7
f. 2, 8, 5	Phosphorus	15

(**A** – all atomic numbers correct; **M** – **A** plus all elements identified correctly)

2.

1	2						
2, 1	2, 2	2, 3	2, 4	2, 5	2, 6	2, 7	2, 8
2, 8, 1	2, 8, 2	2, 8, 3	2, 8, 4	2, 8, 5	2, 8, 6	2, 8, 7	2, 8, 8
2, 8, 8, 1	2, 8, 8, 2						

(**M** – all correct)

For this answer, the second element is placed in the vertical column that represents elements with two electrons in their outer energy level. In the periodic table, the second element (He) is placed in the vertical column that represents elements with eight electrons in their outer energy level – the inert gases.

Unit 11.2 Activity 1D: Atom models (page 57)

1.

Atom	Mass number	Atomic number	Atom model
H	2	1	 H
Be	9	4	 Be
N	14	7	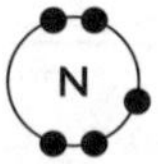 N
S	32	16	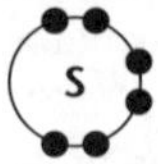 S

(**A** – two atom models correct; **M** – four atom models correct)

2.

	Atom	Atom model showing number of electrons in outer energy level	Description
b.	$^{20}_{10}Ne$ Neon-20		**'Ne'** represents the nucleus of the neon atom containing 10 protons and 10 neutrons, plus 2 electrons in the first energy level.
c.	$^{24}_{12}Mg$ Magnesium-24		**'Mg'** represents the nucleus of the magnesium atom containing 12 protons and 12 neutrons, plus 2 electrons in the first energy level, and 8 electrons in the second energy level.
d.	$^{9}_{5}B$ Boron-9		**'B'** represents the nucleus of the boron atom containing 5 protons and 4 neutrons, plus 2 electrons in the first energy level.
e.	$^{35}_{17}Cl$ Chlorine-35		**'Cl'** represents the nucleus of the chlorine atom containing 17 protons and 18 neutrons, plus 2 electrons in the first energy level, and 8 electrons in the second energy level.

(***A*** – any one element name and any one atom model correct; ***M*** – any four element names and/or atom models and/or descriptions correct; ***E*** – any eight element names and/or atom models and/or descriptions correct)

Unit 11.2 Activity 2A: Covalent bonding (page 64)

1. **a.** and **b.**

H							He
Li	Be	B	C	N	O	F	Ne
Na	Mg	Al	Si	P	S	Cl	Ar
K	Ca						

Elements whose atoms have a complete outer energy level of electrons

Elements whose atoms form covalent bonds with atoms of the same element, or with atoms of other elements

(***A*** – part **a.** correct; ***M*** – parts **a.** and **b.** correct)

2. *Hydrogen* is an element that has an atom with one electron in its outer *energy* level. By *sharing* this electron with another *hydrogen* atom, both atoms obtain a stable outer energy level of *two* electrons. The particle formed is called a *molecule*. All atoms involved in forming a covalent bond have a *complete* outer energy level once the bond is formed. (***A***)

3. a. The hydrogen molecule, H_2

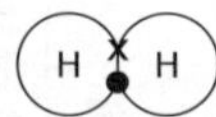

Two separate hydrogen atoms, each with one electron in its outer energy level.

A molecule of hydrogen in which each atom shares one electron with the other atom – both atoms achieve a stable outer energy level of two electrons.

b. The fluorine molecule, F_2

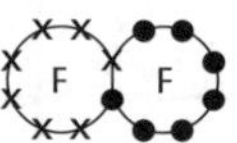

Two separate fluorine atoms, each with seven electrons in their outer energy level.

A molecule of fluorine in which each atom shares one electron with the other atom – both atoms achieve a stable outer energy level of eight electrons.

c. The hydrogen sulfide molecule, H_2S

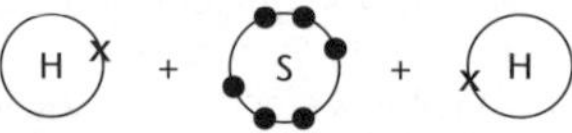

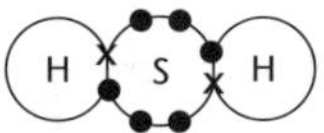

Two separate hydrogen atoms, each with one electron in its outer energy level *and* one sulfur atom with six electrons in its outer level.

A molecule of hydrogen sulfide in which each hydrogen atom shares one electron with the sulfur atom – both hydrogen atoms achieve a stable outer energy level of two electrons *and* the sulfur atom achieves a stable outer energy level of eight electrons.

d. The nitrogen trichloride molecule, NCl_3

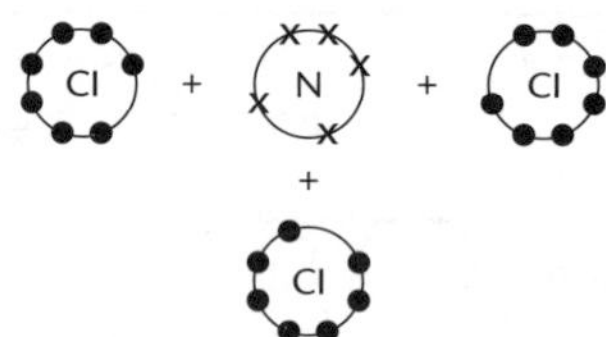

Three separate chlorine atoms, each with seven electrons in their outer energy level, *and* one nitrogen atom with five electrons in its outer energy level.

A molecule of nitrogen trichloride in which each chlorine atom shares one electron with the nitrogen atom – the three chlorine atoms achieve a stable outer energy level of eight electrons *and* the nitrogen atom achieves a stable outer energy level of eight electrons.

e. The silicon tetrahydride molecule, SiH_4

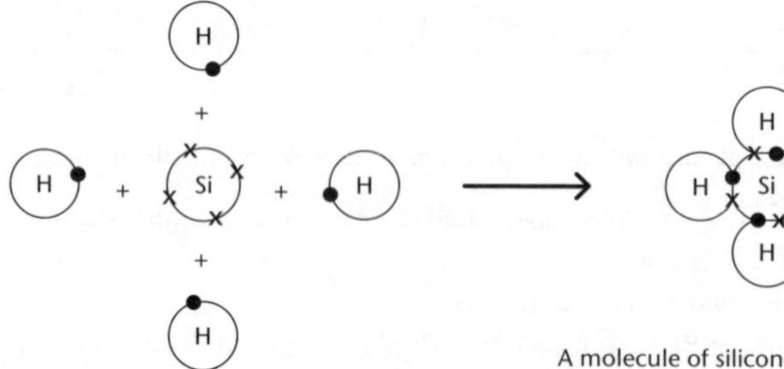

Four separate hydrogen atoms, each with one electron in their outer energy level, *and* one silicon atom with four electrons in its outer energy level.

A molecule of silicon hydride in which each hydrogen atom shares one electron with the silicon atom – the four hydrogen atoms achieve a stable outer energy level of two electrons *and* the silicon atom achieves a stable outer energy level of eight electrons.

(***A*** – one completely correct diagram; ***M*** – three completely correct diagrams; ***E*** – four completely correct diagrams)

4. a. **The ethene molecule, C_2H_4 – one double bond, and four single bonds**

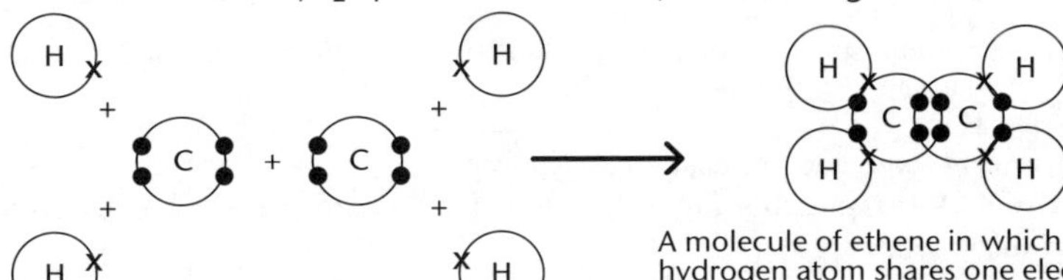

Four separate hydrogen atoms each with one electron in their outer energy level, and two separate carbon atoms each with four electrons in their outer energy levels.

A molecule of ethene in which each hydrogen atom shares one electron with one electron from a carbon atom – hydrogen atom achieves a stable outer energy level of two electrons.
Each carbon atom shares two of its four electrons with a hydrogen atom and the other two electrons with the other carbon atom – each carbon atom achieves a stable outer energy level of eight electrons.

b. **The carbon disulfide molecule, CS_2– two double bonds**

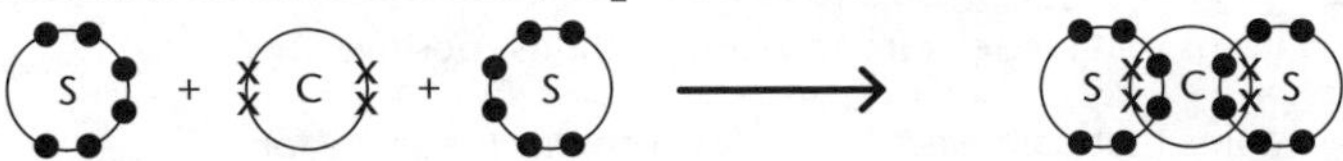

Two separate sulfur atoms each with six electrons in their outer energy level, and one separate carbon atom with four electrons in its outer energy level.

A molecule of carbon disulfide in which each sulfur atom shares two electrons with a carbon atom – each sulfur atom achieves a stable outer energy level of eight electrons.
The carbon atom shares two of its four electrons with a sulfur atom – the carbon atom achieves a stable outer energy level of eight electrons.

c. **The ethyne molecule, C_2H_2 – one triple bond, and two single bonds**

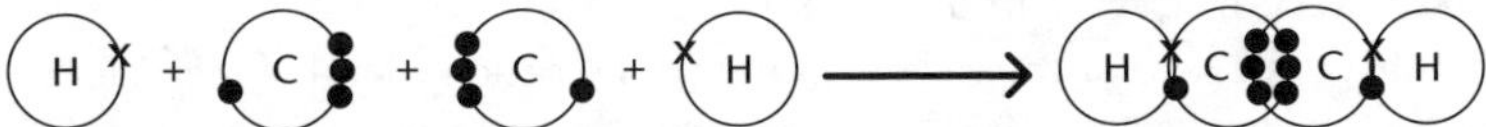

Two separate hydrogen atoms each with one electron in their outer energy level and two separate carbon atoms each with four electrons in their outer energy levels.

A molecule of ethyne in which each hydrogen atom shares one electron with a carbon atom – each hydrogen atom achieves a stable outer energy level of two electrons.
Each carbon atom shares one of its four electrons with a hydrogen atom and three electrons with the other carbon atom – each carbon atom achieves a stable outer energy level of eight electrons.

(**A** – one correct diagram; **M** – two correct diagrams; **E** – three correct diagrams)

5. a. H–F represents a molecule of hydrogen fluoride which contains a *single covalent bond.*

b. S=C=S represents a molecule of carbon disulfide which contains *two double covalent bonds.*

c. H — P — H represents a molecule of phosphorus trihydride which
       |
       H
contains *three single covalent bonds.*

d. $\begin{matrix} H & & H \\ | & & | \\ C & = & C \\ | & & | \\ H & & H \end{matrix}$ represents a molecule of ethene which contains *four single covalent bonds* and *one double covalent bond.*

e. H–O–O–H represents a molecule of hydrogen peroxide which contains *three single covalent bonds*

f. $H-C\equiv C-H$ represents a *molecule* of ethyne which contains *two single covalent bonds* and *one triple covalent bond.*

(**A** – any one of answers **a.** to **f.** completely correct; **M** – any three of answers **a.** to **f.** completely correct; **E** – all answers completely correct)

Unit 11.2 Activity 3A: Ions, ionic compounds and ionic bonding (page 70)

1. **a.** and **b.**

Key	
(light shading)	Atoms that lose electrons to form cations (positive ions)
(dark shading)	Atoms that gain electrons to form anions (negative ions)

H (light)							He
Li (light)	Be (light)	B	C	N (dark)	O (dark)	F (dark)	Ne
Na (light)	Mg (light)	Al (light)	Si	P (dark)	S (dark)	Cl (dark)	Ar
K (light)	Ca (light)						

(**A** – part **a.** correct; **A** – part **b.** correct)

2. Metals have one, two or three *electrons* in the outer energy level of their *atoms*. By *donating* this/these electron(s) to another *atom*, the metal atom becomes stable because it has *eight* (or none, or two in a few cases) electrons in its outer energy level. The atom that *accepts* the electron(s) is from a non-metallic element. The atom accepting the electron(s) gains a *negative* charge. All *ions* are charged and attract each other *strongly*. An ionic compound is electrically neutral – thus the ions must *balance* their charges. The ions are arranged in a three-dimensional lattice where a particular positive ion is surrounded by negative ions and *vice versa*. The overall ratio of ions is *simple*, e.g., 1 : 1, 1 : 2, 2 : 1, etc. The term *molecule* should not be used when referring to ionic compounds – the term *particle* is preferred. (**A** – all correct)

3.

Question	Particle	Electron configuration
b.	Al^{3+}	2, 8
c.	O^{2-}	2, 8
e.	H^{-}	2
f.	Be^{2+}	2
h.	Ca^{2+}	2, 8, 8

O_2^{2-} contains two atoms – each atom has the electron configuration of 2, 8 because of one electron gained and one electron shared.

(**A** – two answers correct; **M** – three or four correct; **E** – five correct)

4. P^{3-} ion; S^{2-} ion; Cl^{-} ion; Ar atom; K^{+} ion; Ca^{2+} ion.
(***A*** – one answer correct; ***M*** – three correct; ***E*** – five correct)

5. **a.**

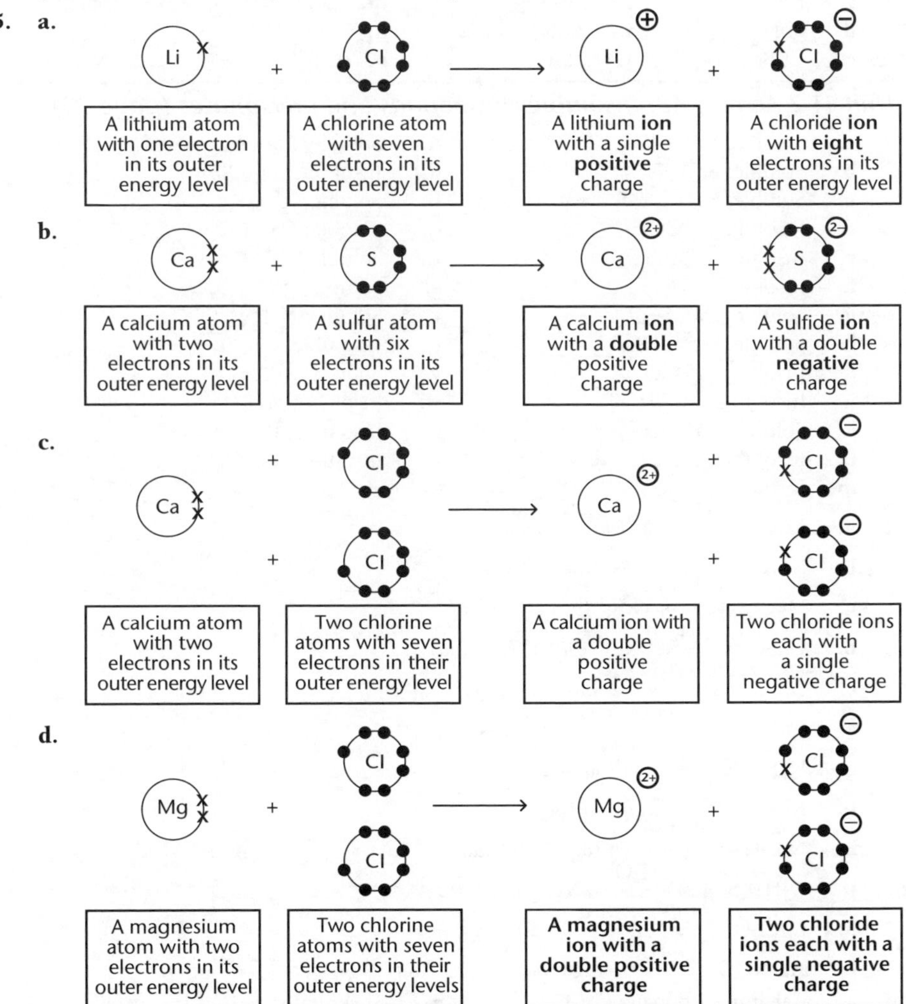

(***A*** – correct words in **a.** or **b.**; ***M*** – **a.** and **b.** correct and charges in **c.** correct; ***E*** – ***M*** plus all **d.** correct)

Unit 11.2 Activity 4A: Formulae of ionic compounds (page 74)

1. **a.** Zinc carbonate
   b. Potassium phosphate
   c. Calcium hydrogen carbonate
   d. Copper(I) oxide
   e. Ammonium dichromate
   f. Potassium thiocyanate
   g. Sodium hypochlorite
   h. Cobalt nitrate
   i. Iron(III) sulfate
   j. Lead hydroxide
   k. Chromium chloride
   l. Sodium permanganate

2. a. $Mg(OH)_2$
b. $Hg(NO_3)_2$
c. FeO
d. PbI_2
e. $CuBr_2$
f. $LiHCO_3$
g. $(NH_4)CO_3$
h. $AgNO_2$
i. Ba_3N_2
j. Ca_3P_2
k. K_2SO_3
l. $Al_2(SO_4)_3$
m. $CoBr_2$
n. Na_2CrO_4
o. $CuCl$
p. FeF_3
q. $KMnO_4$
r. SnO
s. Ag_2S
t. NH_4Cl

Unit 11.2 Activity 4B: Insoluble compounds and precipitates (page 79)

1. a. Soluble; Na^+, Cl^-
b. Soluble; Ba^{2+}, Cl^-
c. Soluble; NH_4^+, NO_3^-
d. Insoluble; $PbSO_4$
e. Insoluble; $Pb(OH)_2$
f. Insoluble; $AgCl$
g. Soluble; Na^+, SO_4^{2-}
h. Soluble; Mg^{2+}, NO_3^-
i. Insoluble; $PbCO_3$
j. Insoluble; $Al(OH)_3$
k. Soluble; Zn^{2+}, Cl^-
l. Insoluble; $CuCO_3$
m. Insoluble; $BaSO_4$
n. Soluble; Fe^{3+}, Cl^-
o. Insoluble; $Zn(OH)_2$
p. Soluble; K^+, Cl^-
q. Soluble; Na^+, CO_3^{2-}
r. Soluble; Na^+, I^-
s. Soluble; Fe^{2+}, SO_4^{2-}
t. Soluble; NH_4^+, CO_3^{2-}
u. Insoluble; $CaCO_3$
v. Soluble; Al^{3+}, Cl^-
w. Insoluble; AgI
x. Insoluble; PbI_2
y. Insoluble; $Cu(OH)_2$

2. a. $KOH(s) \xrightarrow{H_2O} K^+(aq) + OH^-(aq)$
b. $NaNO_3(s) \xrightarrow{H_2O} Na^+(aq) + NO_3^-(aq)$
c. $MgCl_2(s) \xrightarrow{H_2O} Mg^{2+}(aq) + 2Cl^-(aq)$
d. $CuSO_4(s) \xrightarrow{H_2O} Cu^{2+}(aq) + SO_4^{2-}(aq)$
e. $Na_2CO_3(s) \xrightarrow{H_2O} 2Na^+(aq) + CO_3^{2-}$
f. $Al(NO_3)_3(s) \xrightarrow{H_2O} Al^{3+}(aq) + 3NO_3^-(aq)$
g. $NaBr(s) \xrightarrow{H_2O} Na^+(aq) + Br^-(aq)$
h. $KI(s) \xrightarrow{H_2O} K^+(aq) + I^-(aq)$
i. $NaOH(s) \xrightarrow{H_2O} Na^+(aq) + OH^-(aq)$
j. $(NH_4)_2SO_4(s) \xrightarrow{H_2O} 2NH_4^+(aq) + SO_4^{2-}(aq)$
k. $ZnCl_2(s) \xrightarrow{H_2O} Zn^{2+}(aq) + 2Cl^-(aq)$
l. $Mg(NO_3)_2(s) \xrightarrow{H_2O} Mg^{2+}(aq) + 2NO_3^-(aq)$

3. a. Calcium carbonate, $CaCO_3$.
b. Lead iodide, PbI_2.
c. Copper hydroxide, $Cu(OH)_2$.
d. None.
e. Silver chloride, $AgCl$.
f. Barium sulfate, $BaSO_4$.
g. None.
h. Iron(II) hydroxide, $Fe(OH)_2$.

Unit 11.2 Activity 4C: Ionic equations (page 80)

1. $Fe^{2+}(aq) + 2OH^-(aq) \rightarrow Fe(OH)_2(s)$ (**A**)
2. a. $Cu^{2+}(aq) + 2OH^-(aq) \rightarrow Cu(OH)_2(s)$
b. $Cu^{2+}(aq) + CO_3^{2-}(aq) \rightarrow CuCO_3(s)$ (**M**)

c. $Ba^{2+}(aq) + SO_4^{2-}(aq) \rightarrow BaSO_4(s)$
d. $Ag^{+}(aq) + Cl^{-}(aq) \rightarrow AgCl(s)$ (**M**)

3. a. $Ca^{2+}(aq) + SO_4^{2-}(aq) \rightarrow CaSO_4(s)$
b. $Mg^{2+}(aq) + 2OH^{-}(aq) \rightarrow Mg(OH)_2(s)$
c. $Pb^{2+}(aq) + 2Cl^{-}(aq) \rightarrow PbCl_2(s)$ (**M**)
d. $Al^{3+}(aq) + 3OH^{-}(aq) \rightarrow Al(OH)_3(s)$
e. $Mg^{2+}(aq) + CO_3^{2-}(aq) \rightarrow MgCO_3(s)$
f. $Ag^{+}(aq) + I^{-}(aq) \rightarrow AgI(s)$

4. a. $Ca^{2+}(aq) + CO_3^{2-}(aq) \rightarrow CaCO_3(s)$ (**M**)
b. $Pb^{2+}(aq) + 2I^{-}(aq) \rightarrow PbI_2(s)$ (**M**)
c. $Cu^{2+}(aq) + 2OH^{-}(aq) \rightarrow Cu(OH)_2(s)$ (**M**)
d. $Ag^{+}(aq) + Cl^{-}(aq) \rightarrow AgCl(s)$ (**M**)
e. $Ba^{2+}(aq) + SO_4^{2-}(aq) \rightarrow BaSO_4(s)$ (**M**)
f. $Fe^{3+}(aq) + 3OH^{-}(aq) \rightarrow Fe(OH)_3(s)$ (**M**)

Unit 11.2 Activity 4D: Complex ions (page 81)

1. a. $Cu^{2+} + 4NH_3$
b. $[Pb(OH)_4]^{2-}$
c. $[Zn(NH_3)_4]^{2+}$
d. $Al^{3+} + 4OH^{-}$ *or* $Al(OH)_3 + OH^{-}$
e. $[FeSCN]^{2+}$
f. $Ag^{+} + 2NH_3$

2. a. $Zn(OH)_2 + 2OH^{-} \rightarrow [Zn(OH)_4]^{2-}$
b. $Cu^{2+} + 2OH^{-} \rightarrow Cu(OH)_2(s)$
$Cu^{2+} + 4NH_3 \rightarrow [Cu(NH_3)_4]^{2+}$
or $Cu(OH)_2 + 4NH_3 \rightarrow [Cu(NH_3)_4]^{2+} + 2OH^{-}$
c. $Pb^{2+} + 2OH^{-} \rightarrow Pb(OH)_2$
$Pb(OH)_2 + 2OH^{-} \rightarrow Pb(OH)_4^{2-}$
d. $Ag^{+} + Cl^{-} \rightarrow AgCl$
$AgCl^{-} + 2NH_3 \rightarrow [Ag(NH_3)_2]^{+} + Cl^{-}$
or $Ag^{+} + 2NH_3 \rightarrow [Ag(NH_3)_2]^{+}$

Unit 11.2 Activity 5A: Chemical and physical changes (page 84)

1. a. Physical; easily reversed.
b. Chemical; impossible to reverse.
c. Chemical; large quantities of energy released.
d. Physical; properties of reactants retained.
e. Physical; mixture has properties of salt and sand.
f. Chemical; product has new properties (taste).
g. Chemical; products have constant composition.
h. Physical; no new product.

2. The white powder is magnesium oxide, MgO. A compound has formed since the properties of MgO are different from those of the elements it was made from: Mg conducts electricity, (solid) MgO does not. O_2 has a low melting point, MgO has a high melting point.

3.

Physical change	Chemical change
a, d, e	b, c, f, g

Unit 11.2 Activity 5B: Structure of solids (page 95)

1. B, F (***A***)
2. **a.** **ii.** **b.** **iii.** **c.** **iv.** **d.** **i.**, **ii.**, **v.** (***A***)
3. **a.** Metallic – high m.p., good conductor, malleable and ductile.
   b. Molecular – low m.p., gas at room temperature, non-conductor.
   c. Molecular – low m.p., liquid at room temperature, non-conductor.
   d. Ionic – high m.p., solid does not conduct, liquid does, aqueous solution conducts.
   e. Ionic – high m.p., solid does not conduct, liquid does.
   f. Metallic – high m.p., good conductor, malleable and ductile.
   g. Molecular – low m.p., non-conducting.
   h. Ionic – high m.p., conducts when liquid.
   (***A*** – classification; ***M*** – reason)
4.

Substance	Type of solid	Particles present in the solid	Conductivity	Explanation for conductivity/ non-conductivity
Sodium bromide	Ionic	Ions	Conducts when molten or dissolved	Ions free to move
Graphite	2-D covalent network	Atoms	Conducts as solid	Delocalised electrons free to move
Aluminium	Metallic	Positive ions surrounded by moving electrons	Conducts as a solid	Delocalised electrons free to move
Carbon dioxide	Molecular	Molecules	Non-conducting	No mobile ions or delocalised electrons
Silica	3-D covalent network	Atoms	Non-conducting	No mobile ions or delocalised electrons

(***A*** – six out of eight types of solid, types of particles and conductivity correct; ***M*** – also three explanations correct)

5. **a.** SiO_2 consists of Si and O atoms held together by covalent bonds in a 3-D network. As all the bonds in the network are strong, they are difficult to break, so large amounts of energy are required to separate the atoms, and the structure has a high m.p. CO_2 is a molecular solid made up of molecules with weak forces between the molecules. As these attractions are weak, little energy is needed to separate the molecules and the melting point is low. (***E***)
   b. Sodium atoms are held together in a 3-D lattice by metallic bonding, in which valence electrons are 'pooled' among neighbouring atoms. These electrons are able to move freely through the solid, allowing it to conduct electricity. Solid I_2 is made up of molecules and there are no free-moving charged particles, so I_2 cannot conduct electricity. (***E***)

c. KCl is an ionic compound. In the solid there is a regular 3-D array of K^+ and Cl^- ions. These are held in fixed positions and cannot move around, so the solid does not conduct electricity. In the liquid, the ions separate and are free to move around and so conduct electricity. (***E***)

d. Ag is a metallic solid with the Ag atoms in a 3-D array and the valence electrons 'pooled' among the neighbouring atoms. The atoms are held together by the flow of valence electrons around neighbouring atoms. This is a non-directional force, so layers of atoms can slide over each other without disrupting the lattice and breaking the bond. Sodium chloride consists of a 3-D array of positive and negative ions. If the lattice is distorted, like charges will line up next to each other. Since these repel, the layers will separate and the structure will break up. (***E***)

e. Graphite has a 2-D arrangement of C atoms with covalent bonds between them. The layers are held together by weak intermolecular forces. These are easily broken, allowing the layers to slide over each other and a layer can be rubbed onto paper. In diamond, all the bonds are covalent and the atoms are arranged in a 3-D tetrahedral arrangement. This makes the substance very hard, and hence it will cut paper. (***E***)

f. Ice is a molecular solid consisting of water molecules held together by weak intermolecular forces. These are easily broken with little energy input, so the melting point is low. Magnesium oxide is an ionic solid made up of Mg^{2+} and O^{2-} ions held together by a strong ionic bond. These ionic attractions are strong, so more energy is needed to separate the ions than the water molecules, so the melting point of MgO is high. (***E***)

6. a. Metallic bonds. **b.** Weak intermolecular forces.

c. Weak intermolecular forces. (***A***)

7. a. Ionic solids.

b. Metallic bonds.

c. Covalent bonds within molecules – weak forces between molecules.

d. Conduct in solid and molten state.

e. Conduct when molten or dissolved.

f. Graphite.

g. High melting and boiling points.

h. Low melting and boiling points.

8.

Substance:	a. Type of particle found in solid substance:	b. Attractive force broken when solid substance melts:	c. Attractive force existing between particles:
i. Phosphorus, P_4	Molecules	Intermolecular forces	Weak
ii. Silicon dioxide, SiO_2	Atoms	Covalent bonds	Strong
iii. Magnesium chloride	Ions	Ionic bonds	Strong
iv. Zinc, Zn	Atoms/ions (with delocalised electrons)	Metallic bonds	Strong

(***M***)

9.

Substance:	a. Type of particle found in solid substance:	b. Attractive force:	c. Relative melting point:
i. Al	Atoms/ions (with delocalised electrons)	Metallic bonds	High
ii. C	Atoms	Covalent bonds	High
iii. H_2S	Molecules	(Weak) intermolecular forces	Low
iv. CaO	Ions	Ionic bonds	High

(***M***)

10. a. Lead atoms are held together in a 3-D lattice by metallic bonding, in which valence electrons are 'pooled' and attracted to the nuclei of neighbouring atoms. This is a non-directional force; layers of atoms can slide over each other without breaking the metallic bond and disrupting the structure and breaking the metal. (***E***)

b. Zinc chloride is an ionic compound made up of positive Zn^{2+} and negative Cl^- ions. In the solid state, the ions are held in fixed positions by strong electrostatic attractions, or ionic bonds. When molten, the charged ions are separated and free to move and conduct electricity. (***E***)

c. Oxygen consists of O_2 molecules, and weak intermolecular forces exist between the molecules. As these attractions are weak, the molecules are easily separated and the melting point is low. (***E***)

d. Diamond consists of carbon atoms held together by covalent bonds in a tetrahedral arrangement, so that a 3-D network exists. Since the covalent bonds are strong, they are difficult to break, meaning the atoms are held firmly in the solid, so the structure is very hard. (***E***)

e. Copper atoms are held together in a 3-D lattice by metallic bonding, in which valence electrons are 'pooled' and are attracted to the nuclei of neighbouring atoms. Conduction of electricity requires free-moving charges – which are the moving valence electrons; hence, copper can be used for electrical wiring. (***E***)

11. a. Sulfur, S_8, consists of molecules held together by weak intermolecular forces. Since there are no charged particles that are free to move, sulfur does not conduct electricity.

b. Iron atoms are held together in a 3-D lattice by metallic bonding, in which valence electrons are 'pooled' and are attracted to the nuclei of neighbouring atoms. This bond is very strong, so large amounts of energy are needed to separate the atoms. Hence, iron has a high melting point.

c. Silicon carbide consists of silicon and carbon atoms held together by covalent bonds in a tetrahedral arrangement, so that a 3-D network exists. As the covalent bonds are strong, they are difficult to break, meaning the atoms are held firmly in the solid, so the structure is very hard and has a high melting point.

d. NaCl is an ionic compound made up of a regular arrangement of alternating Na^+ and Cl^- ions held together by strong ionic bonds, resulting in a hard substance. When pressure is applied to the solid, ions of like charge will line up next to each other. Since like charges repel, the ions push apart and the layers separate, breaking up the crystal lattice. (***E***)

12. a. Dry ice consists of CO_2 molecules with weak intermolecular forces between the CO_2 molecules. As these attractions are weak, the molecules are easily separated and the sublimation ('melting') point is low and the dry ice sublimes to a gas at low temperatures.

b. CO_2 molecules are non-polar, therefore dry ice is soluble in a non-polar solvent such as cyclohexane. Both molecules have similar weak intermolecular forces, and, on mixing, similar weak forces will exist between the two different molecules.

c. Dry ice does not conduct electricity because it is a molecular solid and there are no freely-moving charged particles in the solid. (***E***)

13. Iodine consists of I_2 molecules, and weak intermolecular forces exist between the molecules. As these attractions are weak, the molecules are easily separated and the 'melting' (sublimation) point is low. Iodine molecules are non-polar, therefore iodine is soluble in a non-polar solvent such as cyclohexane. Both molecules have similar weak intermolecular forces, and similar weak forces will exist between the two different molecules. This allows them to mix and form a solution

Potassium iodide is an ionic solid, consisting of ions held together by strong ionic bonds/electrostatic attractions. As these attractions are strong, a lot of energy is needed to separate the ions (more than for the molecules of iodine), so potassium iodide has a high melting point. When ionic solids are soluble in water, the ions can be separated from the lattice due to the attraction towards the polar water molecules. (***E***)

14. CO_2 exists as molecules. Weak forces exist between the molecules. As all valence electrons are involved in forming covalent bonds, there are no free-moving charges and so there is no electrical conduction. Since the forces between molecules are weak, they are easily overcome – little energy is required to separate the molecules (therefore CO_2 has a low sublimation point, –78 °C). Since the molecules are easily separated, solid CO_2 is easy to break up so is weak/brittle.

SiO_2 exists as a 3-D covalent network solid. Strong covalent bonds hold the Si and O atoms together in a 3-D arrangement. Since all valence electrons are involved in forming covalent bonds, there are no free-moving charges and so there is no electrical conduction. The covalent bonds between atoms are strong and a lot of energy is needed to overcome the bonds and separate the atoms, so the melting point is very high. Also, since the strong covalent bonds hold the atoms firmly in the 3-D structure, the solid is very hard. (***E***)

15. Diamond is used to make saw teeth because it is very hard, due to each C atom being firmly held in a 3-D lattice by covalent bonds to four other C atoms, forming a tetrahedral arrangement. Graphite, however, is a lubricant because it has a 2-D hexagonal arrangement of C atoms, with the layers held together by only weak intermolecular forces. Since these weak intermolecular forces are easily broken, the layers can slide over each other. Graphite is used to make electrodes because there are delocalised electrons between the layers, and these are free to move and conduct electricity. (***E***)

Unit 11.2 Activity 6A: Trends in reactivity of metals and non-metals (page 101)

1. **a.** Metals and non-metals. (**A**)
   b. (**A** – all three correct)
   Metals Non-metals Inert gases
   c. **i.** Periods.
   ii. Groups. (**A** – both correct)
2. **a.** Inert gas.
   b. Metal.
   c. Non-metal.
   d. Inert gas.
   e. Metal.
   (**A** – three correct; **M** – five correct)
3. Elements whose atoms have *eight* electrons in their outer energy levels show no tendency to react. These elements are known as the *inert gases*. If an element has an atom with one, two or three electrons in its outer energy level then these electrons can be *donated* to an atom of an element that has *five*, *six* or *seven* electrons in its outer energy level. In this way, *ions* are formed. Atoms that donate electrons form *positive* ions and negative ions are formed by atoms *accepting* electrons. *Metals* form positive ions and *non-metals* form negative ions.

   The reactivity of a metal is related to the *ease* with which electron(s) in the atoms of the element are donated. Sodium is more reactive than magnesium, which is more reactive than *aluminium*. There is *one* electron in the outer energy level of an atom of sodium, two electrons in the outer energy level of an atom of *magnesium* and three electrons in the outer energy level of an atom of *aluminium*. The *energy* needed to remove one electron from an atom is less than that needed to remove two, which is less than that needed to remove three.

   Similarly one electron is attracted more *strongly* to an atom of a non-metal than are two or three electrons. Hence, fluorine is more reactive than *oxygen*, which is more reactive than *nitrogen*.

   (**A** – ten words correctly placed; **M** – twenty words correctly placed)
4. **a.** Sodium > magnesium > aluminium. (**A**)
   b. Nitrogen > carbon > helium = neon. (**A** – correct order; **M** – helium and neon are stated to be the same reactivity)

Unit 11.2 Activity 6B: Comparison of covalently bonded and ionically bonded compounds (page 104)

1. **a.** **i.**, **ii.**, **iv.**
   b. **iii.**, **v.**, **vi.**
   (**A** – two correct; **M** – four correct; **E** – six correct)
2. **a.** **ii.**, **iii.**, **iv.**, **vii.**
   b. **i.**, **v.**, **vi.**, **viii.**
   (**A** – three correct; **M** – five correct; **E** – seven correct)

3. **a.** **i.** The low boiling point of the compound indicates weak forces between the particles. The fact that it does not conduct electricity as a liquid indicates no charged particles are present in the compound. These particles will be molecules with covalent bonding between the atoms in the molecule.

   b. **i.** The low melting point of the compound and the fact that it does not conduct electricity when molten indicates weak forces between the particles. These particles will be molecules with covalent bonding between the atoms in the molecule.

   c. **ii.** The high melting point of the compound indicates strong forces between the particles. The fact that the compound conducts electricity, when molten, indicates charged particles. These particles will be ions. The insolubility in water is due to the strong attraction between the ions, which cannot be overcome by water molecules attracting the ions away from each other.

   d. **i.** The low boiling point of the compound and the fact that it does not conduct electricity when in the liquid state indicates weak forces between the particles and a lack of charged particles. The particles will be molecules with covalent bonding between the atoms in the molecule. The slight conductivity of the aqueous solution is due to the molecules reacting with the water molecules, to a small extent, to produce a few ions.

   (***A*** – three identifications correct; ***M*** – one explanation linking property to the type of bonding; ***E*** – three explanations linking property to the type of bonding)

4.

	i. Melting point	**ii.** Solubility in water	**iii.** Electrical conductivity of liquid form of substance
a. Nitrogen trichloride	Low	Low	Low
b. Carbon disulfide	Low	Low	Low
c. Sodium sulfide	High	High	High
d. Potassium fluoride	High	High	High

(***A*** – three correct answers for each of **a.** – **d.**)

Unit 11.2 Activity 7A: Features of the periodic table (page 107)

1. **a.** C **b.** A, D **c.** C **d.** E **e.** A, B
2. **a.**, **b.** and **c.**

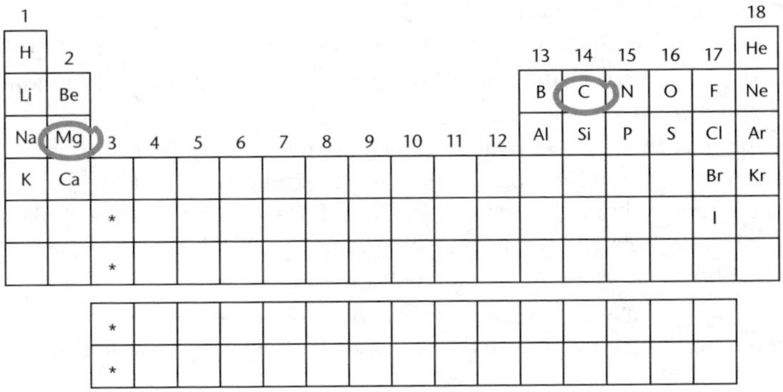

1																	18
H	2											13	14	15	16	17	He
Li	Be											B	C	N	O	F	Ne
Na	Mg	3	4	5	6	7	8	9	10	11	12	Al	Si	P	S	Cl	Ar
K	Ca															Br	Kr
		*														I	
		*															

*															
*															

3. **a.** Hydrogen atoms can lose one electron to achieve a stable electron configuration and form a positive ion with a +1 charge, as can atoms of group 1 elements such as sodium.

   b. Hydrogen atoms can gain one electron to achieve a stable full valence level, as can atoms of group 17 elements such as chlorine. (***M***)

Unit 11.2 Activity 7B: Periodic trends for the halogens (page 109)

1. Halogen atoms have seven electrons in their outer energy level. Two halogen atoms share one electron each to form a single covalent bond and both atoms achieve an octet outer energy level. Two atoms join to make a diatomic molecule.

Two halogen atoms with seven electrons in their outer energy level.

A diatomic molecule – two atoms joined by a single covalent bond.

(***A*** – same number of electrons in outer energy level; ***M*** – sharing of electrons produces octet for both atoms; ***E*** – diagram)

2. **a.** Covalent. **b.** ICl (or ClI) **c.** Red-brown liquid.

   d. **i.** <50 °C **ii.** ~100 °C (***M***)

3. **a.** Each astatine atom has seven electrons in its outer energy level. By forming a diatomic molecule, each atom achieves a noble gas configuration in its outer energy level. (***M***)

   b. The trend in the melting points of the halogens is to increase as the molecular masses of the halogens increase. (***M***)

   An atom of astatine has more electrons than an iodine atom. The molecule of astatine will have stronger intermolecular forces than an iodine molecule because it contains more protons and electrons than does an iodine molecule. (***E***)

Unit 11.2 Activity 7C: Chlorides and oxides of the third period (page 114)

1. **a.** *Trend 1:* Melting point decreases across the period from sodium chloride to chlorine.

   Trend 2: The number of electrons in the valence shell of the atom increases from one to four and the number of chlorine atoms increases to match this increase.

   Discussion: The type of bonding present in the chlorides of the third period explains the trends demonstrated. Melting points decrease, as there is a change from ionic to covalent bonding. (***M***)

   b. Sodium chloride is an ionic compound, formed by the electrostatic attraction of oppositely charged ions. The ionic bond is a very strong bond and a high temperature is required to separate the ions. Phosphorus chloride is a molecular compound and the molecules are held together only by weak intermolecular forces – it is these that break easily, so the melting point is low. (***E***)

   c. The melting point of magnesium chloride is high (712 °C) – this represents a very strong bond – bonding type predicted to be ionic, because metal and non-metal involved.

Sulfur dichloride has a low melting point (–80 °C) – this suggests only weak forces between particles in the compound; compound is thus molecular, with weak intermolecular forces and strong covalent bonds between the S and Cl atoms in the molecule. (***E***)

2. a.

Group	1	2	13	14	15	16	17
Element	Na	Mg	Al	Si	P	S	Cl
Oxides	Na_2O	MgO	Al_2O_3	SiO_2	P_4O_6 P_4O_{10}	SO_2 SO_3	ClO_2 Cl_2O
Bonding	Ionic	Ionic	Ionic	Covalent	Covalent, weak intermolecular forces	Covalent, weak intermolecular forces	Covalent, weak intermolecular forces

(***M***)

b. i. SO_2, ClO_2, Cl_2O

ii. SiO_2 (***M***)

3. NaCl and $MgCl_2$ both have high melting points, so strong bonds must be broken in order to separate the particles (when the compounds melt). NaCl has ionic bonds between Na^+ and Cl^- ions. $MgCl_2$ has ionic bonds between Mg^{2+} and Cl^- ions. Ionic bonds are strong. $SiCl_4$ and SCl_2 have low melting points, so there are weak bonds broken when these compounds melt. $SiCl_4$ and SCl_2 are molecular substances with weak forces of attraction between the molecules. These forces are easily broken, allowing the solids to melt at low temperatures. (***E***)

4. a. Melting and boiling points increase from Na_2O to SiO_2, then drop sharply to SO_2.

b. Na_2O and MgO are ionic compounds.When the solid melts, the strong ionic bonds between the Na^+ and O^{2-} or Mg^{2+} and O^{2-} ions are broken. It takes a lot of energy to separate the ions, so the melting points are relatively high. SiO_2 forms a giant network solid, with strong covalent bonds between the Si and O atoms. A large amount of energy is needed to break all these bonds, so the melting and boiling points are high.

SO_2 is a molecular compound. When the solid melts, the weak forces between the molecules break. Since only a small amount of energy is needed to break these forces, the melting point is low. (***E***)

Unit 11.3 Activity 1A: Valency, formulae of compounds (page 118)

1. **b.**, **c.**, **e.**, **f.** (***A*** – two correct; ***M*** – four correct)

2. a. P_2O_3

b. PF_3

c. OF_2

(***A*** – one correct; ***M*** – two correct; ***E*** – three correct)

3. a. $SeAt_2$

b. $BiAt_3$

c. Bi_2Se_3

(***A*** – one correct; ***M*** – two correct; ***E*** – three correct)

4. a. 7

b. 6

c. 5

(***A*** – two correct; ***M*** – three correct)

5. A^+, B^{2+}, C^{3+}, X^-, Y^{2-}, Z^{3-}
   (***A*** – three correct; ***M*** – six correct)
6. **a.** NaCl
   b. MgS
   c. AlN
   (***A*** – two correct; ***M*** – three correct)
7. **a.** SkO_2
   b. $SkCl_4$
   c. Sk_3N_4
   (***M*** – two correct; ***E*** – three correct)
8. **a.** Li^+
   b. Ca^{2+}
   c. Sc^{3+}
   (***A*** – two correct; ***M*** – three correct)
9. **a.** Na_3PO_4
   b. $Mg(CH_3COO)_2$
   c. $Al_2[(COO)_2]_3$
   (***M*** – one correct; ***E*** – two correct)
10. **a.** Ft^+, Hk^{2+}, Sq^{2-}, Vo^- (***A*** – two correct; ***M*** – four correct)
    b. **i.** Ft_2Sq
    ii. $HkVo_2$
    (***M*** – one correct; ***E*** – two correct)
11. **a.** **i.** Na_2CO_3
    ii. $Al(NO_3)_3$
    iii. NH_4Cl
    iv. $Ca(HCO_3)_2$
    v. $(NH_4)_2SO_4$.
    (***A*** – two correct; ***M*** – three correct; ***E*** – five correct)
    b. **i.** 2 × Na, 1 × C, 3 × O.
    ii. 1 × Al, 3 × N, 9 × O.
    iii. 1 × N, 4 × H, 1 × Cl.
    iv. 1 × Ca, 2 × H, 2 × C, 6 × O.
    v. 2 × N, 8 × H, 1 × S, 4 × O.
    (***A*** – two correct; ***M*** – three correct; ***E*** – five correct)

Unit 11.3 Activity 1B: Writing equations (page 122)

1. **a.** Lithium + oxygen → lithium oxide
   b. Magnesium + water → magnesium hydroxide + hydrogen
   c. Iron + steam → iron oxide + hydrogen
   d. Zinc + sulfuric acid → zinc sulfate + hydrogen
   e. Aluminium + hydrochloric acid → aluminium chloride + hydrogen
   (***A*** – three equations correct; ***M*** – five equations correct)

2. **a.** *Sodium* + oxygen → sodium oxide
   b. Calcium + *sulfur* → calcium sulfide
   c. *Aluminium* + *chlorine* → aluminium chloride
   d. *Sodium* + water → sodium hydroxide + hydrogen
   e. *Magnesium* + steam → magnesium oxide + *hydrogen*
   f. *Magnesium* + sulfuric acid → magnesium sulfate + hydrogen
   g. *Zinc* + *hydrochloric acid* → zinc chloride + hydrogen

   (**A** – three equations correct; **M** – six equations correct)
3. **a.** $\mathbf{2}Li(s) + S(s) \rightarrow Li_2S(s)$
   b. $\mathbf{2}Na(s) + Cl_2(g) \rightarrow \mathbf{2}NaCl(s)$
   c. $\mathbf{4}Al(s) + \mathbf{3}O_2(g) \rightarrow \mathbf{2}Al_2O_3(s)$
   d. $\mathbf{2}Ca(s) + \mathbf{2}H_2O(\ell) \rightarrow \mathbf{2}Ca(OH)_2(aq) + H_2(g)$
   e. $Mg(s) + \mathbf{2}HCl(aq) \rightarrow MgCl_2(aq) + H_2(g)$

   (**A** – two equations correct; **M** – four equations correct; **E** – five equations correct)
4. **a.** $2Mg(s) + O_2(g) \rightarrow 2MgO(s)$
   b. $4Na(s) + O_2(g) \rightarrow 2Na_2O(s)$
   c. $2Na(s) + 2H_2O(\ell) \rightarrow 2NaOH(aq) + H_2(g)$
   d. $2Mg(s) + 2H_2O(\ell) \rightarrow 2Mg(OH)_2(aq) + H_2(g)$
   e. $Zn(s) + H_2O(\ell) \rightarrow ZnO(s) + H_2(g)$
   f. $Mg(s) + H_2SO_4(aq) \rightarrow MgSO_4(aq) + H_2(g)$
   g. $Zn(s) + 2HCl(aq) \rightarrow ZnCl_2(aq) + H_2(g)$
   h. $2Al(s) + 3H_2SO_4(aq) \rightarrow Al_2(SO_4)_3(aq) + 3H_2(g)$

   (For each equation: **A** – all formulae correct; **M** – **A** plus equation balanced correctly)

Unit 11.3 Activity 1C: The language of chemistry (page 123)

1. **b.**, **c.**, **f.** (**A** – one correct; **M** – two correct; **E** – three correct)
2. **a.** Steam is composed of molecules that can be represented as $H_2O(g)$.
   d. Glucose, $C_6H_{12}O_6$, is a sugar whose molecule contains hydrogen atoms, oxygen atoms and carbon atoms.
   e. When you are taking exercise, you need more oxygen in your blood than when you are sitting still.

   (**A** – one sentence correct: **M** – two sentences correct; **E** – three sentences correct)
3. Ammonia is a gaseous chemical with a strong smell. The molecule of ammonia has a formula $NH_3(g)$. There are three times as many *hydrogen atoms* in *a molecule of* ammonia than there are nitrogen *atoms*. *Ammonia* is very soluble in water. In water, the particles present can be represented as $NH_3(aq)$. The solution smells of ammonia because some of the *ammonia* can escape from the solution easily.

   (**A** – two sentences correct; **M** – three sentences correct)
4. **a.** H
   b. H_2
   c. 2O
   d. $2O_2$
   e. $H_2O(\ell)$
   f. $H_2O(g)$.

   (**A** – two correct; **M** – four correct; **E** – all six correct)

5. Sample ***A***, ***M*** and ***E*** answers are:

   Alum is a compound that contains the elements potassium, sulfur, oxygen, aluminium and hydrogen. (***A***)

   Alum is a substance whose smallest particle contains atoms of potassium, sulfur, oxygen, aluminium and hydrogen. (***M***) *Or*: Alum is a substance whose smallest particle contains atoms of potassium, sulfur, oxygen, aluminium and molecules of water. (***M***)

   Alum is a substance whose smallest particle contains two atoms of potassium, four atoms of sulfur, forty atoms of oxygen, two atoms of aluminium and forty-eight atoms of hydrogen. (***E***) *Or*: Alum is a substance whose smallest particle contains two atoms of potassium, four atoms of sulfur, sixteen atoms of oxygen, two atoms of aluminium and twenty-four molecules of water. (***E***)

Unit 11.3 Activity 2A: Oxidation and reduction (page 127)

1. **a.** Oxidation is the addition of oxygen to an element or compound.
   b. Reduction is the addition of hydrogen to an element or compound.
   c. The removal of an electron or electrons from a particle in a reaction.
   d. The addition of an electron or electrons to a particle in a reaction.

   (***A*** – two definitions include key word 'addition' or 'removal' as appropriate; ***M*** – all definitions include key word 'addition' or 'removal' as appropriate; ***E*** – all definitions include key words 'addition' or 'removal' and 'element' or 'particle' as appropriate)
2. **a.** Charcoal. Oxygen has been added to produce carbon dioxide.
   b. Iron. Oxygen has been added to produce iron oxide.
   c. Magnesium. Oxygen has been added to produce magnesium oxide.

   (***A*** – all three substances correctly identified; ***M*** – all three explanations correct)
3. **a.** Vegetable oil. Hydrogen has been added to produce margarine.
   b. Oxygen. Hydrogen has been added to produce steam.
   c. Sulfur. Hydrogen has been added to produce hydrogen sulfide.

   (***A*** – all three substances correctly identified; ***M*** – all three explanations correct)
4. **a.** Mg(s). Electrons have been removed to form the Mg^{2+}(aq) particle.
   b. I^-(aq). Electrons have been removed to form the I_2(s) particle.

   (***A*** – one particle correctly identified; ***M*** – both particles correctly identified; ***E*** – both explanations correct).
5. **a.** Cl_2(aq). Electrons have been added to form the Cl^-(aq) particles.
   b. Fe^{3+}(aq). An electron has been added to form the Fe^{2+}(aq) particle.

   (***A*** – one particle correctly identified; ***M*** – both particles correctly identified; ***E*** – both explanations correct).

Unit 11.3 Activity 2B: Oxidants and reductants (page 131)

1.

Question	i. Oxidant	ii. Reductant	iii. Product of oxidation	iv. Product of reduction
a.	Copper oxide	Charcoal	Carbon dioxide	Copper
b.	Steam	Iron	Iron oxide	Hydrogen
c.	Oxygen	Magnesium	Magnesium oxide	Magnesium oxide

(***A*** – four identifications correct; ***M*** – seven identifications correct; ***E*** – twelve identifications correct)

2. **a.** Aluminium.
b. Aluminium oxide.
(***A*** – one correct; ***M*** – two correct)

3. **a.** $O_2(g)$.
b. $PbSO_4(s)$.
c. $HI(g)$.
d. $Mg(s)$.
(***A*** – two correct; ***M*** – four correct)

4. **a.** **L**oss of **E**lectron is **O**xidation.
b. **G**ain of **E**lectron is **R**eduction.
(***A*** – both correct)

5. **a.** Oxidant (electron acceptor), $Cl_2 + 2e^- \rightarrow 2Cl^-$.
b. Reductant (electron donor), $Fe^{2+} \rightarrow Fe^{3+} + e^-$.
c. Oxidant (electron acceptor), $Fe^{3+} + e^- \rightarrow Fe^{2+}$.
d. Oxidant (electron acceptor), $Cu^{2+} + 2e^- \rightarrow Cu$.
e. Reductant (electron donor), $Mg \rightarrow Mg^{2+} + 2e^-$.
f. Oxidant (electron acceptor), $O_2 + 4e^- \rightarrow 2O^{2-}$.
(***A*** – any two correct; ***M*** – any four correct; ***E*** – all six correct)

6. **a.**

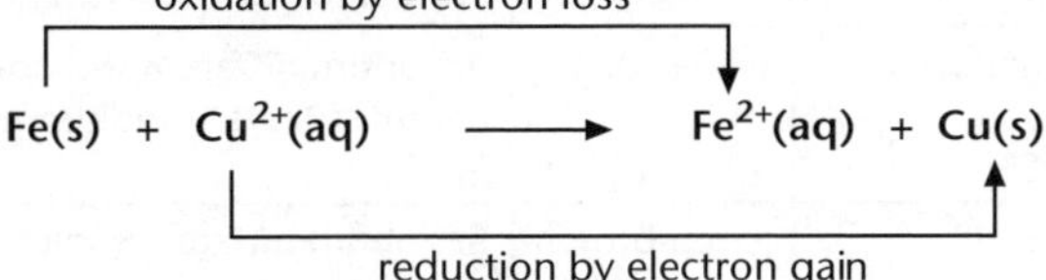

b.

reduction by electron gain

$$2Fe^{3+}(aq) + 2I^-(aq) \longrightarrow 2Fe^{2+}(aq) + I_2(aq)$$

oxidation by electron loss

c.

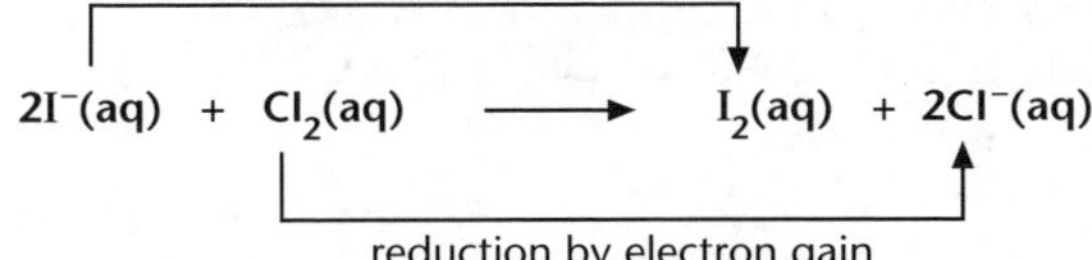

d.

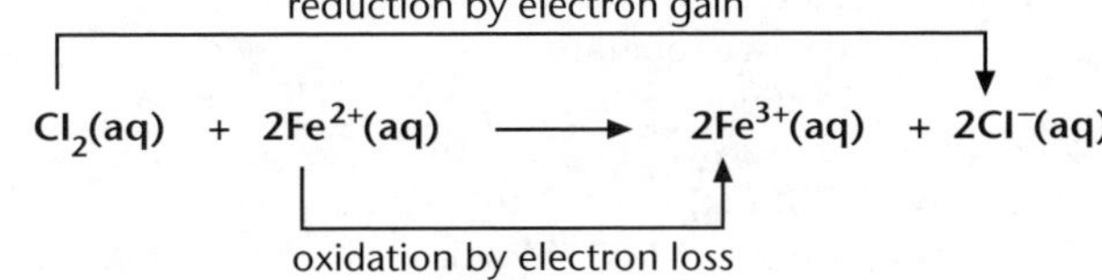

(***A*** – two equations correct for both oxidation and reduction; ***M*** – four equations correct for both oxidation and reduction)

Unit 11.3 Activity 3A: Soluble and insoluble substances, preparation of an insoluble compound (page 137)

1. a. Insoluble. (**A**)
   b. Soluble. (**A**)
   c. Sparingly soluble. (**A**)
2. a. *Any three of:* sodium nitrate, magnesium nitrate, calcium nitrate, aluminium nitrate, zinc nitrate, iron(III) nitrate, lead nitrate, copper nitrate, silver nitrate. (**A**)
   b. Lead chloride or silver chloride. (**A**)
   c. *Any two of:* magnesium carbonate, calcium carbonate, zinc carbonate, iron(II) carbonate, lead carbonate, copper carbonate. (**A**)
   d. Sodium hydroxide, potassium hydroxide. (**A**)
3. A precipitate is an insoluble solid formed when two solutions are mixed. (**A**)
4.

Solution A	Solution B	Precipitate		
		Name and formula	Appearance	Equation in word form
silver nitrate	sodium chloride	silver chloride AgCl	**a.** white solid	**b.** silver nitrate + sodium chloride → silver chloride + sodium nitrate
barium nitrate	sodium sulfate	**c.** barium sulfate $BaSO_4$	white solid	**d.** barium nitrate + sodium sulfate → barium sulfate + sodium nitrate
e. iron(III) nitrate	**f.** sodium hydroxide	iron(III) hydroxide $Fe(OH)_3$	**g.** red-brown solid	**h.** iron(III) nitrate + sodium hydroxide → iron(III) hydroxide + sodium nitrate

(**A** – each correct answer)

5. a. Copper(II) nitrate + sodium hydroxide → copper(II) hydroxide + sodium nitrate. (**A**)
   b. Calcium nitrate + sodium hydroxide → calcium hydroxide + sodium nitrate. (**A**)
   c. Zinc nitrate + sodium carbonate → zinc carbonate + sodium nitrate. (**A**)
   d. $Mg(NO_3)_2(aq) + 2NaOH(aq) \rightarrow Mg(OH)_2(s) + 2NaNO_3(aq)$. (**A**)
   e. $FeSO_4(aq) + Na_2CO_3\,(aq) \rightarrow FeCO_3(s) + Na_2SO_4(aq)$. (**A**)
   f. $Ag^+(aq) + Cl^-(aq) \rightarrow AgCl(s)$. (**A**)
   g. $Ba^{2+}(aq) + SO_4^{2-}(aq) \rightarrow BaSO_4(s)$. (**A**)
6. *Word description:*
   i. Prepare lead nitrate solution and sodium chloride solution.
   ii. Mix equal volumes of the two solutions.
   iii. Lead nitrate + sodium chloride → lead chloride + sodium nitrate, or $Pb(NO_3)_2(aq) + 2NaCl(aq) \rightarrow PbCl_2(s) + 2NaNO_3(aq)$, or $Pb^{2+}(aq) + 2Cl^-(aq) \rightarrow PbCl_2(s)$.
   iv. Filter off the white precipitate of lead chloride, retaining the solid on a filter paper held in a funnel.
   v. Wash the precipitate with distilled water.
   vi. Open up the filter paper and allow the lead chloride to dry by the water evaporating from the solid.

Alternative answer using diagram:

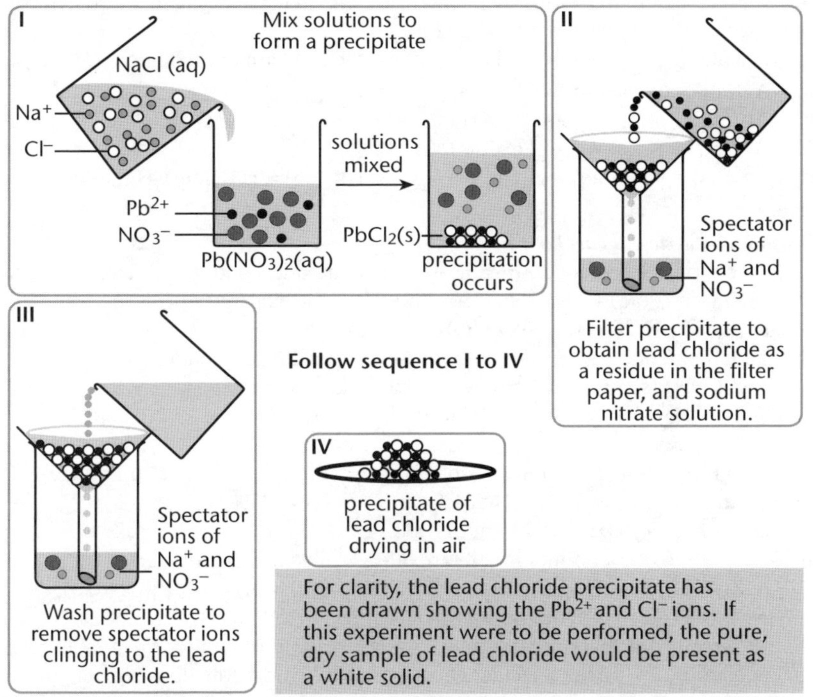

Either word description or diagram acceptable. (***A*** – reagents correct, purification process mentioned, word equation; ***M*** – filtration, washing and drying of ppt, molecular equation; ***E*** – complete description/diagram, ionic equation)

Unit 11.3 Activity 3B: Identifying positive (metal) ions (page 141)

1. **a.** *Any three of:* silver chloride, lead chloride, barium sulfate, zinc hydroxide, aluminium hydroxide, calcium hydroxide, magnesium hydroxide, zinc carbonate, calcium carbonate (or any other metal compound known to be white). (***A***)
b. Iron(II) hydroxide or iron(II) sulfate. (***A***)
c. Copper(II) hydroxide, copper(II) sulfate, copper(II) carbonate or copper(II) nitrate. (***A***)
d. Iron(III) hydroxide or silver oxide. (***A***)

2. **a.** *Any three of:* lead nitrate, zinc nitrate, aluminium nitrate, calcium nitrate, magnesium nitrate. (***A***)
b. $Pb(NO_3)_2$, $Zn(NO_3)_2$, $Al(NO_3)_3$, $Ca(NO_3)_2$, $Mg(NO_3)_2$. (***A***)
c. $Pb(OH)_2$, $Zn(OH)_2$, $Al(OH)_3$, $Ca(OH)_2$, $Mg(OH)_2$. (***A***)
d. Lead hydroxide, zinc hydroxide, aluminium hydroxide. (***A***)

3. **a.** Na^+.
b. Fe^{2+}.
c. Al^{3+}.

(***A*** – all three answers correct)

4. **a.** A white precipitate forms. (***A***)
   b. A green precipitate forms (***A***), which goes brown on standing in air (***M***), due to oxidation of iron(II) to iron(III). (***E***)
   c. A white precipitate forms (***A***), which dissolves in ammonia solution. (***M***)
5. **a.** $Ag^+(aq) + Cl^-(aq) \rightarrow AgCl(s)$. (***A***)
   b. $Al^{3+}(aq) + 3OH^-(aq) \rightarrow Al(OH)_3(s)$. (***A***)
   c. $Ba^{2+}(aq) + SO_4^{2-}(aq) \rightarrow BaSO_4(s)$. (***A***)
   d. $Ag_2O(s) + 4NH_3(aq) + H_2O(\ell) \rightarrow 2Ag(NH_3)_2^+(aq) + 2OH^-(aq)$. (***E***)
6. Zinc.
   Reactions occurring in Part I:
   i. Zinc nitrate + sodium hydroxide → zinc hydroxide + sodium nitrate,
   or $Zn(NO_3)_2(aq) + 2NaOH(aq) \rightarrow Zn(OH)_2(s) + 2NaNO_3(aq)$,
   or $Zn^{2+}(aq) + 2OH^-(aq) \rightarrow Zn(OH)_2(s)$.
   ii. Zinc hydroxide + sodium hydroxide → sodium zincate solution,
   or $Zn(OH)_2(s) + 2NaOH(aq) \rightarrow Na_2ZnO_2\,(aq) + 2H_2O(\ell)$,
   or $Zn(OH)_2(s) + 2OH^-(aq) \rightarrow ZnO_2^{2-}(aq) + 2H_2O(\ell)$.
   Reactions occurring in Part II:
   i. Zinc nitrate + ammonia hydroxide → zinc hydroxide + ammonium nitrate,
   or $Zn(NO_3)_2(aq) + 2NH_4OH(aq) \rightarrow Zn(OH)_2(s) + 2NH_4NO_3(aq)$,
   or $Zn^{2+}(aq) + 2OH^-(aq) \rightarrow Zn(OH)_2(s)$.
   ii. Zinc hydroxide + ammonium hydroxide → zinc ammine solution,
   or $Zn(OH)_2(s) + 4NH_4OH(aq) \rightarrow Zn(NH_3)_4^{2+}(aq) + 4H_2O(\ell) + 2OH^-(aq)$,
   or $Zn(OH)_2(s) + 4NH_3(aq) \rightarrow Zn(NH_3)_4^{2+}(aq) + 2OH^-(aq)$.
   (***A*** – correct metal (zinc) plus one correct equation; ***M*** – two correct molecular equations; ***E*** – three molecular equations or two ionic equations correct)

Unit 11.3 Activity 3C: Identifying negative (non-metal) ions (page 144)

1. **a.** Any of the carbonates of lead, zinc, magnesium, calcium or sodium. (***A***)
   b. Any one of the sulfate solutions of sodium, magnesium, aluminium, zinc, iron(II), iron(III), copper; *or* the carbonate solution of sodium. (***A***)
   c. **i.** Any of the chloride solutions of: sodium, magnesium, calcium, aluminium, zinc, iron(II), iron(III), copper. (***A***)
   ii. *Either* sodium hydroxide solution *or* ammonium hydroxide solution. (***A***)
   d. *Either*: Silver nitrate solution and a solution of any one of the soluble iodides, ie magnesium iodide, calcium iodide, zinc iodide, iron(II) iodide.
   or: Lead nitrate solution and a solution of any one of the soluble iodides, ie magnesium iodide, calcium iodide, zinc iodide, iron(II) iodide.
   (***A*** – two solutions correct)
2. Effervescence; green solid dissolves; blue solution remains. (***A*** – two observations correct; ***M*** – three observations correct)

3. Calcium carbonate (moderate amount of ppt suggests calcium).

Equation	Observation		
	Effervescence	Limewater turns milky	Formation of white ppt
Word equation	calcium carbonate + nitric acid → calcium nitrate + water + carbon dioxide	carbon dioxide gas + limewater → calcium carbonate + water	calcium nitrate + sodium hydroxide → calcium hydroxide + sodium nitrate
Molecular equation	$CaCO_3(s) + 2HNO_3(aq) \rightarrow Ca(NO_3)_2(aq) + H_2O(\ell) + CO_2(g)$	$CO_2(g) + Ca(OH)_2(aq) \rightarrow CaCO_3(s) + H_2O(\ell)$	$Ca(NO_3)_2(aq) + 2NaOH(aq) \rightarrow Ca(OH)_2(s) + 2NaNO_3(aq)$
Ionic equation	$CO_3^{2-}(s) + 2H^+(aq) \rightarrow H_2O(\ell) + CO_2(g)$	$CO_2(g) + 2OH^-(aq) \rightarrow CO_3^{2-}(aq) + H_2O(\ell)$, followed by $CO_3^{2-}(aq) + Ca^{2+}(aq) \rightarrow CaCO_3(s)$	$Ca^{2+}(aq) + 2OH^-(aq) \rightarrow Ca(OH)_2(s)$

(**A** – identification of calcium carbonate and two word equations correct;
M – identification of calcium carbonate and two molecular or ionic equations correct)

4. **a.**

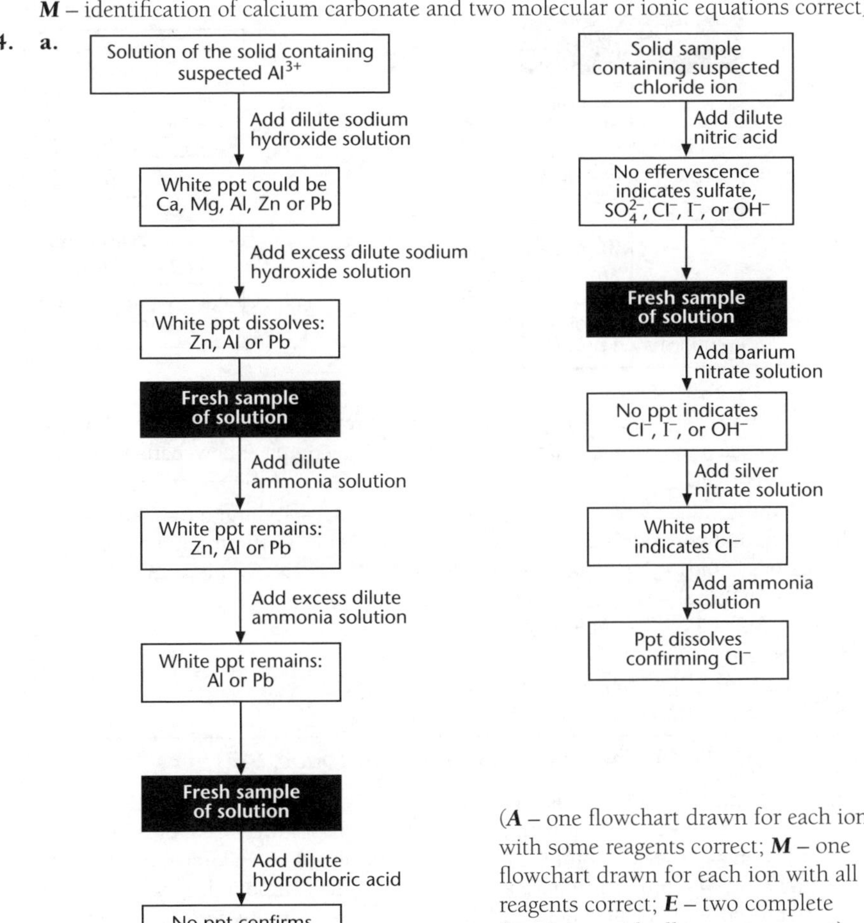

(**A** – one flowchart drawn for each ion with some reagents correct; **M** – one flowchart drawn for each ion with all reagents correct; **E** – two complete flowcharts with all reagents correct)

b.

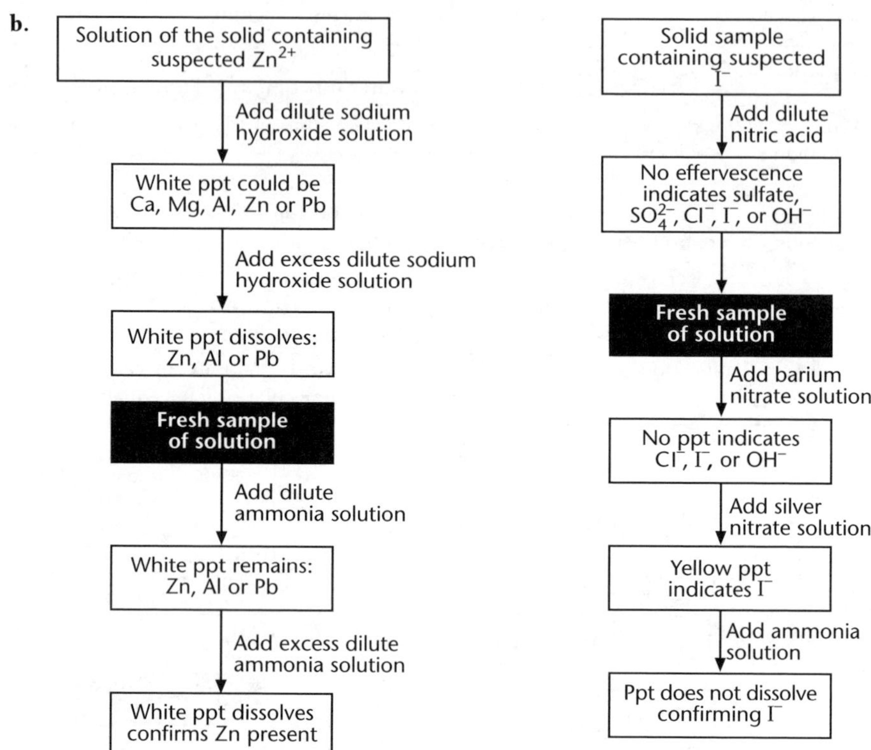

(**A** – one flowchart drawn for each ion with some reagents correct; **M** – one flowchart drawn for each ion with all reagents correct; **E** – two complete flowcharts with all reagents correct)

5. a. Sodium sulfate + barium nitrate → barium sulfate + sodium nitrate. (**A**)
   b. Sodium hydroxide + zinc nitrate → zinc hydroxide + sodium nitrate. (**A**)
   c. $2AgNO_3(aq) + 2NaOH(aq) \rightarrow Ag_2O(s) + H_2O(\ell) + 2NaNO_3(aq)$. (**A**)
   d. $Ca(OH)_2(aq) + CO_2(g) \rightarrow CaCO_3(s) + H_2O(\ell)$. (**A**)
   e. $Ag^+(aq) + I^-(aq) \rightarrow AgI(s)$. (**A**)
   f. $2H^+(aq) + CO_3^{2-}(s) \rightarrow H_2O(\ell) + CO_2(g)$. (**A**)
   g. $2Ag^+(aq) + 2OH^-(aq) \rightarrow Ag_2O(s) + H_2O(\ell)$. (**M**)
   h. $Ba^{2+}(aq) + 2OH^-(aq) \rightarrow Ba(OH)_2(s)$. (**M**)

Unit 11.3 Activity 4A: Thermal decomposition (page 149)

1. a. In chemistry, stable means unreactive (**A**) in the conditions stated. (**M**)
   b. The breakdown of a compound into compounds of lower mass, or into the elements of the compound, by the action of heat. (**A** – breakdown of a compound by heat; **M** – product(s) have lower mass than the compound, or are elements)
   c. Unreactive (**A**) to all chemical reagents and changes in physical conditions. (**M**)
2. a. *Examples*: Calcium carbonate, $CaCO_3$ or copper carbonate, $CuCO_3$. (**A** – name correct; **M** – name and formula correct)

   b. Sodium hydrogen carbonate, $NaHCO_3$. (***A*** – name correct; **M** – name and formula correct)

3. Copper carbonate is green and becomes black on heating. (***A***)

4. Place some limewater (a colourless solution) in the gas jar and shake the jar. If the limewater goes milky then the gas present is carbon dioxide. (***A*** – use of limewater; **M** – colour change from colourless to milky)

5. Calcium is a more reactive metal than copper *or* calcium is higher in the activity series for metals. (***A*** – either answer) Greater activity of metal creates a more stable compound (***M***), which requires more energy for decomposition (***E***).

6. **a.** Copper carbonate → copper oxide + carbon dioxide. (***A***)
   b. Sodium hydrogen carbonate → sodium carbonate + water + carbon dioxide. (***A***)

7. $CaCO_3(s) \rightarrow CaO(s) + CO_2(s)$. (***A*** – two formulae correct; **M** – equation correct)

8. **a.** CO_3^{2-}. (***A***)
   b. HCO_3^{-}. (***A***)

9. **a.** Sodium carbonate. (***A***)
   b. Zinc carbonate. (***A***)
   c. Lead carbonate. (***A***)

10. **a.** Sodium hydrogen carbonate → sodium carbonate + carbon dioxide + water. (***A***)
   b. Lithium carbonate → lithium oxide + carbon dioxide. (***A***)
   c. $ZnCO_3(s) \rightarrow ZnO(s) + CO_2(g)$. (***A***)
   d. $Ca(HCO_3)_2(aq) \rightarrow CaCO_3(s) + H_2O(\ell) + CO_2(g)$. (***A***)

Unit 11.3 Activity 5A: Balanced equations (page 154)

1. $3C_3H_8(g) + 15O_2(g) \rightarrow 9CO_2(g) + 12H_2O(l)$ + heat

2. Using the equation: $zC_xH_y + z(\frac{x}{2} + \frac{y}{4})O_2 \rightarrow zxCO + \frac{zy}{2} H_2O$ + heat

   $4CH_4(g) + 6O_2(g) \rightarrow 4CO(g) + 8H_2O(l)$ + heat

3. $C_3H_8(g) + 5O_2(g) \rightarrow 3CO_2(g) + 4H_2O(l)$ + heat

Unit 11.3 Activity 5B: Combustion reactions – multiple choice (page 157)

1. B
2. A
3. E
4. B
5. D
6. C
7. C

Unit 11.3 Activity 6A: Calculating M_r values (page 160)

1. **a.** $M_r(C_2H_2) = 2 \times 12.0 + 2 \times 1.0 = 26.0$
   b. $M_r(C_3H_6) = 3 \times 12.0 + 6 \times 1.0 = 42.0$
   c. $M_r(C_4H_{10}) = 4 \times 12.0 + 10 \times 1.0 = 58.0$
   (***A*** – one correct; **M** – all three correct)

2. **a.** $M_r(CO) = 12.0 + 16.0 = 28.0$
b. $M_r(SO_3) = 32.0 + 3 \times 16.0 = 80.0$
c. $M_r(N_2H_4) = 2 \times 14.0 + 4 \times 1.0 = 32.0$
d. $M_r(CH_3OH) = 12.0 + 3 \times 1.0 + 16.0 + 1.0 = 32.0$
e. $M_r(C_2H_5OH)= 2 \times 12.0 + 5 \times 1.0 + 16.0 + 1.0 = 46.0$
f. $M_r(NH_4Cl) = 14.0 + 4 \times 1.0 + 35.5 = 53.5$
g. $M_r((NH_4)_2SO_4) = 2 \times (14.0 + 4 \times 1.0) + (32.0 + 4 \times 16.0) = 132.0$
h. $M_r(AlPO_4) = 27.0 + 31.0 + 4 \times 16.0 = 122.0$
i. $M_r(Al_2(SO_4)_3) = 2 \times 27.0 + 3 \times (32.0 + 4 \times 16.0) = 342.0$
j. $M_r(Al(NO_3)_3) = 27.0 + 3 \times (14.0 + 3 \times 16.0) = 213.0$
(***A*** – three correct; ***M*** – six correct; ***E*** – all ten correct)

3. **a.** $M_r(C_{12}H_{22}O_{11}) = 12 \times 12.0 + 22 \times 1.0 + 11 \times 16.0= 342.0$
b. $M_r(C_{10}H_{14}N_2) = 10 \times 12.0 + 14 \times 1.0 + 2 \times 14.0= 162.0$
c. $M_r(C_{17}H_{17}NO(C_2H_3O_2)_2) = 17 \times 12.0 + 17 \times 1.0 + 1 \times 14.0 + 1 \times 16.0 + 2 \times (2 \times 12.0 + 3 \times 1.0 + 2 \times 16.0) = 369.0$
(***A*** – one correct; ***M*** – two correct; ***E*** – three correct)

Unit 11.3 Activity 6B: Calculating composition and yields of chemicals (page 162)

1. **a.** $M_r(C_2H_6) = 2 \times 12.0 + 6 \times 1.0 = 30.0$; percentage carbon = $\frac{24}{30} \times 100 = 80.0\%$
b. $M_r(C_2H_4) = 2 \times 12.0 + 4 \times 1.0 = 28.0$; percentage carbon = $\frac{24}{28} \times 100 = 85.7\%$
c. $M_r(C_2H_2) = 2 \times 12.0 + 2 \times 1.0 = 26.0$; percentage carbon = $\frac{24}{26} \times 100 = 92.3\%$
d. $M_r(C_2H_5OH) = 2 \times 12.0 + 5 \times 1.0 + 16.0 + 1.0 = 46.0$; percentage carbon = $\frac{24}{46} \times 100 = 52.2\%$
e. $M_r(C_6H_{12}O_6) = 6 \times 12.0 + 12 \times 1.0 + 6 \times 16.0 = 180.0$; percentage carbon = $\frac{72}{180} \times 100 = 40.0\%$
(***A*** – one correct; ***M*** – three correct; ***E*** – five correct)

2. **a.** $M_r(CaCO_3) = 40.1 + 12.0 + 3 \times 16.0 = 100.1$; percentage carbon = $\frac{12}{100.1} \times 100 = 12.0\%$
b. $M_r(FeS_2) = 55.9 + 2 \times 32.0 = 119.9$; percentage sulfur = $\frac{64}{119.9} \times 100 = 53.4\%$
c. $M_r(CaCO_3.MgCO_3) = 40.1 + 12.0 + 3 \times 16.0 + 24.3 + 12.0 + 3 \times 16.0 = 184.4$; percentage magnesium = $\frac{24}{184.4} \times 100 = 13.0\%$
d. $M_r(Fe_3O_4.TiO_2) = 3 \times 55.9 + 4 \times 16.0 + 47.9 + 2 \times 16.0 = 311.6$; percentage titanium = $\frac{47.9}{311.6} \times 100 = 15.4\%$

e. $M_r(CaSO_4.2H_2O) = 40.1 + 32.0 + 4 \times 16.0 + 2 \times (2 \times 1.0 + 16.0) = 172.1$;

percentage oxygen = $\frac{96}{172.1} \times 100 = 55.8\%$

(***A*** – three M_r values and one percentage value correct; ***M*** – three percentage values correct; ***E*** – five percentage values correct)

3. **a.** $M_r(H_2O_2) = 2 \times 1.0 + 2 \times 16.0 = 34$; 100g of hydrogen peroxide contains $\frac{32}{34} \times 100 = 94.1$ g of oxygen

b. $M_r(HgO) = 200.6 + 16.0 = 216.6$; 100g of mercury oxide contains $\frac{16}{216.6} \times 100 = 7.4$ g of oxygen

c. $M_r(CO(NH_2)_2) = 12.0 + 16.0 + 2 \times (14.0 + 2 \times 1.0) = 60$; 100 g of urea contains $\frac{28}{60} \times 100 = 46.7$ g of nitrogen

d. $M_r(Al_2S_3) = 2 \times 27.0 + 3 \times 32.0 = 150.0$; 100 g of sulfur produces $\frac{150}{96} \times 100 = 156.3$ g of aluminium sulfide

(***A*** – two M_r values and one calculation correct; ***M*** – three calculations correct; ***E*** – four calculations correct)

4. **a.** $M_r(ZnS) = 65.4 + 32.0 = 97.4$; 1 kg of zinc sulfide contains $\frac{65.4}{97.4} \times 1 = 0.671$ kg *or* 671 g of zinc.

b. $M_r(ZnCO_3) = 65.4 + 12.0 + 3 \times 16.0 = 125.4$; to make 1 kg of calamine requires $\frac{65.4}{125.4} \times 1 = 0.522$ kg *or* 522 g of zinc

c. $M_r(NH_4NO_3) = 2 \times 14.0 + 4 \times 1.0 + 3 \times 16.0 = 80.0$; to make 1 kg of ammonium nitrate requires $\frac{28}{80} \times 1 = 0.35$ kg *or* 350 g of nitrogen

d. $M_r(Ca_3(PO_4)_2) = 3 \times 40.1 + 2 \times (31.0 + 4 \times 16.0) = 310.3$; 1 kg of rock phosphate contains $\frac{62}{310.3} \times 1 = 0.20$ kg *or* 200 g of phosphorus

(***A*** – two M_r values and one calculation correct; ***M*** – three calculations correct; ***E*** – four calculations correct)

Unit 11.3 Activity 6C: Empirical formulae (page 163)

1. Following compounds have different empirical and molecular formulae:

b. Ethene, empirical formula CH_2.

c. Ethyne, empirical formula, CH.

e. Glucose, empirical formula, CH_2O.

f. Ethanoic acid, empirical formula, CH_2O.

(***A*** – all four compounds correctly identified; ***M*** – all four empirical formulae correct)

2. a.

Element	Fe	S
Composition as a percentage	63.6	36.4
Divide composition by atomic mass	$\frac{63.6}{55.9} = 1.14$	$\frac{36.4}{32.0} = 1.14$
Look for the simplest ratio: divide all numbers by the smallest number	$\frac{1.14}{1.14} = 1$	$\frac{1.14}{1.14} = 1$
Empirical formula is **FeS**		

b.

Element	Fe	S
Composition as a percentage	46.6	53.4
Divide composition by atomic mass	$\frac{46.6}{55.9} = 0.83$	$\frac{53.4}{32.0} = 1.67$
Look for the simplest ratio: divide all numbers by the smallest number	$\frac{0.83}{0.83} = 1$	$\frac{1.67}{0.83} = 2$
Empirical formula is $\mathbf{FeS_2}$		

c.

Element	N	H
Composition as a percentage	87.5	12.5
Divide composition by atomic mass	$\frac{87.5}{14.0} = 6.25$	$\frac{12.5}{1.0} = 12.5$
Look for the simplest ratio: divide all numbers by the smallest number	$\frac{6.25}{6.25} = 1$	$\frac{12.5}{6.25} = 2$
Empirical formula is $\mathbf{NH_2}$		

d.

Element	C	O	H
Composition as a percentage	37.5	50.0	12.5
Divide composition by atomic mass	$\frac{37.5}{12.0} = 3.125$	$\frac{50.0}{16.0} = 3.125$	$\frac{12.5}{1.0} = 12.5$
Look for the simplest ratio: divide all numbers by the smallest number	$\frac{3.125}{3.125} = 1$	$\frac{3.125}{3.125} = 1$	$\frac{12.5}{3.125} = 4$
Empirical formula is $\mathbf{COH_4}$			

e.

Element	C	O	H
Composition as a percentage	52.2	34.8	13.0
Divide composition by atomic mass	$\frac{52.2}{12.0} = 4.35$	$\frac{34.8}{16.0} = 2.175$	$\frac{13.0}{1.0} = 13.0$
Look for the simplest ratio: divide all numbers by the smallest number	$\frac{4.35}{2.175} = 2$	$\frac{2.175}{2.175} = 1$	$\frac{13.0}{2.175} = 6$
Empirical formula is $\mathbf{C_2OH_6}$			

f.

Element	C	O	H
Composition as a percentage	40.0	53.3	6.7
Divide composition by atomic mass	$\frac{40.0}{12.0} = 3.33$	$\frac{53.3}{16.0} = 3.33$	$\frac{6.7}{1.0} = 6.7$
Look for the simplest ratio: divide all numbers by the smallest number	$\frac{3.33}{3.33} = 1$	$\frac{3.33}{3.33} = 1$	$\frac{6.7}{3.33} = 2$
Empirical formula is $\mathbf{COH_2}$			

(**A** – correct division of percentage composition by atomic masses for three compounds; **M** – atom ratios correct for three formulae; **E** – atom ratios correct and empirical formulae correct for six compounds)

Unit 11.3 Activity 6D: Calculating masses of chemicals in a reaction (page 165)

1. $M_r(Na_2CO_3) = 2 \times 23.0 + 12.0 + 3 \times 16.0 = 106.0$; $M_r(CO_2) = 12.0 + 2 \times 16.0 = 44.0$. Since 106 g of sodium carbonate produces 44 g of carbon dioxide, then 100 g of sodium carbonate produces $44 \times \frac{100}{106} = 41.5$ g carbon dioxide.

(**A** – correct M_r values; **M** – recognition that carbon dioxide and sodium carbonate are connected in the reaction by the mass ratio of 44 : 106; **E** – correct answer with all working shown)

2. $A_r(Zn) = 65.4$; $M_r(H_2) = 2 \times 1.0 = 2$. Since 65.4 g of zinc produces 2 g of hydrogen, then $\frac{10}{2} \times 65.4 = 327$ g of zinc produces 10 g of hydrogen.

(**A** – correct M_r values; **M** – recognition that zinc and hydrogen are connected in the reaction by the mass ratio of 65.4 : 2; **E** – correct answer with all working shown)

3. $M_r(C_8H_{18}) = 8 \times 12.0 + 18 \times 1.0 = 114.0$; $M_r(O_2) = 2 \times 16.0 = 32.0$. Since $2 \times 114 = 228$ g of octane requires $25 \times 32 = 800$ g of oxygen for complete combustion, then $\frac{800}{228} \times 700 = 2456.1$ g of oxygen is needed for the complete combustion of 1 litre of petrol.

(**A** – correct M_r values; **M** – recognition that octane and oxygen are connected in the reaction by the mass ratio of 114 : 400; **E** – correct answer with all working shown)

4. $A_r(Cu) = 63.6$; $M_r(Cu(NO_3)_2.3H_2O) = 63.6 + 2 \times (14.0 + 3 \times 16.0) + 3 \times (2 \times 1.0 + 16.0) = 241.6$. Since 63.6 g of copper produces 241.6 g of copper nitrate crystals, then $\frac{63.6}{241.6} \times 10 = 2.63$ g of copper produces 10 g of copper nitrate crystals.

One particle of Cu produces one particle of $Cu(NO_3)_2.3H_2O$.

(**A** – correct M_r values; **M** – recognition that copper and copper nitrate crystals are connected in the reaction by the mass ratio of 63.6 : 241.6; **E** – correct answer with all working shown.)

5. $M_r(NaHCO_3) = 23.0 + 1.0 + 12.0 + 3 \times 16.0 = 84$; $M_r(CO_2) = 12.0 + 2 \times 16.0 = 44$. Since $2 \times 84 = 168$ g of baking soda produces 44 g of carbon dioxide, then $\frac{168}{44} \times 10 = 38.2$ g of baking soda produces 10 g of carbon dioxide.
(**A** – correct M_r values; **M** – recognition that baking soda and carbon dioxide are connected in the reaction by the mass ratio of 168 : 44; **E** – correct answer with all working shown)

6. $M_r(NH_3) = 14.0 + 3 \times 1.0 = 17.0$; $M_r(H_2SO_4) = 2 \times 1.0 + 32.0 + 4 \times 16.0 = 98.0$; $M_r((NH_4)_2SO_4) = 2 \times (14.0 + 4 \times 1.0) + 32.0 + 4 \times 16.0 = 132.0$. Since $2 \times 17 = 34$ tonnes of ammonia react with 98 tonnes of sulfuric acid to produce 132 tonnes of sulfate of ammonia, then $\frac{34}{132} \times 1 = 0.26$ tonnes or 260 kg of ammonia produces 1 tonne of sulfate of ammonia. Since 98 tonnes of sulfuric acid react with 34 tonnes of ammonia to produce 132 tonnes of sulfate of ammonia, then $\frac{98}{132} \times 1 = 0.74$ tonnes or 740 kg of sulfuric acid produces 1 tonne of sulfate of ammonia.

If 0.26 tonnes of ammonia produces 1 tonne of sulfate of ammonia, then $1 - 0.26 = 0.74$ tonnes of sulfuric acid is needed, ie there is no need to do the second calculation.

(**A** – correct M_r values; **M** – recognition that ammonia and ammonium sulfate are connected in the reaction by the mass ratio of 34 : 132; **E** – correct answer with all working shown)

Unit 11.3 Activity 7A: pH (page 170)

1. **a.** **i.** More basic. (**A**) **ii.** More basic. (**A**) **iii.** More acidic. (**A**)
   b. **i.** Smaller. (**A**) **ii.** Smaller. (**A**) **iii.** Larger. (**A**)
2. **a.** 10^{-6} mol L^{-1} (**A**) **b.** OH^- (**A**) **c.** Basic. (**A**)
3. **a.** 10^{-3} mol L^{-1} (**A**) **b.** H_3O^+ (**A**) **c.** Acidic. (**A**)
4. **a.** 5×10^{-12} mol L^{-1} (**A**) **b.** 10^{-7} mol L^{-1} (**A**)
5. **a.** Basic. (**A**) **b.** Basic. (**A**) **c.** Acidic. (**A**) **d.** Acidic. (**A**)
6.

$[H_3O^+]$	10^{-1}	10^{-2}	10^{-7}	10^{-10}	10^{-14}
$[OH^-]$	10^{-13}	10^{-12}	10^{-7}	10^{-4}	1 (or 10^0)
pH	1	2	7	10	14
solution	acidic	acidic	neutral	basic	basic

(**A**)

7. **a.** pH = 4 (**A**) **b.** pH = 3 (**A**) **c.** pH = 12 (**M**)
   d. pH = 3 (**M**) **e.** pH = 8.3 (**M**) **f.** pH = 5.4 (**A**)
8. **a.** 10^{-2} mol L^{-1} (**A**) **b.** 10^{-10} mol L^{-1} (**A**)
   c. 4.0×10^{-7} mol L^{-1} (**A**) **d.** 1.3×10^{-3} mol L^{-1} (**A**)
   e. 4.0×10^{-8} mol L^{-1} (**A**) **f.** 6.3×10^{-11} mol L^{-1} (**A**)

9. a. $n(HNO_3) = 0.1 \text{ mol L}^{-1} \times 0.005 \text{ L} = 5 \times 10^{-4} \text{ mol}$

$$c(HNO_3) = \frac{n}{V} = \frac{5 \times 10^{-4} \text{ mol}}{0.1 \text{ L}} = 5 \times 10^{-3} \text{ mol L}^{-1}$$

$[H_3O^+] = 5 \times 10^{-3} \text{ mol L}^{-1}$

b. $pH = -\log(5 \times 10^{-3}) = 2.3$

10. a. 0.01 mol L^{-1} (***A***) **b.** 0.001 mol L^{-1} (***A***) **c.** $10^{-11} \text{ mol L}^{-1}$ (***A***) **d.** pH = 11 (***M***)

11. a. $[H_3O^+] = 0.0850 \text{ mol L}^{-1}$

$pH = -\log(0.0850) = 1.07$

b. pH = 9.22 $[H_3O^+] = 10^{-9.22} = 6.03 \times 10^{-10} \text{ mol L}^{-1}$

$$[OH^-] = \frac{1 \times 10^{-4}}{6.03 \times 10^{-10}} = 1.66 \times 10^{-5} \text{ mol L}^{-1}$$

c. pH = 2.22 $[H_3O^+] = 10^{-2.22} = 6.03 \times 10^{-5} \text{ mol L}^{-1}$

$$[OH^-] = \frac{1 \times 10^{-14}}{[H_3O^+]} = \frac{1 \times 10^{-14}}{6.03 \times 10^{-3}}$$

$$= 1.66 \times 10^{-12} \text{ mol L}^{-1}$$

d. $[H_3O^+] = 8.86 \times 10^{-8} \text{ mol L}^{-1}$

$pH = -\log(8.86 \times 10^{-8}) = 7.05$

e. $[OH^-] = 0.00667 \text{ mol L}^{-1}$

$$[H_3O^+] = \frac{1 \times 10^{-14}}{0.00667} = 1.5 \times 10^{-12} \text{ mol L}^{-1}$$

$pH = -\log(1.5 \times 10^{12}) = 11.8$

12.

Solution	$[H_3O^+]$ / mol L^{-1}	$[OH^-]$ / mol L^{-1}	pH
A	*0.0672*	1.49×10^{-13}	1.17
B	2.51×10^{-11}	3.98×10^{-4}	*10.60*
C	5.18×10^{-4}	*1.93×10^{-11}*	3.28
D	1.26×10^{-6}	*7.94×10^{-9}*	*5.90*
E	*6.35×10^{-10}*	1.57×10^{-5}	9.20
F	1.82×10^{-12}	*0.0055*	11.74
G	*0.00150*	6.67×10^{-12}	2.82
H	1.14×10^{-6}	*8.77×10^{-9}*	5.94
I	3.98×10^{-8}	2.51×10^{-7}	*7.40*

13. a. The slaked lime reacts with the hydrogen ions in the acidic soil to produce a calcium salt and water. Protons are removed from the soil in this process so the pH is raised.

$Ca(OH)_2 + 2H^+ \rightarrow Ca^{2+} + 2H_2O$ (***E***)

b. Indigestion is due to excess acid in the stomach. Both $NaHCO_3$ and Na_2CO_3 react with the acid and neutralise it.

i. $NaHCO_3(aq) + H^+(aq) \rightarrow Na^+(aq) + CO_2(g) + H_2O(\ell)$

ii. $Na_2CO_3(s) + 2H^+(aq) \rightarrow 2Na^+(aq) + CO_2(g) + H_2O(\ell)$ (***E***)

Unit 11.3 Activity 7B: Strong and weak acids and bases (page 175)

1. a. $[OH^-] = 0.01 \text{ mol L}^{-1}$; $\therefore [H_3O^+] = \dfrac{K_w}{[OH^-]} = \dfrac{1 \times 10^{-14}}{0.01} = 1 \times 10^{-12}$
   $pH = -\log[H_3O^+] = -\log(1 \times 10^{-12}) = 12$ (**M**)
   b. $NaOH + HCl \rightarrow NaCl + H_2O$. Since the same amounts of NaOH and HCl are added together (50 mL of 0.01 mol L^{-1} of each), and since the solution is thus neutral, pH = 7. (***A***)
2. Nitric acid is a strong acid and donates all of its protons to water, so $[H_3O^+]$ is high.
   $HNO_3 + H_2O \rightleftharpoons H_3O^+ + NO_3^-$
   Ethanoic acid is a weak acid and doesn't completely dissociate, so $[H_3O^+]$ is lower.
   $CH_3COOH + H_2O \rightarrow CH_3COO^- + H_3O^+$ (**E**)
3. a. $n(H_3O^+) = cV = 0.01 \text{ mol L}^{-1} \times \dfrac{20}{1000} \text{ L}$
   $= 2 \times 10^{-4} \text{ mol}$
   b. 2×10^{-4} mol pH = 7 (***A***)
4. Weak acid – few of the acid particles transfer a proton.
   Dilute acid – few of the acid particles per litre of solution. (**M**)
5. a. Bubbles of gas will form, but the rate of production will be slower for CH_3COOH.
   $2HCl + CaCO_3 \rightarrow CaCl_2 + H_2O + CO_2$
   $2CH_3COOH + CaCO_3 \rightarrow Ca(CH_3COO)_2 + H_2O + CO_2$
   CH_3COOH is a weak acid, so only partially dissociates in water. HCl is a strong acid, so completely dissociates.
   $CH_3COOH + H_2O \rightleftharpoons CH_3COO^- + H_3O^+$
   $HCl + H_2O \rightarrow Cl^- + H_3O^+$
   Hence, there are more H_3O^+ ions present in the HCl of the same concentration. The greater concentration of H_3O^+ will cause the reaction to be faster, as there are more frequent and hence more successful collisions. (**E**)
   b. Measure pH. Since there are fewer H_3O^+ ions in the same volume of the CH_3COOH solution, it will have a higher pH. (**M**)
6. a. HY. Both acids are 0.10 mol L^{-1}. HY has a pH = 1, so $[H_3O^+] = 0.10 \text{ mol L}^{-1}$, which implies that it has fully dissociated – a strong acid. For HX, pH = 3, so $[H_3O^+] = 0.001 \text{ mol L}^{-1}$, so it has only partially dissociated. (**M**)
   b. Measure conductivity. HY will be a better conductor of electricity than HX, as HY has more ions in the solution. *or* Add same-size piece of metal (Zn, Mg) or carbonate ($CaCO_3$). The strong acid, HY, will react faster, so there will be more bubbles produced in the same time; *or* the solid will disappear quicker. (**M**)
7. a. Nitric acid is a strong acid, so completely dissociates in water:
   $HNO_3 + H_2O \rightarrow NO_3^- + H_3O^+$
   Methanoic acid is a weak acid, so only some of the acid molecules dissociate in water, meaning fewer H_3O^+ than for nitric acid. $HCOOH + H_2O \rightleftharpoons HCOO^- + H_3O^+$. Hence $[H_3O^+]$ is less for methanoic acid and the pH is higher.
   b. HNO_3. Since there are more ions present in the solution (because HNO_3 completely dissociates), HNO_3 will be a better conductor. (**E**)

8. HA Acid is a strong acid. Since pH = 1, all of the acid has dissociated.
   HB $[H_3O^+] = 10^{-2.5}$, but acid concentration is higher than this, so only some of the acid is dissociated – so is a weak acid.
   HC Acid is a dilute solution of a strong acid. Since pH = 3, $[H_3O^+] = 0.001$, so all of the acid has dissociated. (***E***)

Unit 11.3 Activity 8A: Equilibrium (page 180)

1. **a.** Reversible; adding water to $CuSO_4$ forms $CuSO_4.5H_2O$. (***A***)
   b. Reversible; add base and blue colour reappears. (***A***)
   c. Irreversible; copper (below zinc on the activity series) won't displace zinc ions from solution. (***A***)
   d. Reversible; haemoglobin transports oxygen to cells by forming oxyhaemogolobin. At the cells, oxyhaemoglobin loses oxygen and haemoglobin is regained. (***A***)
   e. Irreversible; magnesium is above hydrogen on the activity series. (***A***)
   f. Reversible; addition of OH^- ions will reverse this reaction. (***A***)
2. In **c.**, the person is walking up as fast as the escalator is moving down. In **a.**, the seesaw is balanced but there is no movement, and in **b.**, there is no movement of sand in the reverse direction.

Unit 11.3 Activity 8B: Equilibrium expressions (page 181)

1. **a.** $K_c = \dfrac{[HI(g)]^2}{[H_2(g)][I_2(g)]}$ (***A***)
   b. $K_c = \dfrac{[NO_2(g)]^2}{[NO(g)]^2[O_2(g)]}$ (***A***)
   c. $K_c = \dfrac{[FeSCN^{2+}(aq)]}{[Fe^{3+}(aq)][SCN^-(aq)]}$ (***A***)
   d. $K_c = \dfrac{[SO_2(g)]^2[O_2(g)]}{[SO_3(g)]^2}$ (***A***)
   e. $K_c = \dfrac{[CO(g)][H_2(g)]}{[H_2O(g)]}$ (***A***)

2. **a.** $C_2H_4(g) + 3O_2(g) \rightarrow 2CO_2(g) + 2H_2O(g)$
   b. $CH_4(g) + 2H_2S(g) \rightarrow CS_2(g) + 4H_2(g)$
   c. $2NOBr(g) \rightarrow 2NO(g) + Br_2(g)$ (***A***)
3. **a.** $K_c = \dfrac{[NO(g)]^2}{[N_2(g)][O_2(g)]}$
   b. N_2 and O_2
   c. None. (***A***)
4. **a.** **i.** $K_c = \dfrac{[SO_3]^2}{[SO_2]^2[O_2]}$ (***A***)
   ii. SO_3. K_c is very large, so concentration of product is high compared with that of reactants (product concentration is at the top of the ratio). (***M***)
   b. **i.** $K_c = \dfrac{[NO]^2[Cl_2]}{[NOCl]^2}$ (***A***)
   ii. NOCl. K_c is very small, so concentration of reactants is high compared with that of products. (***M***)

5. **a.** **i.** $K_c = \dfrac{[NO_2]^2}{[N_2O_4]}$ (***A***)

ii. N_2O_4. K_c is small (< 1), so concentration of reactants will be more than of products (reactant concentration is on the bottom of the ratio). (***M***)

b. **i.** $\dfrac{[H_2S]^2}{[H_2]^2}$ (***A***)

ii. H_2S. K_c is very high, so concentration of products is much greater than of reactants. (***M***)

Unit 11.3 Activity 8C: Equilibrium changes (page 187)

1. **a.** **i.** Increase the reverse reaction; more PCl_3 and Cl_2 form.

ii. Increase the forward reaction; more PCl_5 forms.

iii. No change, but equilibrium reached more quickly. (***A***)

b. **i.** As pressure decreases, the system acts to oppose the change by increasing the number of gas particles so that the original pressure of the system is maintained. This favours the reverse reaction, since there are 2 mol of reactant gas compared with 1 mol of product gas. (***M***)

ii. As $Cl_2(g)$ concentration increases, the system acts to oppose the change by consuming the $Cl_2(g)$ to form more product, PCl_5. (***M***)

iii. No change – catalysts only increase the speed at which the reaction attains equilibrium. (***M***)

2. **a.** Increase the reverse (endothermic) reaction to use up heat; less ammonia forms. (***M***)

b. Increase the forward (endothermic) reaction; more nitrogen oxide forms. (***M***)

c. Increase the forward (endothermic) reaction; more methanol forms. (***M***)

3. • *Heating the system*. Reverse reaction is endothermic. Heating the system favours the endothermic reaction, which will use up the $N_2O_4(g)$, thus decreasing the amount. (***M***)

• *Decreasing the pressure*. This will favour the formation of more moles of gas, which will shift the equilibrium to the left, where there are 2 mol of gas compared to 1mol on the right. This will reduce the amount of product, N_2O_4. (***M***)

• *Removing some $NO_2(g)$*. This will shift the equilibrium to the left to replace the $NO_2(g)$ removed, and this will use up some N_2O_4, thus decreasing the amount present.

4. **a.** Decrease the number of NO_2 molecules. Increase in pressure favours product, which has fewer moles of gas compared to reactants.

b. Colour becomes lighter brown.
Cooling favours the exothermic forward reaction, so more N_2O_4 forms, thus diluting the colour. (***E***)

5. Equilibrium moves to the right, so more NH_3 decomposes and more N_2 and H_2 are formed. Forward reaction is endothermic. Increasing the temperature favours this reaction to use up added heat energy.

6. a. $K_c = \frac{[FeSCN^{2+}]}{[Fe^{3+}][SCN^-]}$ (***A***)

b. Red colour lightens. Removing a reactant favours the reverse reaction, as the system moves to replace the Fe^{3+} removed. This will use up some $FeSCN^{2+}$, so the red colour intensity is reduced. (***M***)

c. Red colour deepens because adding $FeCl_3$ increases the Fe^{3+} in solution. This favours the forward reaction, as the system moves to use up the Fe^{3+}. More $FeSCN^{2+}$ will be made, so the red colour will be more intense.

7. a. Reverse reaction is endothermic. Increasing temperature favours this reaction. This *reduces* CH_3OH in the system and so reduces the concentration of CH_3OH.

b. Increasing pressure will favour the formation of fewer moles of gas. There are less moles of gas in the product than reactants, so more products will form. This *increases* the amount of CH_3OH and so increases the concentration of CH_3OH.

c. Adding a reactant will favour the forward reaction, so more product is formed. This *increases* the amount of CH_3OH and thus increases the concentration of CH_3OH.

8. a. Forward reaction is endothermic. Increasing temperature favours this reaction. This increases the concentration of H_2.

b. Decreasing pressure favours the formation of more moles of gas. There are 2 mol of product gas compared with one mol of reactant gas, so more products will form. This *increases* the concentration of H_2.

c. Removing a product will favour the forward reaction. This *increases* the concentration of H_2.

d. Since C is a solid, the amount present will not affect the equilibrium position, so *no change* in the concentration of H_2.

9. a. Reverse reaction is endothermic. Increasing temperature favours this reaction. This will *increase* the concentration of chlorine gas.

b. Decreasing pressure favours the formation of more moles of gas. There are 10 moles of reactant gas compared with 4 of product, so more reactants will form. This means an *increase* in the concentration of chlorine gas.

c. Removing a product favours the forward reaction, so reactants will be used up. This will *reduce* the concentration of chlorine gas.

d. Catalysts do not change the equilibrium position, so *no change* in the concentration of chlorine gas.

10. a. i. Fizziness would increase; extra HCO_3^- would increase $[HCO_3^-]$, so equilibrium moves to left, producing more $CO_2(g)$. (***M***)

ii. Fizziness would increase; adding an acid increases $[H^+]$, equilibrium moves to left, more $CO_2(g)$ produced. (***M***)

b. i. Endothermic. Decrease in pH means increase in $[H^+]$, so forward reaction favoured by heating, so is endothermic. (***E***)

ii. Value of K_c will increase.
Forward reaction is favoured by heating, so more products formed ($K_c = \frac{[\text{products}]}{[\text{reactants}]}$), so K_c will increase. (***E***)

c. Opening a can decreases pressure; equilibrium moves to produce more moles of gas, so moves to the left-hand side, $[H^+]$ decreases, so pH increases. (***E***)

11. High pressure – favours forward reaction, since there is 2 mol of gas in products and 3 mol in reactants.
Low temperature – forward reaction is exothermic, so will be favoured by reduced temperatures. (**M**)

12. Forward reaction is exothermic. As temperature increases, the reverse reaction, which is endothermic, is favoured. Less product will be formed, so percentage yield is reduced. (**M**)

13. Increasing pressure – causes more ethanol to form (increase in yield). There are fewer gas particles in the products than in the reactants, so increasing pressure favours the forward reaction. Increasing temperature causes less ethanol to form. The formation rection is exothermic, so it will not be favoured at high temperatures.

14. a. $K_c = \frac{[NH_3]^2}{[N_2][H_2]^3}$ (**A**)

b. More NH_3 is formed. Increasing pressure favours formation of fewer moles of gas. Product has 2 moles of gas, while reactant has 4 moles. NH_3 is the product, so more is formed at higher pressure. (**M**)

c. More $NH_3(g)$ is formed. Liquefying the NH_3 removes a product ($NH_3(g)$) from the system, which favours the forward reaction, so more NH_3 is produced. (**M**)

d. Less NH_3 produced at higher temperatures. Reverse reaction is favoured by heating, so this reaction is endothermic. Reaction for formation of NH_3 is *exothermic*. (**E**)

15. Use a catalyst. (**A**)

Unit 11.4 Activity 1A: Rates of reaction (page 194)

1. a. Surface area of solid reagent; concentration of reagent; temperature of the reagents. (**A** – each correct answer)

b. Surface area of solid reagent – reactions are due to collisions; powders have a greater surface area than a single lump, enabling more collisions to occur.
Concentration of reagent – reactions are due to collisions; high concentration of reagent means more particles in a given volume and therefore more collisions are possible.
Temperature of the reagents – reactions are due to collisions; changing the temperature of a reaction changes the energy of the particles; particles with high energy move faster, creating more collisions in a given time; the energy involved in the collision is greater.

(**A** – recognition that particles need to collide; **M** – **A** plus one explanation of why more/fewer collisions can occur if factor is changed; **E** – **A** plus three explanations of why more/fewer collisions can occur if factor is changed)

2. a. One or both of the chemicals have been used up. (**E**)

b. i. 23 mL (**A**)

ii. 40 – 23 = 17 mL (**A**)

iii. As the reaction proceeds, the reagents are used up. (**A**) There is less surface area of the solid and fewer particles of acid in the second minute than in the first minute. Fewer collisions occur and therefore, less reaction, producing less gas. (**M**)

c. **Rate of reaction of metal + acid**

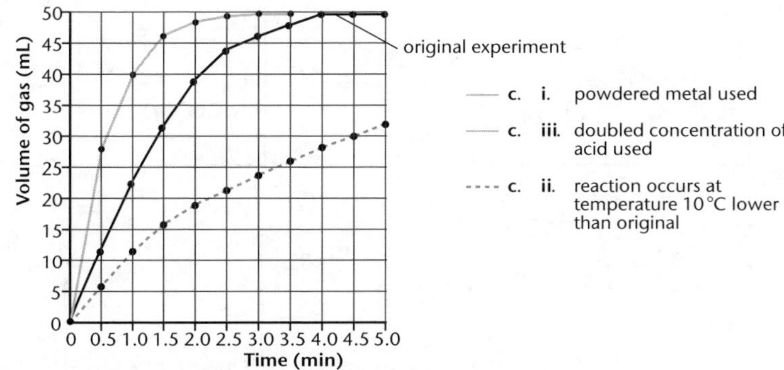

Two of the new graphs are identical, or nearly so.

(***A*** – one new graph reasonably correct; ***M*** – two new graphs reasonably correct; ***E*** – three new graphs reasonably correct.

3. **a.** Calcium carbonate + hydrochloric acid → calcium chloride + water + carbon dioxide. (***A***)

b. Carbon dioxide gas. (***A***)

c. The readings (negative) would be greater at a given time interval. (***A***)

d. The concentration of the acid could be decreased, *or* the temperature of the reaction lowered.

Both these factors would decrease the rate of the reaction to counter the increase in the rate of reaction due to powdered marble being used.

(***M*** – either factor stated)

Unit 11.4 Activity 2A: Enthalpy changes (page 200)

1. **a.** Exothermic. (***A***) **b.** Endothermic. (***A***) **c.** Exothermic. (***A***)
d. Endothermic. (***A***) **e.** Endothermic. (***A***) **f.** Exothermic. (***A***)
g. Exothermic. (***A***) **h.** Endothermic. (***A***) **i.** Exothermic. (***A***)
j. Exothermic. (***A***) **k.** Exothermic. (***A***)

2. **a.** Negative. (***A***)

b.

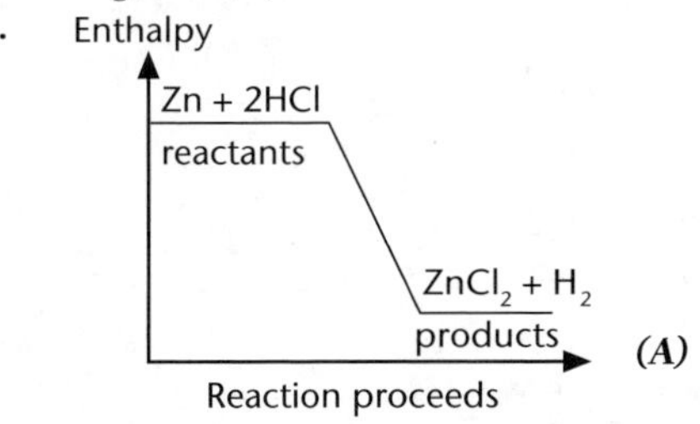

(***A***)

3.

Enthalpy

$CO_2(g)$

$\Delta_r H > 0$

$CO_2(s)$

Reaction proceeds

(***M***)

4. **a.** Exothermic.

b. Endothermic.

c. Endothermic.

d. Exothermic. (***A***)

5. a.

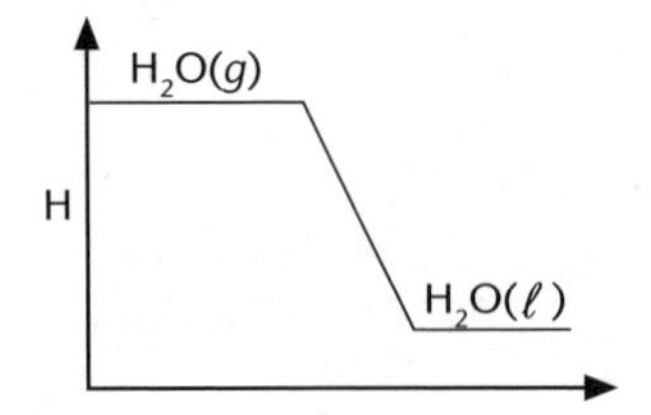

b.

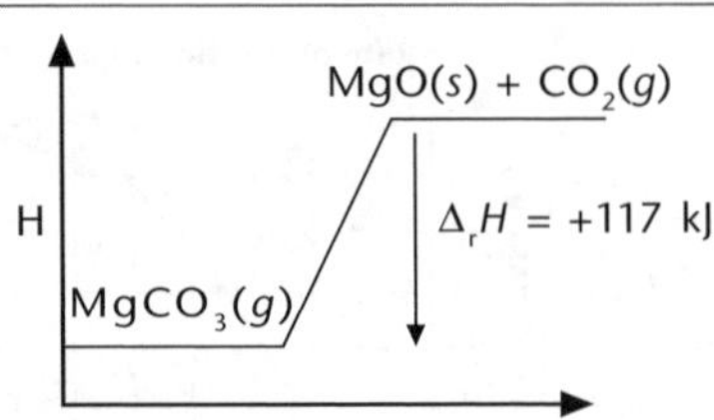

c.

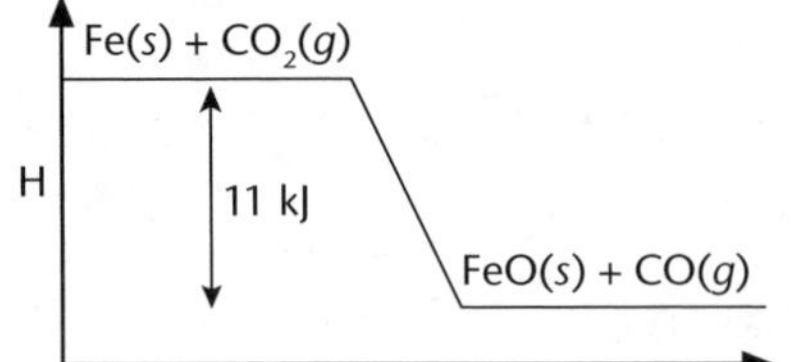

d.

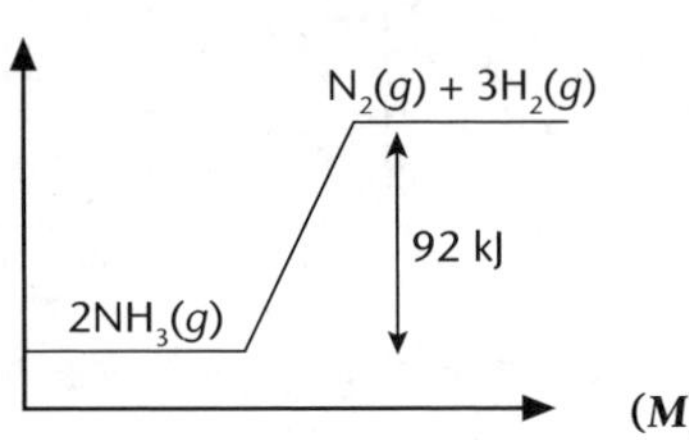

(M)

Unit 11.4 Activity 2B: Calculating enthalpy changes (page 202)

1. $n(\text{Mg}) = \dfrac{m}{M} = \dfrac{6.0\text{ g}}{24.3\text{ g mol}^{-1}} = 0.25\text{ mol}$

0.25 mol releases 150 kJ

2 mol releases $\dfrac{150}{0.25} \times 2 = 1200$ kJ (**M**)

2. 1 mol of C produces 400 kJ, 1 mol = 12 g

6 g produces 200 kJ

$\Delta_rH = -200$ kJ (**M**)

3. a. Δ_rH is negative, so reaction is exothermic. (**A**)

b. 1 mol of reactants releases 57.4 kJ

0.5 mol will release $\dfrac{57.4}{2} = 28.7$ kJ

$\Delta_rH = -28.7$ kJ (**A**)

4. a. 184 kJ **b.** 23 kJ (**A**)

c. 92 kJ produced from 1 mol

1840 produced from $\dfrac{1}{92} \times 1840 = 20$ mol (**M**)

5. a. 1 mol O_2 releases 230 kJ

8 mol O_2 releases 8 × 230 kJ = 1840 kJ

b. 2 mol of C burned when 230 kJ released

2 mol is 24 g

When 6 g is burned, $\dfrac{230}{24} \times 6 = 57.5$ kJ released (**M**)

6. a. Temperature goes up so heat released. Therefore, reaction is exothermic. Also, Δ_rH is –ve so reaction is exothermic. (**A**)

b. $n(Cu^{2+}) = cV = 0.2\text{ mol L}^{-1} \times 0.05\text{ L} = 0.01\text{ mol}$

1 mol releases 216 kJ

0.01 mol releases 216 × 0.01 = 2.16 kJ (**M**)

7. a. $\Delta_r H = -98.2\ \text{kJ mol}^{-1}$

If 5 mol used, *E* released = 5 × 98.2 kJ

= 491 kJ (***A***)

b. $\frac{1}{2}$ mol O_2 = 16 g

16 g formed when 98.2 kJ released

1 g formed when $\frac{98.2}{16} \times 1 = 6.14$ kJ released (***M***)

c. 98.2 kJ produced from 34 g H_2O_2

600 kJ produced from $\frac{34}{98.2} \times 600 = 208$ g (***E***)

8. a.

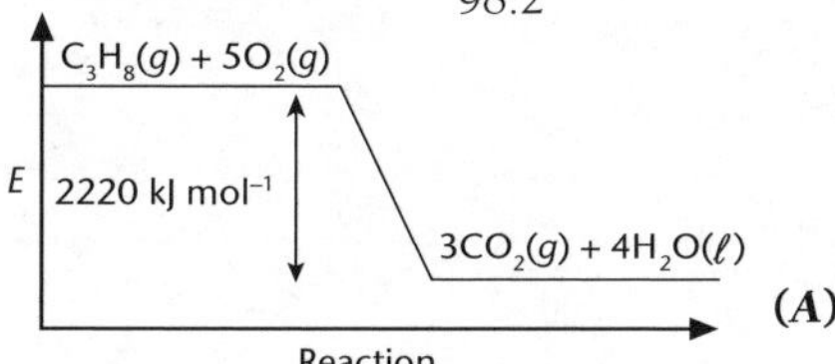

(***A***)

b. 1 mol C_3H_8 = 44 g

44 g burned produces 2220 kJ

11 g burned produces $\frac{2220}{44} \times 11 = 555$ kJ

or $n(C_3H_8) = \frac{11\ \text{g}}{44\ \text{g mol}^{-1}}$ 0.25 mol

So, *E* produced = 0.25 mol × 2220 kJ mol^{-1}

=555 kJ (***M***)

c. 44 g produces 2220 kJ

10 000 kJ produced from $\frac{44}{2220} \times 10\,000 = 198$ g

or 2220 kJ from 1 mol

10 000 kJ from $\frac{2220\ \text{mol}}{10000} = 4.50$ mol

$m(C_3H_8) = nM = 4.50\ \text{mol} \times 44\ \text{g mol}^{-1} = 198$ g (***M***)

9. a. $C_2H_6O(\ell) + 3O_2(g) \rightarrow 2CO_2(g) + 3H_2O(g) \quad \Delta_r H = -1367\ \text{kJ mol}^{-1}$ (***A***)

b. $n(C_2H_6O) = \frac{m}{M} = \frac{4.71\ \text{g}}{46.0\ \text{g mol}^{-1}} = 0.102$ mol

Heat released = 1367 kJ mol^{-1} × 0.102 mol

= 139 kJ (***M***)

10. a. $CH_4(g) + 2O_2(g) \rightarrow CO_2(g) + 2H_2O(g) \quad \Delta_r H = -890\ \text{kJ mol}^{-1}$

$C_4H_{10}(g) + \frac{13}{2}O_2(g) \rightarrow 4CO_2(g) + 5H_2O(g) \quad \Delta_r H = -3\,000\ \text{kJ mol}^{-1}$ (***A***)

b. $m(CH4) = 1$ g $\quad n(CH4) = \frac{m}{M} = \frac{1\ \text{g}}{16.0\text{g mol}^{-1}} = 0.0625$ mol

890 kJ from 1 mol

890 × 0.0625 = 55.6 kJ from 1 g

$m(C_4H_{10}) = 1$ g $\quad n(C_4H_{10}) = \frac{1\ \text{g}}{58.0\ \text{g mol}^{-1}} = 0.0172$ mol

3000 from 1 mol

3000 × 0.0172 = 51.6 kJ from 1 g

ie, CH_4 releases more energy per gram. (***E***)

11. a. $n(\text{HgO}) = \dfrac{m}{M} = \dfrac{277\text{ g}}{217\text{ g mol}^{-1}} = 1.28\text{ mol}$

2 mol needs 182 kJ

1.28 mol needs $\dfrac{182}{2} \times 1.28\text{ kJ} = 116\text{ kJ}$ (***M***)

b. 182 kJ produces 2 mol Hg, so 91 kJ produces 1 mol

555 kJ produces $\dfrac{555}{91} = 6.1\text{ mol Hg}$

$m(\text{Hg}) = nM = 6.1\text{ mol} \times 201\text{ g mol}^{-1}$

$= 1226\text{ g}$ (***E***)

12. a. 2 mol SO_2 releases 188 kJ

$n(SO_2) = \dfrac{m}{M} = \dfrac{32.0\text{ g}}{64.0\text{ g mol}^{-1}} = 0.5\text{ mol}$

0.5 mol releases $\dfrac{188}{2} \times 0.5 = 47\text{ kJ}$ (***M***)

b. 188 kJ from 2 mol SO_2

94 kJ from 1 mol SO_2

3000 kJ from $\dfrac{3000}{94} = 31.9\text{ mol } SO_2$

$m(SO_2) = nM = 31.9\text{ mol} \times 64.0\text{ g mol}^{-1}$

$= 2\ 043\text{ g}$ (***E***)

13. 1648 kJ produced by 4 mol Fe

$m(\text{Fe}) = 448\text{ g}$ $\quad n(\text{Fe}) = \dfrac{m}{M} = \dfrac{448\text{ g}}{55.9\text{ g mol}^{-1}} = 8.01\text{ mol}$

8.01 mol produces $\dfrac{1648}{4} \times 8.01\text{ kJ} = 3300\text{ kJ}$ (***M***)

14. $\text{mol}(B_5H_9) = \dfrac{1\text{ g}}{63.0\text{ g mol}^{-1}} = 0.0159\text{ mol}$

2 mol releases 9036 kJ

0.0159 mol releases $\dfrac{9036}{2} \times 0.0159\text{ kJ} = 71.7\text{ kJ}$

heat released per gram = 71.7 kJ (***M***)

15. $n(NH_3) = \dfrac{600000\text{ g}}{17\text{ g mol}^{-1}} = 35\ 294\text{ mol}$

4 mol releases 912 kJ

35 294 mol releases $\dfrac{912}{4} \times 35\ 294\text{ kJ} = 8.05 \times 10^6\text{ kJ}$

16. a. $n(NO_2) = \dfrac{m}{M} = \dfrac{500\text{ g}}{46\text{ g mol}^{-1}} = 10.9\text{ mol}$

2 mol NO_2 needs 68 kJ

10.9 mol needs $\dfrac{68}{2} \times 10.9 = 371\text{ kJ}$ (***M***)

b. 68 kJ uses 1 mol N_2

350 kJ uses $\dfrac{350}{68}\text{ mol} = 5.15\text{ mol } N_2$

$m(N_2) = nM = 5.15\text{ mol} \times 28.0\text{ g mol}^{-1}$

$= 144\text{ g}$ (***E***)

17. 2 mol N_2H_4 releases 1048 kJ

$$n(N_2H_4) = \frac{m}{M} = \frac{100000\text{ g}}{32.0\text{ g mol}^{-1}} = 3125\text{ mol}$$

$$3125\text{ mol releases } \frac{1048}{2} \times 3125\text{ kJ} = 1.64 \times 10^6\text{ kJ}$$ (***M***)

18. a. 1 mol glucose releases 2803 kJ

$$n(\text{glucose}) = \frac{m}{M} = \frac{10\text{ g}}{180\text{ g mol}^{-1}} = 0.0556\text{ mol}$$

Energy produced = 2803 × 0.0556 = 156 kJ (***M***)

b. 1 g glucose releases 15.6 kJ

For 1 g sucrose:

$$n(\text{sucrose}) = \frac{1\text{ g}}{342\text{ g mol}^{-1}} = 0.00292\text{ kJ}$$

Energy produced per gram = 5640 × 0.00292 kJ
= 16.5 kJ

ie, sucrose produces more energy per gram. (***E***)

Unit 11.4 Activity 3A: Reaction rates and energy diagrams (page 208)

1. a.

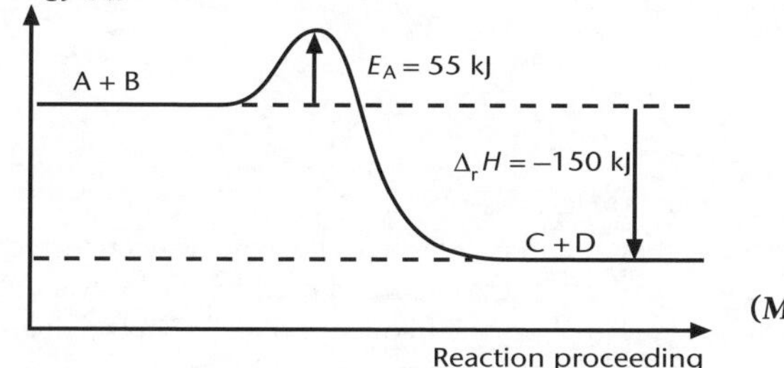

(***M***)

b.

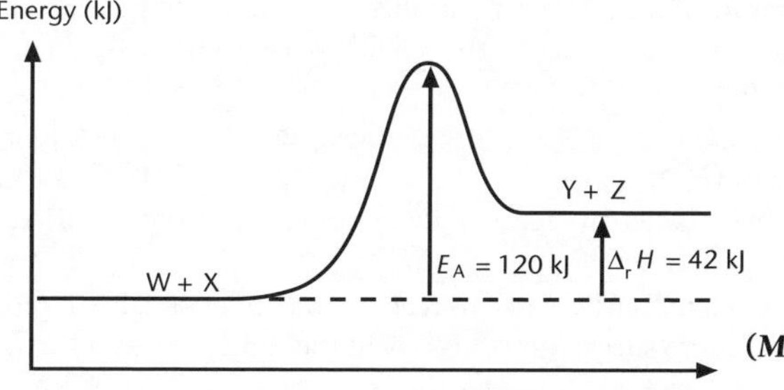

(***M***)

2. a. Exothermic. **b.** B **c.** A (***A***)

3. a. Endothermic. **b.** B (***A***)

Unit 11.4 Activity 3B: Collision theory and reaction rates (page 212)

1. a.

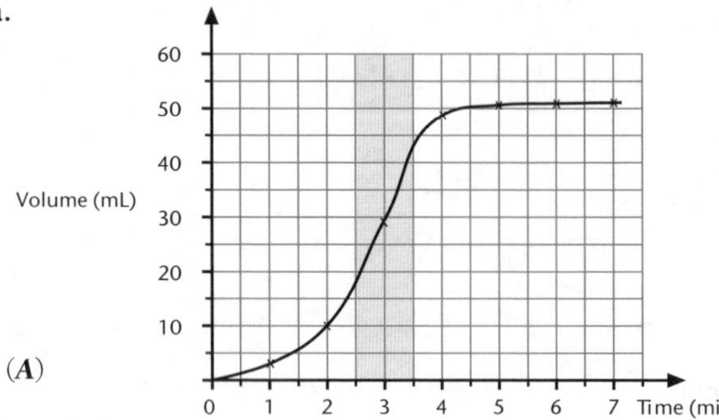

(*A*)

b. *Shaded area of graph* (ie steepest area). (***A***)

c. After about 5 minutes. (***A***)

d.

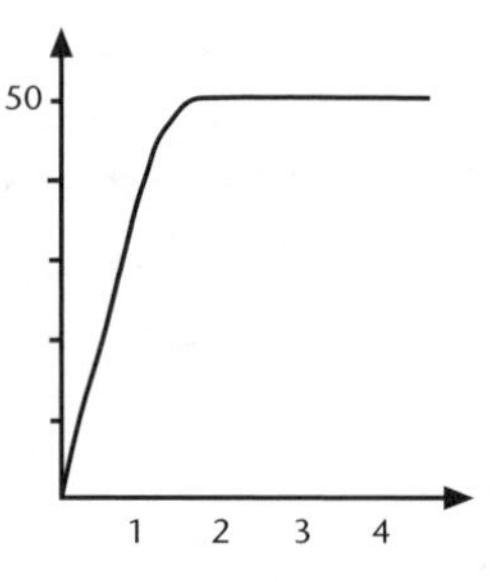

(***M***)

Note: Same volume of gas produced in a shorter time.

e. *Lower temperature* means that the molecules have less kinetic energy, so collisions will be less effective – since there is less chance of reaching the activation energy, the rate of reaction will be reduced. Also, the frequency of collision is reduced, so there will be fewer successful collisions in the same time. (***E***)

More dilute acid means there are fewer particles in a given volume, so collisions will be less frequent – meaning fewer successful collisions in the same time. This decreases the reaction rate. (***M***)

2. a. Coal dust has a large surface area so more particles are exposed, therefore greater chance of collisions. (***M***)

b. Splintered wood has a larger surface area to combine with oxygen during combustion, so more frequent collisions. (***M***)

c. Powdered zinc has a large surface area to react with copper sulfate, so more frequent collisions, meaning more successful collisions in the same time as for zinc granules. This will increase the reaction rate. (***M***)

d. The 0.1 mol L^{-1} acid is less concentrated than the 0.2 mol L^{-1} acid, so fewer acid particles; smaller number of collisions. (See explanation for **1. e.**) (***M***)

e. There are more oxygen molecules in pure oxygen than in the same volume of air, so more frequent collisions occur, meaning more successful collisions in the same time and so a faster reaction rate. (***M***)

f. Protective oxide layer on the aluminium surface prevents any effective collisions. (***M***)

g. Iron wool has a large surface area, so more frequent collisions. (***M***)

3. a.
- Use larger lumps of zinc.
- Use more dilute HCl.
- Cool the reaction mixture. (***A***)

b.
- With larger lumps of zinc, there is less surface area for collisions to occur on, so there will be lower frequency of collisions. (***M***)
- Fewer HCl particles to react with zinc, so lower frequency of collisions. (***M***)
- Lowering temperature reduces the energy of the particles. Collisions become less frequent and since particles collide with less energy, they are less likely to reach activation energy. Both these factors slow the reaction down. (***E***)

4. a. No change (***A***); same frequency of collisions with the same energy, since concentration is unchanged. (***M***)

b. Decrease (***A***); less frequent collisions since solution is less concentrated, so fewer effective collisions and rate is reduced.

c. Increase (***A***); larger surface area, so greater frequency of collisions and more effective collisions, resulting in faster rate. (***M***).

5. a. 3 (***A***)

b. 1; smallest concentration of acid present, so limited number of particles to react. (***M***)

c. Raise the temperature; use powdered $CaCO_3$; use a catalyst. (***A***)

6. a. *Increase* in rate. Reacting particles have more kinetic energy, so when molecules collide there is a greater chance of reaching the activation energy and the reaction rate is increased. Also, there is an increase in the frequency of collisions, causing more successful collisions in the same time. (***E***)

b. *Decrease* in rate. Fewer reacting particles in a given volume, meaning collisions will be less frequent. This means fewer successful collisions in the same time and the rate is reduced. (***M***)

c. *Increase*. Increasing the surface area means more particles are available for collisions. More frequent collisions will occur and so more successful collisions in the same time, so rate increases. (***M***)

d. *No change*. (***A***)

7. a. **i.** More time.

ii. More time.

iii. Less time.

iv. More time.

b. **i.** Reduced temperature (see answer to **1. e.**). (***E***)

ii. Reduced concentration (see answer to **1. e.**). (***M***)

iii. Increased concentration – more reacting particles in a given volume, so more frequent collisions. Greater chance of successful collisions in the same time, so rate is increased and reaction takes less time. (***M***)

iv. This reduces the overall concentration (see **1. e.** for explanation). (***M***)

8. V_2O_5 is a catalyst for the reaction. The catalyst provides an alternative pathway of lower activation energy (E_A). Molecules which did not previously have enough energy to react may now reach the lowered E_A during a collision. The successful collision rate increases and reaction proceeds faster. Since V_2O_5 is not used up in the reaction, only a small amount is required. (***E***)

Unit 11.5 Activity 1A: Physical properties of metals (page 217)

1. Behaviours dependent upon physical properties are: **b.**, **c.**, **d.**, **f.**
   (**A** – two correct; **M** – three correct; **E** – four correct)
2. **a.** Grey solid.
   b. Silvery liquid.
   c. Colourless vapour.
   (**A** – each correct answer)
3.

Physical property	Definition	Example	Explanation of the property
Density	Mass per unit volume	Zinc has a density of 7.1 g mL^{-1}.	*Metal atoms have a high mass and are closely packed.*
Lustre	*Ability of a solid to reflect light*	Aluminium can be polished and used as a mirror.	The tightly packed atoms will reflect light.
Electrical conductivity	Capable of allowing an electric current to pass through readily	*Copper wire is used in large quantities to carry electric current.*	*Metal atoms have electrons in the outer energy level that are mobile.*
Thermal conductivity	Capable of allowing heat to pass through readily	Saucepans often have copper bottoms.	The tightly packed copper atoms can pass heat energy from one to another.
Malleability	Ability to be beaten into a sheet	*Aluminium foil is useful in cooking to wrap food in.*	*Metal atoms can slide over each other.*
Ductility	*Ability to be drawn into a wire*	Steel can be used to make No. 8 wire.	Atoms of iron can slide over each other.

(**A** – one definition and one example correct; **M** – two definitions and two examples correct; **E** – two definitions plus two examples plus two explanations correct)

Unit 11.5 Activity 1B: Alloys, and uses of metals and alloys (page 220)

1. **a.** Steel. (**A**)
   b. *Any two of*: bronze, pewter or solder. (**A**)
   c. Bronze, brass, cupronickel. (**A**)
2. The metals that are to form the alloy are melted and then mixed in the correct proportions. The molten alloy is then allowed to cool to produce the solid alloy. (**A** – melting of metals, mixing of liquid metals and cooling to produce solid alloy)
3. Duralumin is stronger and harder than aluminium. (**M**) The alloy does not have the ability to burn that magnesium possesses. (**E**)
4. $9 \div 24 \times 50 = 18.75$ g. (**M** – correct mathematical expression; **E** – answer correct)

5. **a.** Lead. (***M***)

   A cube of side 1 cm has a volume of 1 cm^3, or 1 mL. The density of the solid is 11 g mL^{-1}. Lead is very malleable and is used to weatherproof awkward joins in roofs.

   b. Gold. (***A***)

   Gold is rare and therefore valuable. Gold is a 'noble' metal, ie it will not react with chemicals. Aqua regia (translates from Latin as 'royal water') is a very powerful acid plus oxidising agent and will dissolve gold.

Unit 11.5 Activity 2A: The activity of metals (page 222)

1. **a.** A property of a substance that results in the formation of a new substance. (***A***)
   b. An arrangement of metals with the most active metal at the beginning of the series. (***A***)
2. **a.** Sodium, lithium, calcium. (***A***) Each of these metals reacts vigorously with the air. (***M***)
   b. Gold, silver. (***A***) Each of these metals has no reaction with the air. (***M***)

   In volcanic regions, sulfur compounds in the air, especially hydrogen sulfide gas, tarnish silver.
3. Zinc, iron, lead, gold. (***A***)

Unit 11.5 Activity 2B: Metals reacting with oxygen (page 225)

1. **a.** Calcium. (***A***)
   b. Aluminium. (***A***)
   c. Gold. (***A***)
2. **a.** Magnesium. (***A***)
   b. Aluminium. (***M***)

Unit 11.5 Activity 2C: Metals reacting with water or steam (page 227)

1. **a.** *Any three of*: Sodium, lithium, calcium, magnesium. (***A***)
   b. *Any three of*: Lead, copper, silver, gold. (***A***)
   c. *Either of*: Zinc or iron. (***A***)
2. The sodium was returned to the wrong bottle (the bottle contained phosphorus under water) and was added to water. There is a violent reaction between sodium and water involving the release of hydrogen and heat. Adding more water would intensify the reaction resulting in an explosion.

   (***A*** – sodium added to the wrong bottle; ***M*** – violent reaction between sodium and water; ***E*** – adding more water intensifies the reaction)
3. Water, when heated, becomes steam. (***A***) There is a vigorous reaction between hot magnesium and steam. (***M***) The magnesium burns in steam adding to the intensity of the fire. (***E***)

4.

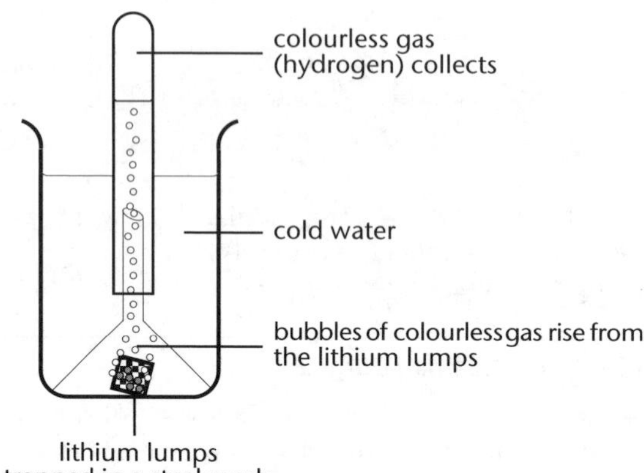

(**A** – apparatus shows a gas can be collected and the lithium below the water surface; **M** – diagram shows the components of the apparatus in reasonable proportions; **E** – complete, correct, clearly labelled diagram)

Unit 11.5 Activity 2D: Metals reacting with acids (page 230)

1. **a.** *Any one of*: Zinc, magnesium, iron. (***A***)
   b. Lead. (***A***)
   c. Aluminium. (***A***)
   d. *Any one of*: Lead, copper, gold, silver. (***A***)
2. **a.** Zinc. (***A***)

   Zinc would produce hydrogen gas at a rate that was fast enough for the experiment to be successful.

   b. Aluminium metal is coated with an oxide that prevents reaction between the metal and the acid. Lead reacts slowly with hydrochloric acid, producing insoluble lead chloride, which coats the lead and prevents further reaction between the acid and the metal. Copper does not react with dilute hydrochloric acid.

   (**M** – explanation why one metal (other than zinc) is not suitable; **E** – explanation why all three metals (other than zinc) are unsuitable)
3. **a.** Magnesium. (***A***)
   b. Either gold or silver. (***A***)
   c. Aluminium. (***A***)
4. **a.** Lithium reacts vigorously with oxygen, water and acids, which are all present in the atmosphere. Storing lithium under oil protects the metal from these chemicals.
   b. Aluminium metal reacts rapidly with oxygen to form a hard coating of aluminium oxide on the metal's surface. This oxide layer protects the metal from further reaction by oxygen or other chemicals in the atmosphere.
   c. Copper does not react with water. Copper does not react with the gases that will be dissolved in hot water, eg oxygen and carbon dioxide from the atmosphere.

   (**A** – one correct statement for each metal; **M** – clear explanation for one metal; **E** – clear explanation for three metals

Unit 11.5 Activity 2E: Displacement of metals (page 232)

1. Magnesium. (***A***)

Magnesium is more reactive than zinc and will displace the zinc from a solution of zinc nitrate. Lithium and calcium would also displace the zinc but would preferentially react with the hydrogen ions in the water. Aluminium, although higher in the activity series than zinc, has a protective layer of oxide on its surface.

2. **a.** No reaction. (***A***)

b. The blue colour of the solution would fade. A brown-red deposit of copper would form. The zinc strip would dissolve.

(***M*** – two observations correct; ***E*** – three observations correct)

3. **a.** Place a strip of lead foil in copper nitrate solution. Filter off the deposited copper from the lead nitrate solution using a funnel and a filter paper. Wash the copper with distilled water and then allow it to dry in air at room temperature by opening up the folded filter paper. (***A***)

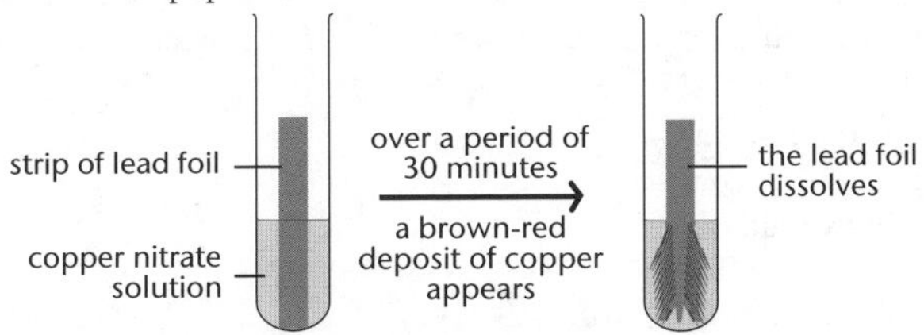

Lead metal displaces copper from copper nitrate solution

The equation for the reaction is $Pb(s) + Cu^{2+}(aq) \rightarrow Pb^{2+}(aq) + Cu(s)$.

(***M*** – ***A*** plus diagram showing reaction between lead metal and copper nitrate solution ***E*** – ***M*** plus equation)

b. Silver metal could be produced in a similar manner. (***M***)

No other metal in the series can be produced using lead metal for displacement. Gold, the least active of the metals in the series, is not obtainable in solution.

4. Lithium is a very reactive metal. (***A***) When placed in any aqueous solution that contains a metal ion (eg zinc nitrate solution contains the zinc ion), the lithium metal displaces hydrogen from the water in preference to displacing the metal (eg zinc). (***M***)

Unit 11.5 Activity 3A: Extraction of metals (page 236)

1. **a.** *Either of*: Gold or silver.

b. *Either of*: Aluminium or iron.

c. *Either of*: Zinc or lead.

(***A*** – any one answer correct; ***M*** – all answers correct)

2. **a.** Bauxite.

b. Zinc blende.

(***A*** – one answer correct; ***M*** – both answers correct)

3. **a.** A process of reduction using heat. (***A***)

b. Using electrical energy to produce chemicals by passing an electric current through an electrolyte. (***A*** – using electrical energy; ***M*** – ***A*** plus passing an electric current through an electrolyte)

4. Copper metal is less active than aluminium metal. (**A**) Higher energy is needed to separate aluminium from oxygen than is needed to separate copper from oxygen. (**M**) More energy is available from electricity than from heat. (**E**)

5. **a.** $C(s) + O_2(g) \rightarrow \mathbf{CO_2(g)}$
   b. $\mathbf{C(s)} + CO_2(g) \rightarrow \mathbf{2}CO(g)$
   c. $Fe_3O_4(s) + \mathbf{4}CO(g) \rightarrow \mathbf{3}Fe(s) + \mathbf{4CO_2(g)}$
   d. $2PbS(s) + \mathbf{3O_2(g)} \rightarrow 2PbO(s) + 2SO_2(g)$
   e. $ZnO(s) + C(s) \rightarrow \mathbf{Zn(s)} + \mathbf{CO(g)}$
   f. $\mathbf{2}Al_2O_3(s) \rightarrow 4Al(s) + \mathbf{3O_2(g)}$

   (**A** – four insertions correct; **M** – eight insertions correct; **E** – all insertions correct)

6. **a.** **iii.** Gold, **ii.** copper, **i.** aluminium. (**A**)
   b. Metals were discovered in the reverse order of their activity, ie the least active metals were discovered first. (**A**)

Unit 11.5 Activity 3B: Corrosion of metals (page 239)

1. Oxygen, water vapour, acid gases (eg carbon dioxide and, in industrial areas, sulfur dioxide and nitrogen dioxide). (**A** – any two chemicals named; **M** – indication of acid gases in air)

2. **a.** Rusting would occur.
   b. No rusting would occur.
   c. No rusting would occur.
   d. Rusting would occur.
   e. No rusting would occur.

   (**A** – two correct; **M** – four correct; **E** – five correct)

3. **a.** No change. Gold has no reaction with the chemicals present in air. It will remain yellow and shiny for thousands of years.
   b. No visible change. As soon as aluminium is in contact with the air, the surface of the metal is converted to aluminium oxide. The coating of the oxide is firmly attached to the remaining aluminium metal beneath. The coating protects the remaining aluminium from further attack. The protection will last permanently.
   c. The shiny surface would become dull. Lead is slowly converted into its oxide, hydroxide, and finally the carbonate, as it remains in contact with air. The lead becomes dull in appearance but will remain protected permanently.
   d. The bright, shiny pink would become dull and slowly develop a green colouration. Copper is converted to its oxide, hydroxide and carbonate on the surface of the metal. The appearance of the copper changes from bright pink to dull brown, and eventually to a green patina (coating) of a mixed hydroxide and carbonate of copper. The patina will protect the remaining copper for hundreds of years.

   (**A** – three observations correct; **M** – two explanations of changes occurring; **E** – four explanations of changes that occur)

4. **a.** Alloyed with aluminium for aeroplane frame construction. The silver shiny alloy would go dull. The oxygen, water vapour and carbon dioxide in the air convert the metal to magnesium oxide, hydroxide and carbonate.
   b. Protection of iron/steel. The silver-grey shiny metal would go dull. The oxygen, water vapour and carbon dioxide in the air convert the metal to zinc hydroxide and carbonate.

c. For jewellery. The bright silver metal would become black. Sulfur compounds, especially hydrogen sulfide, in the atmosphere would convert the silver to silver sulfide.

(***A*** – three uses correct; ***M*** – ***A*** plus list of chemicals in the atmosphere causing changes; ***E*** – ***M*** plus identification of all new chemicals formed)

5. Aluminium has a lower density than steel; aluminium does not corrode in contact with air – no protection is needed. (***A*** – one reason given; ***M*** – two reasons given)

Unit 11.5 Activity 4A: Physical properties of non-metals (page 242)

1. **a.** Green-yellow gas.
   b. Colourless gas.
   c. Colourless gas.
   d. Yellow solid.
   (***A*** – at least three correct)
2. **a.** **i.** Strong smell. (***A***)
   ii. Colourless. (***A***)
   iii. Brittle solid. (***A***)
   b. **i.** Gas at room temperature (cf. oxygen, nitrogen).
   ii. Colourless (cf. nitrogen).
   iii. Very low solubility in water (cf. oxygen, sulfur).
   iv. No smell (cf. oxygen, nitrogen).
   The term 'cf.' means 'compare'. (***A*** – two correct; ***M*** – four correct)
   3. **a.** matches with **ii.**; **b.** matches with **iii.**; **c.** matches with **i.**
   (***A*** – each correct answer)
4. **a.** Brittle – refers to a solid that breaks easily into assorted fragments, ie it shatters when hammered.
   b. Insoluble – refers to a solute that does not dissolve in a solvent.
   c. Shiny – refers to a solid that has a polished surface that reflects light well.
   d. Appearance – usually refers to how a substance is seen by the eyes, but can include the effect on the other senses (ie smell, sound, taste, touch). (***A*** – any two correct; ***M*** – all correct)

Unit 11.5 Activity 4B: Allotropes of oxygen and sulfur (page 247)

1. **a.** Colourless gas with a strong smell.
   b. Yellow crystalline solid with a rhomboid shape,

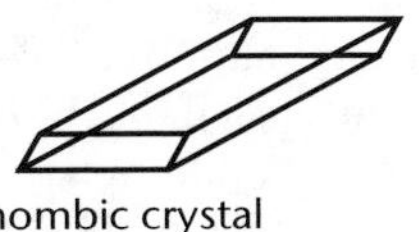

rhombic crystal

   c. Yellow crystalline solid with a needle shape,

monoclinic crystal

   d. Brown-black rubbery solid that changes to a yellow brittle solid over an hour or so.
   (***A*** – three correct; ***M*** – four correct)
2. Ozone is used to freshen air in underground railway stations. (***A***) The ozone destroys harmful bacteria (***M***) and decomposes to produce oxygen. (***E***)

3. **a.** Yellow solid.
   b. Yellow solid.
   c. Dark red and very thick liquid.
   d. Black liquid that is close to boiling.
   (***A*** – two descriptions correct; ***M*** – four descriptions correct)
4. **a.** Sulfur (rhombic). (***A***)
   b. Ozone (oxygen). (***A***)
5. **a.** Rhombic sulfur crystals are brittle. (***A***) The crystal form is easily destroyed and a powder is formed. (***M***)
   b. Ozone gas contains triatomic molecules of oxygen, ie the formula is O_3. Oxygen gas has diatomic molecules, ie O_2. These two substances are the same element in the same physical state but with different arrangements of atoms.
   (***A*** – different arrangement of atoms; ***M*** – in the same physical state)
   c. If liquid sulfur, above 180 °C, is suddenly cooled, long chains are formed (***A***) resulting in a plastic material. The rubbery texture is due to the long chains being able to stretch, and then to contract. (***M***)
6. Isotopes – atoms of the same element with different numbers of neutrons in the nucleus of their atoms. Examples – $^{16}_{8}O$, $^{17}_{8}O$ and $^{18}_{8}O$ are isotopes of oxygen.
   Allotropes – different forms of the same element in the same physical state. Examples – diatomic oxygen, O_2, and triatomic oxygen (ozone, O_3) are allotropes of oxygen.
   (***A*** – one definition correct; ***M*** – one definition plus one set of examples correct; ***E*** – both definitions and both sets of examples correct)

Unit 11.5 Activity 4C: Reactions, uses and preparation of chlorine (page 250)

1. **a.** *Any one of*: sodium chloride, brine or common salt. 'Salt' is not correct. (***A***)
   b. **iii.** Carnallite (potassium magnesium chloride). (***A***)
2. **f.**, **c.**, **b.**, **d.**, **a.**, **e.** (Brine, a sodium chloride solution, is placed in a cell. An electric current is passed through the electrolyte (sodium chloride solution) whereupon chlorine is produced at the titanium anode, and hydrogen is produced at the nickel cathode. The electrolyte becomes a solution of sodium hydroxide.)
   (***A*** – three statements in correct order; ***M*** – six statements in correct order)
3. **a.** Electrolysis is a process of passing an electric current through an electrolyte – a solution or molten solid that conducts electricity. The electrolyte is decomposed by the electric current. (***A*** – using electricity to decompose a chemical; ***M*** – the chemical must be an electrolyte)
   b. The material through which electricity is passed and which undergoes chemical change. (***A***) Electrolytes are either molten ionic substances, or aqueous solutions that contain ion. (***M*** – both parts of explanation)
   c. Brine is a solution of sodium chloride. (***A***)
4. Metals, hydrogen, hydrogen. (***A*** – three correct)
5. **a.** Chlorine + oxygen → chlorine monoxide (***A***)
   b. $H_2(g) + Cl_2(g) \rightarrow 2HCl(g)$ (***A***)
   c. $Cl_2(g) + H_2O(\ell) \rightarrow HCl(aq) + HOCl(aq)$
   (***A*** – one formula correct; ***M*** – formulae correct; ***E*** – formulae and states correct)
6. Purifying water; manufacture of PVC; bleaching wood pulp. (***A*** – one use; ***M*** – two or three uses)

Unit 11.5 Activity 4D: Reactions and preparation of oxygen and nitrogen (page 253)

1. a. i. Air.
   ii. Air. (**A** – both **i.** and **ii.** correct)
   b. iv. Chile saltpetre (sodium nitrate). (**A**)
2. **c.**, **a.**, **e.**, **b.**, **d.** (Air is drawn in from the atmosphere and has dust, water vapour and carbon dioxide removed. The air, without dust, water vapour and carbon dioxide, is compressed. The compressed air is allowed to expand rapidly, which produces a cooling effect sufficient to liquefy the air. The cold liquid is warmed gently to allow nitrogen to boil off leaving liquid oxygen in the vessel. The nitrogen gas can be compressed into cylinders, or turned back into the liquid state for bulk transportation. (**A** – any three statements in correct order; **M** – all five statements in correct order)
3. Nitrogen; nitric oxide; Nitric oxide; nitrogen dioxide. (**A** – any two spaces correctly filled; **M** – four spaces correctly filled)
4. Fuel – a substance that can be used domestically or commercially to react with oxygen and produce heat energy. (**A** – produces heat energy; **M** – produces heat energy in a reaction with oxygen (air))

   Exothermically – a process that proceeds with the release of heat energy. (**A**)

   Product of combustion – the substance(s) formed as the result of a combustion reaction. (**A**).

   Inert – unreactive chemically. 'Unreactive' accepted. (**A**)
5. Atmosphere; lightning; dioxide; acids; nitrite; bacteria; oxidised; roots; protein; animal's; decomposed; returns (**A** – four words in correct position; **M** – eight words in correct position; **E** – twelve words in the correct position)
6. Ammonia can be converted into ammonium compounds which make good nitrogenous fertilisers. (**A**) The fertilisers added to soil can be converted into the nitrite ion, which is part of the nitrogen cycle. (**M**) Crops grown in some countries are not consumed locally and the waste products from the consumption of the crops are not returned to the soil where the crops were grown. For continuous crop production, it is necessary for the soil to be enriched with added man-made fertilisers. (**E**)

Unit 11.5 Activity 4E: Reactions, uses and preparation of sulfur (page 256)

1. a. *Any one of*: fossil fuel, natural gas or petroleum. (**A**)
   b. **i.** Gypsum (calcium sulfate dihydrate). (**A**)
2. hydrogen; poisonous; SO_2; SO_2 (**A** – two answers correct; M – four answers correct)
3. a. Rain that is more acidic than normal due to the presence of sulfur dioxide, nitrogen dioxide and other acidic gases formed in industrial areas. (**A** – rain more acidic than normal; **M** – **A** plus one gas named that is responsible for the increased acidity; **E** – **A** plus two or more gases named that are responsible for the increased acidity)
   b. A chemical capable of carrying out the process of reduction. (**A**) The chemical is capable of either removing oxygen from a substance, or adding hydrogen to a substance, or adding electrons to a particle involved in the reaction. (**M** – one reduction process described; **E** – all three reduction processes described)
   c. Another name for a reductant. (**A**)

4. Sulfur; sulfur dioxide; acid (**A** – three terms correct)
5. **a.** Sulfur + oxygen → sulfur dioxide (**A**)
   b. $\mathbf{H_2S}(g) + 1.5O_2(g) \rightarrow H_2O(\ell) + SO_2(g)$ (**M**)
   c. $\mathbf{2}H_2S(g) + SO_2(g) \rightarrow \mathbf{2}H_2O(\ell) + 3S(s)$ (**M**)
6. To make sulfuric acid; to convert latex into rubber; to combat bacterial/fungal growth on plants. (**A** – each correct use)

Unit 11.5 Activity 5A: Reactions to prepare gases, and the collection of gases (page 259)

1. **a.** **i.** Upward displacement of air.
   ii. Downward displacement of air.
   iii. Displacement of water.
   b. **i.** Sulfur dioxide is denser than air.
   ii. Ammonia is less dense than air.
   iii. Nitrogen monoxide is insoluble in water.
   (**A** – two of part **a.** correct; **M** – two of part **a.** and two of part **b.** correct)
2. **a.** Sulfur dioxide. (**A**)
   b. Ammonia. (**A**)
3. **a.** $2NH_4Cl(s) + Ca(OH)_2(s) \rightarrow CaCl_2(s) + 2H_2O(\ell) + \mathbf{2NH_3(g)}$
   b. $NaCl(s) + H_2SO_4(\ell) \rightarrow NaHSO_4(s) + \mathbf{HCl(g)}$
   c. $2NaCl(s) + H_2SO_4(\ell) \rightarrow Na_2SO_4(s) + \mathbf{2HCl(g)}$
   d. $NH_4Cl(s) + \mathbf{NaOH(aq)} \rightarrow NaCl(aq) + H_2O(\ell) + NH_3(g)$
   e. $Na_2SO_3(s) + 2HNO_3(aq) \rightarrow 2NaNO_3(aq) + H_2O(\ell) + \mathbf{SO_2(g)}$
   (**A** – two equations with the correct formulae; **M** – four equations with the correct formulae; **E** – all equations with the correct formulae, and three equations correctly balanced)
4. **a.** Concentrated sulfuric acid. (**A**)
   b. Water.
   If the nitrogen monoxide is collected by displacement of water then no wash bottle would be needed. (**A**)

Unit 11.5 Activity 5B: Apparatus for preparing and collecting gases (page 264)

1. **a.** **b.** **c.** **d.**

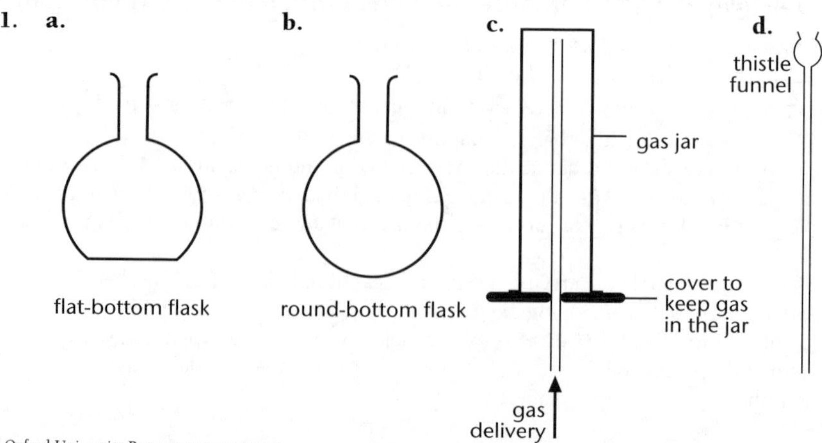

(***A*** – two correct diagrams; ***M*** – four correct diagrams, with at least one of appropriate scale; ***E*** – four correct diagrams, all to appropriate scale)

2. *Either:*

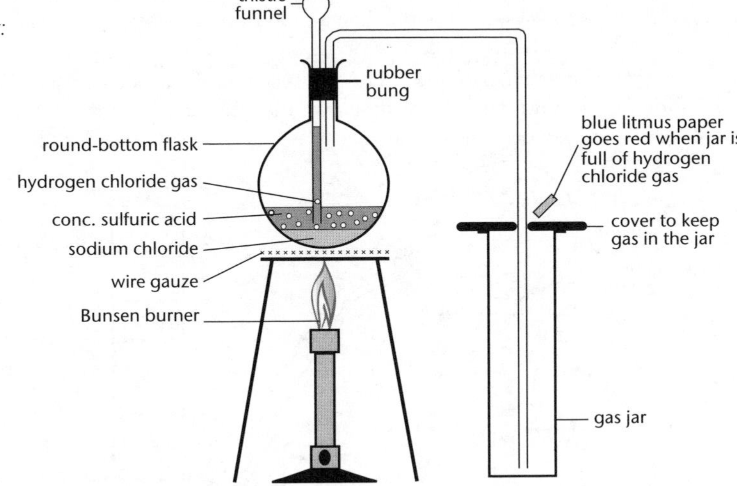

$2NaCl(s) + H_2SO_4(\ell) \rightarrow Na_2SO_4(s) + 2HCl(g)$.

Or:

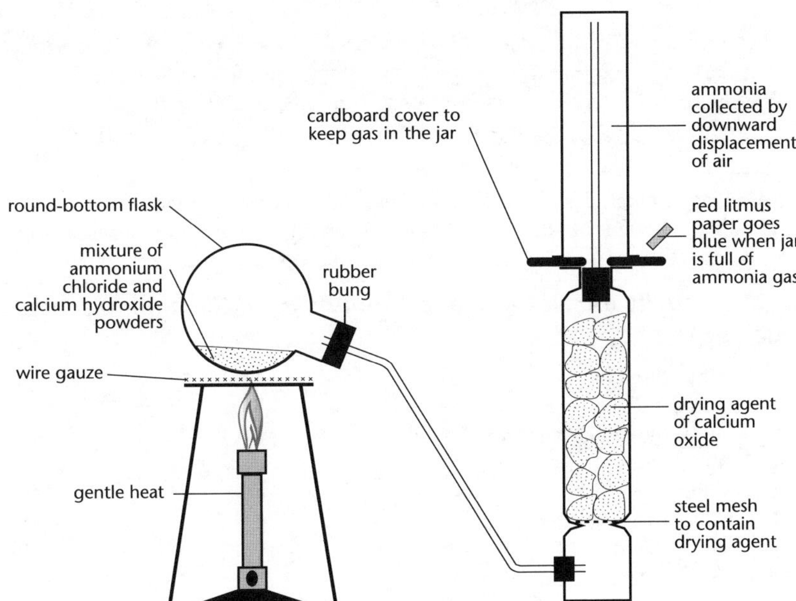

$2NH_4Cl(s) + Ca(OH)_2(s) \rightarrow CaCl_2(s) + 2H_2O(\ell) + 2NH_3(g)$.

(***A*** – diagram showing round-bottom flask, heat (Bunsen burner), wire gauze between flask and heat source; ***M*** – ***A*** plus correct combination of chemicals to produce a gas; ***E*** – ***M*** plus drying agent (if required), connections in place, jar for collection of gas in correct displacement and correct equation)

3. **a.** A round-bottom flask is stronger than a flat-bottom flask – there is no sharp bend in the glass where the heat is being applied. When glass is heated, and expansion of the glass occurs, a sharp bend is a point of weakness where a crack can occur.

b. A flat-bottom flask is more stable than a round-bottom flask – the flask will sit comfortably on a bench without support.

(***A*** – statement of advantage of round-bottom flask and of flat-bottom flask; ***M*** – explanation for the advantage of one flask over the other; ***E*** – both explanations for the advantage of one flask over the other)

4. 

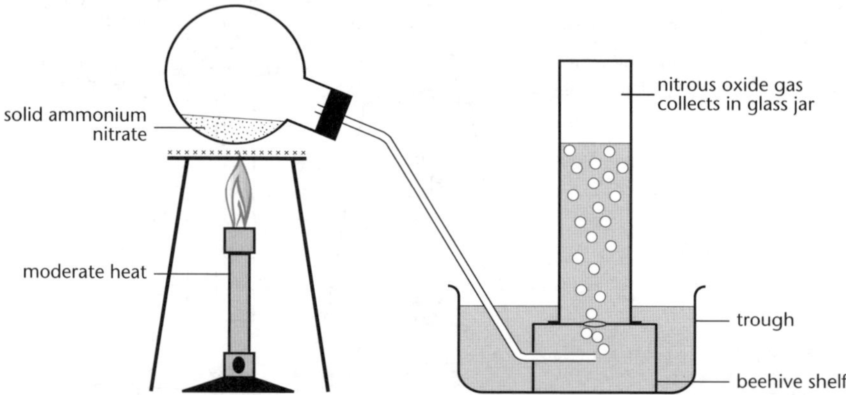

Features of the apparatus are: a round-bottom flask is used when heating is necessary; the flask slopes so that water formed by the condensation of the steam that is produced in the reaction cannot run back onto the hot glass; the gas, being insoluble in water, is collected by displacement of water.

$NH_4NO_3(s) \rightarrow N_2O(g) + 2H_2O(\ell)$.

(***A*** – diagram includes round-bottom flask with tube delivering gas to a water trough, flask being heated; ***M*** – ***A*** plus apparatus is drawn so no gas can escape; ***E*** – complete and correct diagram drawn in reasonable proportions, with correct equation)

Unit 11.5 Activity 6A: Non-metal compounds and the human senses (page 266)

1. **a.** Sulfur dioxide.

b. Nitrogen dioxide (brown).

c. Sulfur dioxide *or* nitrogen dioxide.

d. Nitric acid.

e. Sulfuric acid.

f. Nitric acid.

g. Sulfuric acid.

h. Nitric acid *or* sulfuric acid.

(***A*** – three correct; ***M*** – six correct; ***E*** – eight correct)

2. **a.** Wear safety glasses. Place the gas jar in a fume cupboard with the extractor fan running. Slide the lid off the jar so that about one quarter of the jar is open. Using an open hand, waft the gas towards your nose.

(***A*** – two safety precautions stated; ***M*** – three safety precautions stated; ***E*** – four safety precautions stated)

b. Wear safety glasses. Place the equipment in a fume cupboard with the extractor fan running. Wear protective (rubber) gloves. Remove the stopper from the bottle and place it in a secure position in the fume cupboard. Transfer 10 mL of the liquid to a measuring cylinder by carefully pouring the liquid from the bottle and avoiding any spillage. Transfer the 10 mL of the liquid to the beaker. Carefully wash the measuring cylinder with plenty of water (assume the liquid is soluble in water).
(***A*** – three safety precautions stated; ***M*** – five safety precautions stated; ***E*** – six or more safety precautions stated)

c. Wear safety glasses. Place the equipment in a fume cupboard with the extractor fan running. Wear protective (rubber) gloves. Place the jar of nitrogen dioxide in the fume cupboard with the jar upright. Place the jar of air, upside down, on top of the jar containing the nitrogen dioxide. Slide the lid off the jar containing the nitrogen dioxide, making sure the jars fit without any gaps where the jars meet. Observe that the brown colour of the gas moves (upwards) into the jar of air, despite the fact that the gas is heavier than air.
(***A*** – three safety precautions stated plus the jar of nitrogen dioxide being placed in an upright position; ***M*** – ***A*** plus recognising that the gas should not escape; ***E*** – ***M*** plus recognising that the gas will flow up into the jar of air even though its density is greater than that of air)

Unit 11.5 Activity 6B: Solubility and acidity of non-metal compounds (page 272)

1. Nitrogen dioxide *or* sulfur dioxide. (***A***)

2. $SO_2(aq)$.
(***A*** – formula correct; ***M*** – formula and state correct)

3. Nitrogen dioxide gas is very soluble in water. (***A***) The pressure in the test tube is much lower than the atmospheric pressure above the water in the beaker. (***M***) The greater external pressure pushes the water to the top of the test tube. (***E***)

4. Sulfur dioxide gas is dissolving in water to produce a solution of the gas. (***A***) The gas is dissolving physically in the water. (***M***)

5. **Result X** – gas X is insoluble in water.
Result Y – gas Y is moderately soluble in water.
Result Z – gas Z is very soluble in water.
(***A*** – two results correct; ***M*** – three results correct)

6. **a.** A solvent is usually a liquid which dissolves a solute to produce a solution. Water is a very common solvent. It dissolves common salt (sodium chloride) to produce salt solution.

b. An aqueous solution is a solution using water as the solvent. Sea water is an aqueous solution of sodium chloride in water.

c. An anhydride is a substance that reacts with water to produce a new product that is an acid. Sulfur dioxide is the anhydride of sulfurous acid.

(***A*** – one term explained with a satisfactory example; ***M*** – two terms explained with satisfactory examples; ***E*** – three terms explained with satisfactory examples)

7. The acid would sink below the water. The mixture would become hot. The mixture would boil if the acid were added quickly to the water. When the mixing was complete, there would be one layer of sulfuric acid solution.
(***A*** – one observation; ***M*** – three observations; ***E*** – four observations)

8. **a.** $H_2SO_4(\ell) + \mathbf{H_2O(\ell)} \rightarrow H_3O^+(aq) + HSO_4^-(aq)$
   b. $HSO_4^-(aq) + H_2O(\ell) \rightarrow \mathbf{H_3O^+(\mathit{aq})} + SO_4^{2-}(aq)$
   c. $\mathbf{HNO_3(\mathit{aq})} + H_2O(\ell) \rightarrow H_3O^+(aq) + NO_3^-(aq)$
   d. $HNO_2(aq) + H_2O(\ell) \rightarrow H_3O^+(aq) + \mathbf{NO_2^-(\mathit{aq})}$
   (**A** – two formulae correct; states not necessary; **M** – four formulae correct; states not necessary; **E** – four formulae correct with states)
9. **a.** $H_2SO_3(aq) + H_2O(\ell) \rightarrow H_3O^+(aq) + HSO_3^-(aq)$
   b. $HNO_3(aq) + H_2O(\ell) \rightarrow H_3O^+(aq) + NO_3^-(aq)$
   (**A** – three formulae correct; states not necessary; **M** – five formulae correct; states not necessary; **E** – both equations correct, including the states of the substances)

Unit 11.5 Activity 6C: Photochemical smog and acid rain (page 274)

1. **a.** *Either of*: Combustion of fossil fuels *or* volcanic activity.
   b. *Either of*: From the exhaust of a vehicle *or* during a thunderstorm.
   (**A** – either **a.** or **b.** correct; **M** – both **a.** and **b.** correct)
2. Sulfurous acid. (**A**)
3. **a.** Rain dissolves gases in the atmosphere. In industrial areas, fossil fuels produce sulfur dioxide gas when the fuels are burnt. Sulfur dioxide, dissolved in rain, is strongly acidic. Oxides of nitrogen, formed in car engines and emitted through the exhaust system, dissolve in rain and are strongly acidic. Rain in industrial areas can be highly acidic. (**E**)
   (**A** – two sentences with good information; **M** – four sentences with good information)
   b. Photochemical means chemical reactions involving sunlight. Ozone is formed by oxygen in the air reacting, in the presence of sunlight, with oxides of nitrogen present in the exhaust gases of the internal combustion engine. Some petrol and other fuels remain unburnt in the internal combustion engine and hydrocarbons are released into the atmosphere. Ozone reacts with unburnt hydrocarbons from petrol, etc, to produce other organic molecules. These molecules form an aerosol in the atmosphere which gives the atmosphere a blue-brown haze – a photochemical smog. (**E**)
   (**A** – two sentences with good information; **M** – four sentences with good information)
   c. Ozone is an allotrope of oxygen. An allotrope is a different form of an element in the same physical state. Oxygen forms molecules, O_2. Ozone forms molecules, O_3. Ozone is formed in the lower atmosphere by photochemical reaction between oxygen and oxides of nitrogen. (**E**)
   (**A** – two sentences with good information; **M** – four sentences with good information)
4. **a.** *Either of*: Nitrogen dioxide *or* nitrogen monoxide.
   b. Sulfur dioxide.
   (**A** – one substance correct; **M** – two substances correct)

Unit 11.5 Activity 7A: Metalloids (page 280)

1. Metalloids are usually found combined with non-metals, in which they exist as cations in a positive oxidation state. To obtain them in their pure elemental form then, they need to be reduced.
2. **a.** $2BCl_3(g) + 3H_2(g) \rightarrow 2B(s) + 6HCl(g)$
   b. $SiO_2(s) + C(s) \rightarrow Si(s) + CO_2(g)$
   c. $2As_2O_3(s) + 3H_2(g) \rightarrow 2As(s) + 3H_2O(g)$

3. B_2H_6. Diborane. Three-centre bond arrangement.
4. Catenation (the linkage of atoms of the same element into longer chains) occurs in carbon, which can form long chains or ring molecules. Catenation between Si atoms (–S–Si–) forms silanes (Si_2H_6, Si_6H_{14}).

Unit 11.5 Activity 8A: Commercial preparation, properties and uses of ammonia and sulfuric acid (page 284)

1. **a.** Air, sulfuric acid, water, zinc blende.
   b. Air, natural gas, water.
   (***M*** – one set of chemicals correct; ***E*** – two sets of chemicals correct. If extra chemical(s) are named then ***M*** becomes ***A***, ***E*** becomes ***M***.)
2. **a.** and **b.**
   i. $S(s) + O_2(g) \rightarrow \mathbf{SO_2(g)}$; sulfuric acid.
   ii. $2ZnS(s) + \mathbf{3O_2(g)} \rightarrow 2ZnO(s) + 2SO_2(g)$; sulfuric acid.
   iii. $CH_4(g) + H_2O(\ell) \rightarrow CO(g) + \mathbf{3H_2(g)}$; ammonia gas.
   iv. $K_2CO_3(aq) + \mathbf{CO_2(g)} + \mathbf{H_2O(\ell)} \rightarrow 2KHCO_3(aq)$; ammonia gas.
   (***A*** – two formulae correct plus two products correct; ***M*** – four formulae correct plus three products correct; ***E*** – all formulae correct and all products correct)
3. **a.** Sulfuric acid has wide range of properties and uses.
   b. Nitric acid is used to make many conventional high and low explosives – these substances are necessary agents of war.
   c. In the manufacture of sulfuric acid, the gases sulfur dioxide and oxygen come into contact with vanadium pentoxide catalyst.
   (***A*** – two explanations; ***M*** – three explanations)
4. **a.** Carbon dioxide gas is very soluble in potassium carbonate solution. If a mixture of all four gases is passed through potassium carbonate solution, the carbon dioxide gas will be retained and the other gases will pass through the solution.
   b. Ammonia gas is easily liquefied by cooling and also it is highly soluble in water. A mixture of ammonia, nitrogen and hydrogen gases can be cooled, whereupon the ammonia will liquefy, or the mixture can be passed through water, whereupon the ammonia will dissolve. The nitrogen and hydrogen in both cases will be unchanged.
   (***A*** – statement of the method of separation for **a.** and **b.**; ***M*** – explanation in either part **a.** or part **b.**; ***E*** – complete explanation in both parts)
5. **a.** $H_2SO_4(\ell)$.
   b. $NH_3(g)$.
   (***A*** – both chemical formulae correct; ***M*** – ***A*** plus one physical state correct; ***E*** – ***A*** plus both physical states correct)

Unit 11.5 Activity 8B: Commercial preparation and uses of superphosphate, sodium hypochlorite and sulfur dioxide (page 288)

1. **a.** Rock phosphate, sulfuric acid.
   b. Chlorine gas, sodium hydroxide solution.
   (***M*** – one set of chemicals correct; ***E*** – two sets of chemicals correct. If extra chemical(s) are named then ***M*** becomes ***A***, ***E*** becomes ***M***.)

2. a. Sodium hydroxide is used to make sodium hypochlorite by reaction with chlorine gas.
   b. The 'super' in superphosphate indicates that the compound has superior solubility in water compared with rock phosphate.
   (**A** – one correct explanation; **M** – two correct explanations)
3. Superphosphate is a mixture of calcium hydrogen phosphate and calcium sulfate. (**A**) Both these compounds are valuable plant fertilisers. (**M**)
4. a. $NaOCl(aq)$.
   b. $Ca_3(PO_4)_2(s)$.
   c. $CaSO_4(s)$.
   d. $Ca(H_2PO_4)_2(s)$.
   (**A** – two chemical formulae correct; **M** – four chemical formulae correct; **E** – all formulae, plus all four physical states, correct)
5. a. Sulfur dioxide. (**A**)
   b. Sodium hypochlorite solution. (**A**)
6. a. OCl^-.
   b. $Ca(H_2PO_4)_2 + CaSO_4$ (ie a mixture of $Ca(H_2PO_4)_2$ and $CaSO_4$).
   (**A** – one formula correct; **M** – two formulae correct; **E** – all formulae correct)
7. a. Sulfur dioxide is an antioxidant. It combines with oxygen dissolved in the fruit drink and thus microbes cannot survive. (**A**) Removal of the microbes means that no toxins can be produced by them. (**M**) The solubility of sulfur dioxide in water is sufficient to allow this process to operate for some months. (**E**)
   b. A gymnasium contains human beings who get hot and perspire. (**A**) These are ideal conditions for microbes to multiply, some of which are harmful to the human system. (**M**) Washing the walls of the shower box regularly with a chemical that destroys microbes will lessen the chance of disease spreading. (**E**)
8. Today's newspaper would have white paper. The newspaper that was two years old would be yellow due to the wood pulp combining with oxygen in the air and returning to its original colour.
   (**A** – description of the appearance of the two newspapers; **M** – explanation for the difference related to oxygen in the air restoring the paper to its original colour)
9. The first dish would have a heavy infestation of colonies, randomly spread over the agar jelly. The second dish would have clear patches of no infestation where the sodium hypochlorite solution had been placed. The sodium hypochlorite solution would destroy any microbes and those regions of agar jelly would be clear of colonies.
   (**A** – description of the appearance of the two dishes; **M** – explanation for the difference related to the sodium hypochlorite solution destroying the microbes)

Supplementary Unit Activity A: Practical investigation (page 298)

1. a. To pose a problem that can be, or may be, answered by practical measurement, or by practical observation.
   b. A proposition made as a basis for reasoning, without the assumption of its truth.

(**A** – one explanation correct; **M** – two explanations correct)

2. **a.** Aim.
   b. Prediction. Assuming it is known that it is easier for a human being to float in the Dead Sea than in the Pacific Ocean.

   (***A*** – one term identified correctly; ***M*** – two terms identified correctly)

3.

Investigation	a. Independent variable	b. Dependent variable
1	Concentration of ammonia solutions	Volume of acid used for neutralisation
2	Time	Volume of gas collected
3	Constant mass of different antacid tablets	Volume of alkali used for neutralisation
4	Concentration of ethanoic acid solutions	Volume of alkali used for neutralisation
5	Constant mass or volume of different alcohols	Temperature rise of a fixed mass or volume of water
6	Concentration of reagent	Form of precipitate formed – fine, flocculent or gelatinous

(***A*** – one independent and one dependent variable identified correctly; ***M*** – three independent and three dependent variables identified correctly; ***E*** – five independent and five dependent variables identified correctly)

4.

Volume of acid solution taken (concentration is 2 g L^{-1}) (mL)	Volume of water to add to make 100 mL of solution (mL)	Total volume of solution (mL)	Concentration of prepared solution (g L^{-1})
a. 100	0	100	2.0
b. 80	20	100	1.6
c. 40	60	100	0.8
d. 5	95	100	0.1

(***A*** – two quantities correct; ***M*** – four quantities correct; ***E*** – all seven quantities correct)

5. **a.** 4.4 mL of sodium hydroxide solution in experiment where 10 mL of 2.5 g L^{-1} of ethanoic acid was used. The value is more than 0.2 mL greater than the other results.

   (***A*** – correct identification of 4.4 mL; ***M*** – explanation correct)

   b.

Concentration of ethanoic acid solution (g L^{-1})	Average volume of sodium hydroxide solution used (mL)
5	8.3
2.5	4.05
1	2.17
Diluted vinegar solution	4.57

(***A*** – all averages correct)

c.

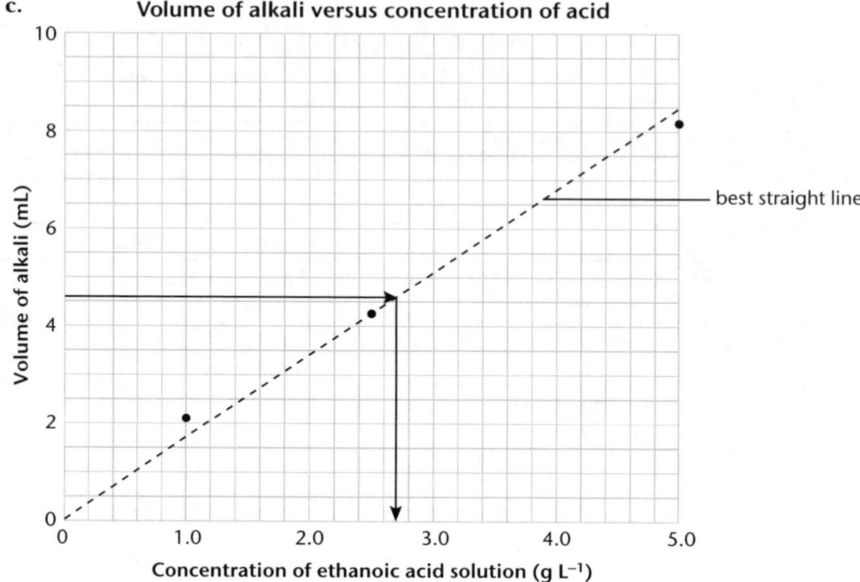

(***A*** – graph drawn with axes and title; ***M*** – points plotted accurately; ***E*** – best straight line drawn)

d. 2.70 g L^{-1}. (***A*** – correct concentration; ***M*** – indication on graph in **c.** of how the figure was obtained)

e. 27.0 g L^{-1}. (***A***).

6.

Similarities	Differences
Both made of glass	Pipette delivers one stated volume
Both measure volumes of liquid accurately	Burette delivers a variety of volumes by control of the tap – from 0.1 mL to 50.0 mL.

(***A*** – one similarity and one difference; ***M*** – two similarities and two differences)

7. The two variables (independent and dependent) are directly proportional (***A***), ie if one variable doubles, so does the other. (***M***)

8. a. The ammonia solution contains a gas, ammonia, that escapes readily from solution. (***A***) If the temperature of different estimations is the same, the amount of ammonia lost will be constant. (***M***)

b. Three drops of indicator is sufficient to see clearly the colour change of the indicator at the point where neutralisation occurs. (***A***) By using the same volume of indicator, the intensity of the colour change will be the same and the readings are consistent. (***M***)

c. Ammonia gas is released from ammonia solution (***A***), and it should not be inhaled. (***M***)

Glossary/Index

***Note:** Numbers in brackets refer to pages on which term is defined or discussed.*

acid (**108**)**:** a substance in solution that will turn blue litmus paper red and has a pH of less than 7.

acid radical (**134**)**:** a group of atoms in a molecule that, when joined to hydrogen atom(s), form an acid.

acid rain (**2**, **255**, **265**, **273**)**:** rain that is more acidic than normal due to the presence of sulfur dioxide, nitrogen dioxide and other acidic gases formed in industrial areas.

activity series (**145**, **221**, **230**)**:** an order for the arrangement of metals, with the most active metal at the top of the series; for Grade 11 Chemistry, the metals listed are, in order of activity: sodium > lithium > calcium > magnesium > aluminium > zinc > iron > lead > copper > silver > gold.

aerosol (**273**)**:** a dispersion of very small drops of liquid in a gas, eg air; forming part of what is commonly referred to as 'smog'.

alcohol (**39**, **153**, **289**)**:** a compound of carbon formed by replacement of a hydrogen atom by a hydroxyl group in the molecule of an alkane; general formula $C_nH_{2n+1}OH$.

alkali (**260**, **269**, **277**, **289**, **292**)**:** a soluble base that will turn red litmus paper blue and has a pH greater than 7.

alkaline (**144**)**:** solution containing more hydroxide ions than hydronium ions.

alkane (**152**, **163**, **278**)**:** a compound of hydrogen and carbon, in a series with the general formula, C_nH_{2n+2}.

alkene (**152**)**:** a compound of hydrogen and carbon, in a series with the general formula, C_nH_{2n}.

allotrope (**90**, **242**, **247**, **273**, **277**)**:** different physical form of an element.

alloy (**15**, **39**, **154**, **218–20**, **239**, **279**)**:** two or more metals dissolved in each other; (*Note:* carbon, a non-metal, can form an alloy with iron (called steel)).

alumina (**235**)**:** aluminium oxide, obtained from the mineral bauxite and used to produce aluminium metal.

amphoteric (**275**, **277**)**:** metal oxide or hydroxide that can be acidic (react with bases) or basic (react with acids).

amu (**159**)**: atomic mass unit**; the unit is $\frac{1}{12}$ of the mass of a carbon-12 atom and is the standard against which all atomic and molecular masses are compared.

analysis (**139**, **142**)**:** separation of compounds/mixtures into their constituents and/or properties of constituents.

anhydride (**271**)**:** a compound that reacts with water to produce an acid.

anhydrous (**200**, **238**, **258**)**:** without water – used to describe crystals that have lost their water of crystallisation.

anion (**67**, **73**, **76**, **87**, **271**, **276**)**:** a negative ion that is attracted to the anode during electrolysis.

anode (**164**, **249**)**:** the positive terminal in an electrolysis process to which anions are attracted.

antioxidant (**255**, **270**, **287**): another name for a reducing agent, or reductant; in terms of food preservation, the antioxidant removes oxygen, and thus prevents bacteria surviving to cause food to decay.

aqueous (**75**, **81**, **103**, **110**, **112**, **121**, **136**, **172**, **179**, **184**, **194**, **249**): dissolved in water.

aqueous solution: a solution formed by dissolving a substance in water.

A_r (**159**, **160**, **275**): symbol used to indicate relative atomic mass; relative atomic mass of an atom is the mass of an atom in amu.

atom (**11**, **49–57**, **61**): the smallest particle of an element that can take part in a chemical reaction; composed of protons, neutrons and electrons.

atomic number (symbol Z) (**49**, **54**, **83**, **99**, **105**, **275**, **301**): the number for an element that indicates its position in the periodic table, as well as the number of protons, and the number of electrons, in one atom of the element.

Atomic Theory (**36**, **49**): theory put forward by John Dalton, between 1803 and 1808, that convinced the world that atoms exist.

Avogadro's law (**28**, **35**): all gases at the same temperature and pressure contain the same number of molecules.

baking soda (**147**): a common (trivial) name for sodium hydrogen carbonate.

balanced (**120**, **185**): the number of each type of atom on both sides of a chemical equation is the same.

base (**108**, **169**, **173**, **260**, **283**): the oxide, hydroxide, carbonate or hydrogen carbonate of a metal; a substance that will neutralise an acid.

basic (**108**, **167**, **169**, **174**, **237**, **277**): can neutralise acids.

bauxite (**235**): a mineral containing oxides of aluminium and iron; used as a source of aluminium metal.

bends (**241**): a decompression sickness suffered by divers ascending too quickly from deep water; caused by dissolved nitrogen rapidly escaping from blood.

bicarbonate (**145**, **147**): another name for the hydrogen carbonate radical.

bleaching agent (**286**): a chemical capable of removing the colour from another chemical substance.

boiling point, bp (**10**, **13**, **18**, **20**, **40**, **88**, **93**, **108**, **215**, **244**, **276**): the temperature at which a liquid changes to a gas or vapour.

bond (**9**, **49–114**): a force between two atoms which creates a molecule, a part of a molecule, or an ionic substance; or a force between two molecules responsible for the liquid or solid state of a substance.

bonding electrons (**278**): shared electrons in a bond.

brittle (**88**, **241**, **275**): a property of a solid that causes it to break easily into irregular-shaped fragments.

buckminsterfullerene (**92**): the first fullerene compound to be discovered, commonly called a buckyball.

buckyball (**92**): common name for buckminsterfullerene.

burette (**170**, **292**, **293**): a piece of glass equipment used to deliver any volume of liquid between 0.1 mL and 50.0 mL accurately.

calibrated (294): set to a specific value/measurement.

capillary attraction (19): the manner by which a liquid can rise up a thin tube to create a meniscus; also, how a liquid can move through the pores of paper or other porous materials.

carbon dating (52): a process for dating dead organic matter, eg charcoal, over periods of tens of thousands of years.

cathode (249): the negative terminal in an electrolysis process to which cations are attracted.

cation (73, 77, 86): a positive ion that is attracted to the cathode in an electrolysis process.

charcoal (152, 155): an impure form of carbon obtained by heating wood in the absence of air.

chemical bonds (59–98, 278): strong attractive forces between atoms, molecules or ions.

chemical changes (84, 148, 179, 191, 249): changes that result in a new substance and that are difficult or impossible to reverse.

chemical equation (119, 201): a brief and accurate way of summarising what happens in a chemical reaction.

chemical property (105, 215, 221–32, 275–80): a property of a substance that, when displayed, produces a new chemical substance.

chromatogram (18–19): the result of a chromatography process showing each substance present in a mixture.

chromatography (18): a process of separation of chemically similar substances by use of a stationary and a moving phase.

cleave (8–9): to split a crystal along one of its axes.

closed system (178, 184): system from which substances involved are not allowed to enter or escape.

coke (233): produced from coal that has been heated in the absence of air; oils and gases are expelled from the coal to leave a porous solid that is mainly carbon.

collision theory (191, 206): describes the conditions necessary for a reaction to occur and explains why the rate at which a reaction occurs might change.

colour (18, 19, 77, 167, 205, 215, 265, 293): a physical property of a substance based on the detection by the eyes of light that is being reflected from the substance.

combustion (151–7, 223, 251): a chemical reaction involving oxygen that produces heat and light and sometimes sound.

complete combustion (152–4, 164): a form of combustion in which the maximum amount of oxygen is used; commonly applied to the combustion of hydrocarbons in which the products are carbon dioxide and water.

compound (4, 11, 68, 73–83, 88, 102, 108, 115, 135–6, 162, 257–74): a substance formed by chemical reaction between two or more elements.

compressible (9): the property of a substance capable of reducing its volume when subjected to the application of pressure; commonly a property of gases.

concentrated (41, 174, 263, 268): a term that applies to a solution containing a large number of solute particles per unit volume of solvent.

concentration (27, 40–6, 167, 173–5, 180, 183, 192, 209, 289–98): a measure of the amount of solute in a given volume of solution.

condensation (178): changing of a substance from gaseous phase to liquid phase.

conical flask (17): a cone-shaped flask used in the laboratory for heating and in titration reactions.

Contact Process (255, 282): a process for the industrial manufacture of sulfuric acid.

corrosion (219, 233–40, 273, 283): a chemical process by which a metal is 'eaten' away or the surface of the metal is changed in appearance.

covalent bond (59–64, 67, 83, 90–3, 99, 102–13, 115, 279): a bond formed between two atoms by the sharing of electrons from the outer energy level of each atom; the bond is due to the nucleus of each atom attracting the outer electrons of the other atom.

covalent network (85, 89): a one-, two- or three-dimensional network of atoms covalently bonded.

cryolite (235): a substance (Na_3AlF_6) added to alumina to reduce the melting point during the electrolytic process for the production of aluminium.

crystal (8, 17, 23, 44, 88, 216, 243, 245, 278): a solid of regular shape, with faces at constant angles to each other.

decanting (15, 24, 77): separating a liquid from a solid in a vessel by pouring the liquid carefully from the vessel to leave the solid at the bottom of the vessel.

decomposers (252): organisms responsible for decay of dead matter.

dehydrate (284): remove water that is chemically bonded in a compound.

delocalised electrons (86, 90, 94): electrons that are free to move in an element, ion or compound.

density (24, 28, 215–19, 257, 276): the ratio of mass : volume for any substance; a common unit of density is g mL^{-1}.

deuterium (52): an isotope of hydrogen with a mass number of 2.

diamond (83, 91): an allotrope of carbon which is a three-dimensional network.

diatomic (61, 63, 108, 242): a molecule that contains two atoms.

diffuse (27–30, 265–6): the process whereby a gas will spread to occupy all available space.

dilute (41, 174, 177, 210, 221, 228, 268): a solution containing a small number of solute particles per unit volume of solvent.

discrete molecular substances (92, 113): substances that exist as individual molecules held together by weak forces of attraction; sometimes referred to as molecular substances.

displacement (221, 230, 257–9): a reaction where one substance takes the place of another in a compound, eg zinc will displace copper from copper sulfate to produce zinc sulfate solution and copper metal.

dissociation constant of water, K_w (168): the product ('multiplication') of the concentrations of hydronium and hydroxide ions in aqueous solutions, constant at a specific temperature; $K_w = 10^{-14}$ mol^2 L^{-2} at 25 °C, also called the ionic product of water.

dissolve (16, 23–5, 39–47, 75, 89, 93, 133): a solute mixing with a solvent to become a solution.

distillation (17–18, 155): the process whereby a liquid, on heating, becomes a vapour which is led away from the heating flask and then condensed back to the liquid state.

distillation flask (17): a glass vessel with a round bottom used in distillation; containing a side tube attached to the neck of the flask so that vapours formed by heating a liquid in the vessel can pass to a condenser.

double bond (60, 63): a covalent bond formed by two atoms, each sharing two of their electrons with another atom – four electrons are used to form the bond.

downward displacement (257): the process for collecting a gas that is less dense than air by delivering the gas to the top of gas jar so that the air is displaced downwards.

drying agent (238, 258, 261): a substance that will absorb water – commonly used to dry gases.

ductile (86, 87, 275): a property of metallic substances – can be drawn out into a thin wire.

dynamic (168, 178): not constant, but rather continuously moving.

effervescence (42, 142, 268): the release of a gas during a reaction involving a liquid; the gas causes the liquid to froth.

electrical conductivity (85, 88, 94, 103, 215–18): the ability of a material to conduct electricity without undergoing chemical change.

electrodes (89, 235): solid bodies that carry current to or from an electrolyte in an electrolytic cell.

electrolysis (164, 235, 249): using an electric current to decompose an electrolyte to produce chemicals at electrodes.

electrolyte (173, 235, 249): a molten substance, or an aqueous solution, that contains ions and which will conduct electricity thereby undergoing chemical change.

electron (27, 49–56, 59–64, 67–8, 73, 86–91, 100, 108, 115, 125, 216): a subatomic particle of negligible mass and carrying a unit negative charge.

electron configuration (54, 99, 106): the arrangement of electrons in an atom, in energy levels, outside the nucleus of the atom.

electron dot diagram (60): also known as a Lewis diagram; a diagram showing bonding between two atoms; the dots represent electrons from the outer energy level of each atom; different-coloured dots, or dots and crosses, may be used to differentiate between the electrons in the outer energy level of each atom.

electronegative (46): having a negative charge.

electronegativity (276, 279): a measure of the pull of an atom on the electrons it shares in a covalent bond.

electrostatic attraction (86, 253): forces of attraction between oppositely charged particles, or of repulsion between similarly charged particles.

electrostatic charge (68): an electrical charge that builds up on an object, caused by an excess or deficiency of electrons on the object.

element (4, 11–15, 49, 52–5, 59, 67, 83, 99, 105, 301): a substance that cannot be broken down into anything simpler by chemical means; contains one type of atom – all the atoms having the same, and equal, numbers of protons and electrons.

empirical formula (87, 162): a formula that gives the simplest whole number ratio of the atoms or ions in a compound.

endothermic (185, 198, 253): refers to a chemical process in which heat energy is absorbed.

endothermic reaction (186, 198, 208): overall, energy released is less than energy absorbed.

energy diagram (197, 207): shows relative energies and energy changes of reactants, products and activation energy during a reaction.

energy level (49, 54–67, 87, 100, 115, 216): a term used to define the status/position of an electron in an atom, outside the nucleus of the atom; the term 'energy shell' is sometimes used.

enthalpy (198–202, 207): the energy contained in substances in the form of either kinetic or potential energy.

enthalpy change (198–202, 207): the energy change during a reaction; is equal to the difference between the enthalpy of products and the enthalpy of reactants.

enzymes (212): protein catalysts common in biological systems.

equilibrium (39, 42, 168, 178–86): the state of reversible reactions where the rate of the forward reaction equals the rate of the backward reaction.

equilibrium constant (180–1): an expression showing the ratio of products to reactants of a system in equilibrium; *K* has a constant value for any particular system at a given temperature.

esters (39): organic compounds containing a –COO– functional group.

evaporation (16–17, 40, 177–8): changing of a substance from liquid phase to gaseous phase.

exothermic (152, 185, 197, 253): energy is produced/given out.

exothermic reaction (186, 197, 207, 271): reaction that involves the release of energy, since, overall, energy released is greater than energy absorbed.

face (8, 191): a side of a crystal.

fermentation (151, 268): the natural process whereby carbohydrates, eg starch or glucose, can be converted, by enzymes, to ethanol and carbon dioxide.

fertiliser (253, 285): a substance that provides essential nutrients to improve plant growth.

filtration (16, 77, 136): a separation technique using filter paper to separate a solid from a solution.

formal equations (80): show all species associated with a reaction – include spectator ions.

formula (60, 115, 159): the simplest unit of an element or compound shown in symbol form with the type and number of atoms of each element indicated.

fossil fuels (156, 255): coal, oil and gas.

fractional distillation (18): the process of separation of two volatile liquids by careful heating of the mixture of liquids so that the more volatile liquid is separated as a vapour, which then condenses into a receiver. The less volatile liquid remains in the distillation flask.

fuel (155, 251): a substance that can be used domestically or commercially to react with oxygen in the air and the reaction releases heat energy.

galvanising (220, 237): a process of covering steel with zinc to protect the iron in steel from rusting.

gas (8): a state of matter in which the particles present try to occupy all available space.

gelatinous (77): jelly-like.

general formula (279): a formula used to represent the atom composition of a molecule of any member of a homologous series.

GER (129): an acronym (a word formed from the initial letters of a phrase) to help students remember that 'gain of electron is reduction'.

gram (156): a unit of mass.

graphite (90): an allotrope of carbon which has a two-dimensional structure.

group (45, 99, 105): vertical column of the periodic table; atoms of the elements in a group have the same number of valence electrons but different numbers of energy levels.

Haber process (184, 253, 281): industrial process for the production of ammonia from nitrogen and hydrogen.

half-equation (130): an equation written to illustrate an oxidising or a reducing process involving electron transfer; adding a half equation (eg illustrating oxidation) to the half equation that is complementary (eg illustrating reduction) produces an overall equation.

halides (109): compounds formed from a halogen (Group 17) element and another element or elements.

heat of reaction (198): *see* enthalpy change.

heavy water (52): water in which the molecule contains the isotope deuterium instead of hydrogen.

hydrocarbon (93, 152): a compound that contains carbon and hydrogen only.

hydrogen carbonate (145): the radical with the composition HCO_3, forming the hydrogen carbonate ion, HCO_3^-.

hydrogen ions (67, 167): H^+; effectively a proton.

hydronium ions, H_3O^+ (108, 167): the form in which hydrogen ions exist in aqueous solution.

incomplete combustion (153): a form of combustion in which less than the maximum amount of oxygen is used; commonly applied to the combustion of hydrocarbons in which the products are carbon monoxide and/or carbon plus water.

incompressible (7, 85): not capable of having the volume reduced by pressure; applies to solids and liquids in which the particles are closely attached and there is no space between the particles.

indicator (167, 296): dye that changes colour with a change in pH.

inert (55, 67, 99, 107, 145): unreactive to all chemical reagents and changes in physical conditions; the term applies to gases such as helium, neon, argon, krypton and xenon, forming group 18 of the periodic table.

insoluble (15, 76, 133, 267): substances that do not readily form solutions/cannot dissolve in water.

insulator (215): a substance that does not conduct electricity.

intermolecular (92): between molecules.

intramolecular (92): within molecules.

ion (67): an atom, or group of atoms, that has a positive or negative charge; the ion is formed by the loss or gain of electrons from the atom or group of atoms.

ionic bond (67, 68, 87): a bond formed between two atoms by electron(s) being donated by one atom and accepted by the other atom.

ionic compounds (68): compounds formed from elements whose atoms combine by loss and gain of electrons, producing charged particles called ions. Ionic compounds have the characteristics of high melting point, electrically conducting when molten, and crystalline in form.

ionic equations (80): equations that include only the ions involved in a reaction; spectator ions are omitted.

ionic product (168): *see* dissociation constant of water.

iron sand (233): an ore containing oxides of iron and titanium.

isotopes (52, 159): atoms with the same number of protons but different numbers of neutrons.

kinetic energy (10): energy possessed by moving objects.

lattice (8, 68, 87, 116): stable structure in which anions and cations attract each other electrostatically so that each is surrounded by a group of oppositely charged ions arranged in a regular order.

Le Chatelier's principle (183): states that when a change is applied to a system in equilibrium, the system will react in such a way as to minimise the effects of the change; allows the behaviour of systems in equilibrium to be predicted when conditions change.

LEO (129): an acronym (a word formed from the initial letters of a phrase) to help students remember that 'loss of electron is oxidation'.

Lewis diagrams (60): diagrams showing covalent bonds between atoms as lines (as opposed to dots in an electron dot diagram).

limewater (270): a solution of calcium hydroxide used for detecting carbon dioxide gas.

liquid (8, 9): a state of matter where the substance has the shape of the container; the particles in a liquid are closely attached but arranged irregularly, with spaces between some particles.

litmus (257): a substance used in solution, or absorbed on paper, that has a red colour in acid and a blue colour in alkali.

lustre (145, 215): the ability of a substance to shine by reflecting visible light.

macromolecular (89): made up of a single molecule consisting of millions of atoms covalently bonded to neighbouring atoms.

malleable (86): a property of metallic substances – they bend easily.

mass (11): the amount of matter in a substance.

mass number (symbol **A**) (**50**): the number of protons plus the number of neutrons in the nucleus of an atom.

matter (**7**, **83**): the stuff around us and of what we are made.

melting point, mp (**10**, **88**): the temperature at which a solid changes to a liquid.

metal (**67**, **99**): an element whose atoms have 1, 2 or 3 electrons in their outer energy level; most metals are silver-grey and shiny in appearance.

metallic bond (**87**): strong electrostatic force of attraction between the positive ions and the delocalised electrons that make up a metal.

metalloids (**106**, **275**): elements with properties similar to both metals and non-metals, also called semi-metals.

mineral (**233**): naturally occurring substance that is mined, or occurs in the sea, and is commercially valuable.

miscible (**41**, **211**): the property of two liquids where they can be shaken together to form a homogeneous mixture.

mixed anhydride (**271**): a compound which, when added to water, produces two acids; eg nitrogen dioxide, when added to water, produces nitric and nitrous acids.

mixture (**12**, **83**): two or more elements and/or compounds, added to each other, in any possible combination of quantities, and retaining the properties of each of the components.

mnemonic (**129**): aid to memory.

molar mass (**28**, **37**): the mass, in grams, of one mol of a substance.

mole/mol (**28**): the number of carbon atoms present in 12 g of carbon-12, and is equal to 6.02×102^3 particles.

molecular formula (**60**): the formula of a substance that indicates the number and type of atoms present in one molecule or unit cell of the substance.

molecule (**31**, **59**): group of atoms bonded covalently together.

molten (**88**): the liquid state.

monatomic (**73**): ion made up of one type of atom only.

monoclinic (**243**): a crystal form having three unequal axes, one pair of which are not at right angles to each other.

monomer (**110**, **249**): a simple molecule that can be bonded to other molecules of the same, or different, types to form a large molecule – the product is called a polymer.

M_r (**159**): a symbol used to represent the relative molecular mass of an element or compound; relative molecular mass of an element or compound is the mass of a molecule of an element or compound in amu.

negative ion (**44**, **67**, **101**, **142**): an atom or a radical that has a negative charge by virtue of the number of electrons in the particle exceeding the number of protons by one or more.

neutral (**169**): having a pH of 7, ie neither acidic nor basic.

neutralisation (**169**): making an acidic solution less acidic or a basic solution less basic.

neutron (**49**, **50**): a subatomic particle with a mass of 1.00 amu and no charge; all atomic nuclei, except hydrogen-1, contain one or more neutrons.

nitrogen cycle (**252**)**:** the natural process whereby nitrogen circulates through the atmosphere, soil, and plants and animals.

non-metal (**67**, **99**, **101**)**:** an element whose atoms have 4, 5, 6, 7 or 8 electrons in their outer energy level; there are no common properties of non-metals in the way that there are for metals.

nucleus (**49**, **50**)**:** the heavy, central part of any atom, which contains protons and neutrons; comprises a very small part of the total size of the atom.

octet rule (**55**, **59**, **63**)**:** the recognition that many atoms are trying to achieve eight electrons in the outer energy level of their atom in order to become stable.

open system (**177**)**:** system from which some or all of the substances involved are allowed to escape or be added.

ore (**22**)**:** naturally occurring compound from which economically valuable constituents, eg metals, can be obtained.

organic acids (**172**)**:** act as acids because the hydrogen at the –COOH group can be transferred to a base; another term for carboxylic acids.

organic compounds (**44**)**:** all compounds containing carbon except carbon monoxide, carbon dioxide and the carbonates.

overall equation (**130**)**:** an equation to illustrate an oxidising process and a complementary reducing process involving electron transfer; formed by adding two half equations, one illustrating oxidation and the other illustrating reduction.

oxidant (**oxidising agent**) (**128**, **151**)**:** a chemical substance that: can add oxygen to another substance, *or* can remove hydrogen from another substance, *or* possesses particles (atoms or ions) that can remove electron(s) from the particles of another substance.

oxidation (**125**)**:** a chemical reaction in which: oxygen is added to a substance, *or* hydrogen is removed from a substance, *or* particles (atoms or ions) remove electron(s) from the particles of another substance.

oxidation numbers (**ONs**) (**277**)**:** a number describing the 'degree' to which an element has been oxidised or reduced.

oxidation state (**276**)**:** the 'degree' to which an element has been oxidised or reduced.

oxidising agent (**128**, **263**)**:** a reactant that oxidises another reactant.

ozone (**242**)**:** a gaseous allotrope of oxygen whose molecule contains three atoms of oxygen.

particle (**7**)**:** an atom, molecule, or ion; a small unit in a chemical substance.

percentage composition (**161**)**:** the mass of each element in a substance expressed as a percentage of the total mass.

period (**99**)**:** horizontal row of the periodic table in which atoms of the elements have the same number of energy levels, but different numbers of valence electrons.

periodic table (**49**, **59**, **105**, **301**)**:** arrangement of all the 110 known elements according to the number of valence electrons and the number of electron shells; increase in atomic number occurs across the table, columns have elements with similar properties down the table.

pH (167): a numerical scale used to indicate acidity or alkalinity in an aqueous solution. The usual range of the scale is 0 – 14. pH of 0 – 6 indicates acidity with 0 very high acidity, and 6 extremely low acidity. pH 7 indicates neutrality. pH 8 – 14 indicates increasing alkalinity.

pH scale (167): a scale formed from the pH of different solutions.

phlogiston theory (125): an early theory of burning such that on combustion any material released phlogiston (in the form of flame, soot, etc) and left a calx as a residue.

photosynthesis (52, 273): a process occurring in the leaves of trees whereby carbon dioxide and water, in the presence of sunlight and chlorophyll, produce glucose and oxygen.

physical change (84): a change that does not result in a new substance and that is easy to reverse.

physical dissolving (270): a process of dissolving in which the solute does not react chemically with the solvent.

physical property (215): a property that, when displayed, does not change the substance chemically, eg colour, electrical conductivity, ductility.

pigment (255): a substance added to impart colour to another substance.

pipette (293): a glass tube calibrated to dispense an exact volume of solution.

plastic (90): the description of a material that is easily moulded, ie the material can be stretched and shaped; the term is also used to describe materials that have these properties, especially if the material is synthetic (man-made).

plastic sulfur (90, 243, 245): an allotrope of sulfur of temporary existence, which is pliable.

polar covalent bond (110): a covalent bond in which the electrons are not uniformly shared between the two atoms, ie the atom at one 'end' of the bond has a small positive charge and the atom at the other end of the bond has a small negative charge; the term also applies to a molecule in which the polar bonds do not cancel, ie the molecule has a positive end and a negative end.

polarity (93): the property of a covalent bond, or a molecule, that has a positive end and a negative end.

polyatomic (73, 117): a particle that has more than two atoms in its molecule.

polyatomic ions (117): a radical that has a negative charge, by virtue of the number of electrons in the particle exceeding the number of protons by one or more; or a radical that has a positive charge, by virtue of the number of electrons in the particle being less than the number of protons by one or less.

polyethene (90): a polymer made from ethene monomers; structure is that of covalently bonded carbon and hydrogen atoms in linear chains.

polymer (278): large organic molecule made up of many monomer units linked together by covalent bonds, also called macromolecule.

polymerisation (197): the process of combining many small molecules of the same type to form one very large molecule.

porous (16): the property of a solid material that allows a liquid to penetrate through it, eg filter paper is porous.

positive ions (67, 138): atoms or radicals that have a positive charge by virtue of the number of electrons in the particle being less than the number of protons in the particle by one or more.

ppt (133): an abbreviation for the word 'precipitate'.

precipitate (76, 77, 133): an insoluble solid formed from mixing two solutions together.

precipitation (76, 184): formation of a solid (precipitate) from the combining of ions in solution.

pressure (10, 184): defined as force per unit area, with the units of N m^{-2}; commonly used to change the volume of a gas by applying a force.

products (12): substances produced by a chemical reaction.

protein (252): a fundamental material of plant and animal structure that is composed of amino acids; most commonly observed as muscle in animals.

proton (49, 50): a subatomic particle with a mass of 1.00 amu, and a positive charge; all atomic nuclei contain one or more protons; in acid/base chemistry – an H^+ ion (*see* hydrogen ions).

pure substance (13, 83): matter made up from a single element or single compound.

quantitative (80): the study of how much is produced.

radioactive (49): the spontaneous emission of radiation from the nuclei of unstable atoms of an isotope of an element.

rate of reaction (191, 205): the speed of a reaction measured in, for example, g s^{-1}, or m L^{-hr}, or in some other quantity of a reagent or product involved in the reaction.

reactants (75): substances that react in a chemical reaction.

reagent (81, 120, 138, 192): a chemical compound or solution used to create a chemical reaction.

redox (129): describes chemical reactions where one substance is reduced while another is oxidised.

reducing agent (128): a reactant that reduces another reactant.

reductant (reducing agent) (128): a chemical substance that: can remove oxygen from another substance, *or* add hydrogen to another substance, *or* possesses particles (atoms or ions) that can add electron(s) to the particles of another substance.

reduction (125): a chemical reaction in which: oxygen is removed from a substance, *or* hydrogen is added to a substance, *or* particles (atoms or ions) add electron(s) to the particles of another substance.

relative atomic mass (159): the mass of an atom in amu; symbol is A_r.

relative molecular mass (159): the mass of a molecule of an element or compound in amu; symbol is M_r.

respiration (273): a chemical reaction occurring in plants and animals involving oxygen and a fuel to produce energy and, most commonly, carbon dioxide and water as products of respiration.

reversible reactions (178): reactions that can proceed in the direction of forming products or in the direction of forming reactants.

rhombic (243): a crystal form having three individually perpendicular, unequal axes.

rust (237): the product of a chemical reaction, involving oxygen, water and carbon dioxide from the air, and iron.

salt (228): an ionic compound; formed when the hydrogen in an acid is replaced by a metal.

saturated (23): compounds in which all the covalent bonds are single bonds, such as the alkanes.

semi-metals (106): elements with properties similar to both metals and non-metals, also called metalloids.

significant figures (171): the digits obtained in a measurement (which are then used to determine the number of figures used in the answer to a calculation).

silica (92, 112): a common name for silicon dioxide, SiO_2; a three-dimensional network solid.

silicon dioxide (92, 112): *see* silica.

solid (8): a state of matter where the substance has its own shape; the particles in a solid are closely attached and arranged in a regular pattern.

solubility (41): the measurement of how much solute will dissolve in a fixed mass, or fixed volume, of a solvent; units are commonly *either* g (of solute) per 100 g (of solvent) *or* g (of solute) per litre (of solvent).

solubility grids (77): grids useful for predicting what precipitates will form when two solutions are mixed.

solubility rules (45, 76): rules that help to decide whether a compound is soluble or not, and whether a precipitate will form when mixing solutions together.

soluble (16, 133): describing the property of substance that will dissolve in a solvent to form a solution.

solute (16, 39): the substance that dissolves in a solvent to produce a solution.

solution (16, 39): a homogeneous mixture formed when a solute dissolves in a solvent.

solvent (16, 39, 133): a liquid that dissolves a solute to make a solution.

sparingly soluble (76, 133): describing a solute that has a low solubility in a solvent; as a rough guide, the solubility is between 0.1 g and 1.0 g of solute per 100 g of solvent.

spectator ion (76, 134): those ions that are part of reagent solutions but do not take part in the reaction.

stable (41, 145): unreactive in the conditions stated, eg in the presence of air, or when a substance is heated.

steel (219, 237): an alloy of iron and carbon; specialist steels also contain other metals.

strong acid (113, 172): an acid for which nearly all molecules present donate H^+ ions to water.

strong base (173): a base that readily accepts protons and/or completely dissociates in water to produce OH^- ions.

structural formula (60): formula that gives the number of each kind of atom in a molecule and shows how the atoms are joined together.

subatomic particle (50): a particle smaller than any other atom, eg protons, neutrons and electrons.

sublimation (93, 200): the transformation of a substance from the solid phase to the gaseous phase without going through the liquid phase, or vice versa.

synthetic (282): human-made.

tarnish (224): the result of the slow corrosion of a metal in air whereby the appearance of the metal becomes dull.

thermal conductivity (215): the ability of a material to conduct heat without undergoing chemical change.

thermal decomposition (145): the action of heat on a compound producing compounds of lower molecular mass and/or elements.

thermochemical equation (201): chemical equation that shows the enthalpy change for a reaction.

titanomagnetite (233): a mineral present in iron sand whose composition is given by the formula $Fe_3O_4.TiO_2$.

tonne (12): an industrial unit of mass, equal to 1000 kg.

transition metal (77, 107): element in groups 3 to 12 of the periodic table.

triple bond (64): a covalent bond formed by two atoms each sharing three of their electrons with another atom – six electrons are used to form the bond.

tritium (54): an isotope of hydrogen with a mass number of 3.

universal indicator (167): an acid–base indicator that can be used to determine the pH of a solution because it produces a range of colours corresponding to different pH values.

unsaturated (152): compound that contains a double (C=C) or triple (C≡C) carbon-carbon bond.

upward displacement (257): the process for collecting a gas that is denser than air by delivering the gas to the bottom of a gas jar so that the air is displaced upwards.

valence (86, 216): a term used to describe electrons in the outer energy level of an atom that are used for bonding.

valence level (107): the energy level that valence electrons are found in, also called the outer level.

valency (115): magnitude or size of the charge on an ion or the number of covalent bonds formed by an atom.

vaporised (18): the property of substance that was a liquid, and by evaporation or boiling, has become a gas/vapour.

vapour (9): another term for gas; usually referring to the product formed by a liquid evaporating or boiling.

volatile (18, 266): describing a liquid with a low boiling point (< 100 °C); the liquid vaporises easily.

wash bottle (260): bottle of deionised water used to wash solid or solution from the sides of a container back into a solution.

weak acids (172): acids that donate protons only to a limited extent; only partly ionise in solution to form hydrogen ions.

weak base (173): accepts protons only to a limited extent; only partly ionised.

word equation (119): a statement of a reaction in words; identifies the reagents and products of a reaction by name only.